Causation with a Human Face

OXFORD STUDIES IN PHILOSOPHY OF SCIENCE

Recently Published in the Series:

Mathematics and Scientific Representation
Christopher Pincock

Simulation and Similarity: Using Models to Understand the World
Michael Weisberg

Systemacity: The Nature of Science
Paul Hoyningen-Huene

Causation and its Basis in Fundamental Physics
Douglas Kutach

The Ant Trap: Rebuilding the Foundations of the Social Sciences
Brian Epstein

Understanding Scientific Understanding
Henk de Regt

The Philosophy of Science: A Companion
Anouk Barberousse, Denis Bonnay, and Mikael Cozic

Calculated Surprises: The Philosophy of Computer Simulation
Johannes Lenhard

Chance in the World: A Skeptic's Guide to Objective Chance
Carl Hoefer

Brownian Motion and Molecular Reality: A Study in Theory-Mediated Measurement
George E. Smith and Raghav Seth

Branching Space-Times: Theory and Applications
Nuel Belnap, Thomas Müller, and Tomasz Placek

Causation with a Human Face: Normative Theory and Descriptive Psychology
James Woodward

Causation with a Human Face

Normative Theory and Descriptive Psychology

JAMES WOODWARD

OXFORD
UNIVERSITY PRESS

Oxford University Press is a department of the University of Oxford. It furthers the University's objective of excellence in research, scholarship, and education by publishing worldwide. Oxford is a registered trade mark of Oxford University Press in the UK and certain other countries.

Published in the United States of America by Oxford University Press
198 Madison Avenue, New York, NY 10016, United States of America.

CIP data is on file at the Library of Congress
ISBN 978-0-19-758541-2

DOI: 10.1093/oso/9780197585412.001.0001

3 5 7 9 8 6 4

Printed by Integrated Books International, United States of America

For my dear wife Julia and my brave daughter Katherine

Contents

Acknowledgments

I have been interested in the empirical psychology of causal cognition for decades, but the topic really began to come into focus for me when I participated in a series of workshops at the Stanford Center for Advanced Study in the Behavioral Sciences (CASBS) in 2003–4. These were funded by the McDonnell Foundation, with Alison Gopnik as the principal investigator under the title "Causal Learning: Computational Learning Mechanisms and Cognitive Development." The workshops brought together philosophers, psychologists, statisticians, and others with a common interest in using computational theories of causal learning to illuminate its empirical psychology. From my point of view, they were paradigms of successful interdisciplinary interaction. I am very grateful to the participants in these workshops for all that I have learned from these interactions. In this connection, I particularly want to mention John Campbell, Patricia Cheng, David Danks, Alison Gopnik, Clark Glymour, Chris Hitchcock, Tamara Kushnir, Andrew Meltzoff, Laura Schulz, Josh Tenenbaum, and Henry Wellman and to apologize to the many other participants whose names I have omitted.

In the intervening years I wrote several papers on the relationship between normative/computational theories of causal cognition and empirical psychology, but I did not begin this book until I was a fellow at the CASBS during 2016–17. My fellowship at the Center (along with support from the University of Pittsburgh) provided me with an uninterrupted nine months that allowed me to complete most of an initial draft of this book. I'm very grateful to the Center and to Margaret Levi, its director, for this support. While at the center I interacted very regularly with several social scientists with an interest in causal inference, including Paul Jargowsky, Ross Matsueda, Teppei Yamamoto, and David Yeager. I am very grateful to them for many helpful conversations.

A number of people have commented on portions of the present manuscript. Thomas Blanchard and Clark Glymour read through the entire manuscript and provided a number of very helpful suggestions. Others who have commented or who have provided me with advice and feedback on issues related to those I discuss include Colin Allen, Wesley Buckwalter, Patricia Cheng, Alison Gopnik, Chris Hitchcock, Josh Knobe, Edouard Machery, Michael Rescorla, Michael Waldmann, and Zina Ward. I taught portions of an early version of this manuscript in a graduate seminar at the University of Pittsburgh in 2018 and have benefited from the input of students in that class. I have also greatly benefited from the comments of an anonymous referee for Oxford University Press. (There was also, inevitably, referee number 2, who did not comment at all on the substance of my manuscript but objected very strongly to its "anti-metaphysical" tone. I've tried in revising to be gentler in my treatment of metaphysics.)

The Department of the History and Philosophy of Science at the University of Pittsburgh has provided a very supportive intellectual environment for my work on this and related projects. I am grateful to my colleagues there, as well as Bob Batterman and Mark Wilson in Philosophy for their friendship and encouragement. Having colleagues at Carnegie-Mellon with a particular interest in causal inference and in some cases empirical psychology has also been a tremendous benefit—here I particularly want to mention David Danks, Clark Glymour, Peter Spirtes, Richard Scheines, and Kun Zhang. Finally, although they have not commented directly on this manuscript, I want to mention Frederick Eberhardt and Jiji Zhang, who have certainly influenced many aspects of my views about causation and causal reasoning.

Causation with a Human Face

Introduction

I.1. Preamble

The title (as well as the cover) of this book reflects my conviction that in understanding causal cognition (and causation itself) it is crucial to appreciate that causal reasoning is an activity carried out by human beings. More specifically, the ways we think about causation are in part shaped by human interests and goals and human abilities, including various epistemic limitations we face. Understood as a claim about causal cognition or the causal notions we employ, it is hard to see how this could be disputed. However, it is somehow easy to lose sight of the fact that when we talk about causal relations "as they are in the world" we are still employing categories and ways of thinking that are human products, reflecting human concerns.

This conviction—that "causation" and "causal cognition" need to be understood together and with reference to the sorts of creatures that we humans (and perhaps other sorts of subjects)[1] are—provides one of the primary motivations for the issues and topics taken up in this book. These issues are various—perhaps they will seem to some readers to be too much so to be appropriately discussed under one cover. I discuss theoretical or normative treatments of causation and causal reasoning (normative in the sense that these are theories about how we *ought* to reason causally).[2] Some of these derive from philosophy, others from disciplines like machine learning and statistics. I discuss how these normative theories interact with descriptive research on the empirical psychology of causal cognition in humans and (to some small extent) in other animals. I try to illustrate how these two enterprises—the theoretical/normative and the empirical—can mutually and beneficially inform one another: normative ideas can guide and structure empirical research, and the results of empirical research can suggest possible normative ideas. I also attempt to connect all of

[1] In subsequent chapters I briefly discuss learning and behavior with cause-like features in non-human animals as well as causal learning in very young children. So when I refer later to human limitations and goals, the reader may think of this as shorthand for limitations and goals for creatures who are broadly like us in not being omnipotent or omniscient.

[2] The idea that accounts of causal reasoning and causation can be understood as "normative" may strike some philosophical readers as odd, but this (and a corresponding contrast with "descriptive" accounts) is standard outside of philosophy. I defend it in what follows. In my view, it better captures what is really at stake when we theorize about causation than alternative approaches.

Causation with a Human Face. James Woodward, Oxford University Press. © Oxford University Press 2021.
DOI: 10.1093/oso/9780197585412.003.0001

this with methodological debates in philosophy concerning the role of appeals to "intuitions" or "judgments about cases," focusing on examples involving judgments about causal relationships. Such appeals figure prominently in many philosophical discussions of causation, but an additional reason for making this connection is that claims that people hold various intuitions or make various judgments are empirical claims, and in this respect arguably not so different in principle from empirical claims about causal cognition resulting from psychological experiments. I advance some suggestions about how such judgments about cases as well as empirical results from psychological experiments can, when appropriate additional background assumptions hold, be relevant to normative theorizing about causation.

My justification for discussing all of these issues together is that however ungainly the result, they belong together but are often not discussed in a connected way.[3] Thus one goal is simply to get readers to see that such connections exist and are worth exploring—hopefully this claim will be persuasive even to readers who do not accept many of the details of what I have to say. What I would like to do is to *open up* discussions of causation and causal reasoning, particularly but not exclusively in philosophy, helping participants in these discussion to appreciate the relevance of considerations drawn from many different sources, both descriptive and normative, that have not always received much philosophical attention. I thus oppose various *strategies of dismissal* (as I call them) that are sometimes taken to show that many of the issues that I discuss together have little or nothing to do with each other. These include the claims that normative issues about causal reasoning have no bearing on empirical studies of causal cognition and conversely and that the "metaphysics" of causation (understood as having to do with causation as it is in the world) has no connection to the epistemology or methodology of causal inference. I discuss some of these strategies in a general way in this introduction and in more detail in subsequent chapters.

I am of course aware that a number of philosophers will think that empirical facts about how humans and others think causally, as well as ideas from disciplines like statistics and machine learning, could not possibly be relevant to philosophical projects connected to causation. What I hope to show is that the results will be richer and more interesting if one does not proceed on such assumptions of irrelevance.

[3] So, to readers who think this is far too much to discuss in one book, I sympathize but also think not discussing these issues together would be a missed opportunity.

I.2. Interventionism

The framework for thinking about causation adopted in this book is broadly interventionist in the sense described in Woodward 2003. As a very rough characterization (there is more detail in Chapter 2), interventionists think of causal claims as claims about the outcomes of hypothetical experiments. Different sorts of causal claims may be distinguished on the basis of the kinds of experiments with which they are associated but a generic version is that *C* causes *E* if and only if there is some possible intervention on *C* (and depending on which causal notion is being characterized, perhaps on other variables) that, if performed, would be associated with a change in *E*. An intervention may be thought of as an idealized experimental manipulation. However, many of the conclusions I reach—about the importance of normative theorizing and its relation to descriptive claims, about the legitimate role of appeals to intuitions, as well as the descriptions of particular experimental results—seem to me to be relatively independent of a commitment to this framework. I hope that they will interest readers who are not persuaded by interventionism.

I.3. Some Misused Dichotomies (and Other Assumptions)

It is not a very deep or original observation that the reception of philosophical ideas often depends upon presuppositions that are regarded by those who hold them as so obvious and uncontroversial as to require little examination and defense. This is a problem for those, like me, who do not share a number of these assumptions: views of the sort I will defend will not seem prima facie plausible or even intelligible unless readers are prepared to consider the possibility that some of their presuppositions may be mistaken.

In what follows I attempt to identify several of these presuppositions and to briefly explain why I do not find them persuasive. I do not expect to change minds, but I hope that this exercise will help some readers to better understand what I am trying to say. I see these presuppositions as connected to entrenched distinctions or dichotomies. I do *not* object to the distinctions themselves, which I take to be uncontroversial, but rather to some of the uses to which the distinctions are put and to certain mistaken inferences they are used to motivate. Roughly speaking, the mistaken inferences involve moving from the correct observation that X is different from Y to the conclusion that X and Y are completely unrelated and can (and should) be discussed entirely independently of each other, so that any attempt to relate them involves "conflating" X and Y. The distinctions thus function rhetorically as devices for dismissal of the sort mentioned earlier.

I.3.1. Normative and Descriptive

I take it to be uncontroversial that the descriptive claim that so-and-so is the case does not (aside perhaps from some trivial exceptions) by itself entail the normative claim that so-and-so ought to be the case. Similarly the normative claim does not entail the corresponding descriptive claim. From the fact that people make certain causal inferences or judgments and not others, it does not (without further assumptions) follow that they are correct or normatively justified in doing so. From the claim that people ought to make certain judgments, it does not follow that, as a descriptive matter, they do make such judgments.

However, although there is no interesting entailment from "is" to "ought" or conversely, it does not follow, at least in the case of causal cognition, that we should think of the descriptive and normative as completely disconnected areas of inquiry: one obvious possibility (but not the only one) is that there are additional true assumptions that might be used to establish connections between these two areas. As an illustration, suppose that there is reason to believe that some causal inferences and judgments we make are the result of selection processes (involving either natural selection or learning or a combination of these) that operate so as to preserve true judgments while discarding false or mistaken ones, so that people are, in this respect, *successful* in many of the causal judgments and inferences they make.[4] Then the descriptive fact that people make such judgments and inferences may, in conjunction with the assumption about truth-conducive selective processes just described, be taken to have some bearing on whether those judgments or inferences are correct or normatively justified. (Again, I say more about standards of normative justification in subsequent chapters.) This need not mean that we should conclude just on the basis of the descriptive facts and the assumption about selection that the judgments

[4] I say more about success in subsequent chapters. But for those who are worried that invocations of success presuppose the correctness of some particular account of causation, I offer the following: First, we do have some uncontroversial cases of causal knowledge—one mark of success in an account of causation or a causal inference procedure is that it reproduces these, that it is well calibrated in this sense. Second, and relatedly, different accounts of causation often agree on what causal relations are present in particular cases, even if they disagree about the bases for such judgments—again, these cases can be used as standards for success or correctness that don't presuppose the correctness of any particular theory. Finally, accounts of causal inference and of other features of causal reasoning can sometimes (depending on the account in question) be assessed "internally"—in terms of whether they accomplish the goals they specify. For example, if one thinks (as interventionists do) that causal discovery has to do with identifying relationships that support manipulation, one can ask whether some particular inference procedure reliably accomplishes that. Of course, even if this is the case, a critic might deny that such manipulation-supporting relationships are causal relationships, but the requirement just described can function as an internal check. A useful analogy, developed in more detail subsequently, is the claim that the human visual system enables us to interact "successfully" with our environment. One can have good grounds for thinking that this is correct without presupposing any particular theory of how the visual system works. Similarly for human causal cognition.

and inferences are normatively justified. A weaker possible conclusion is that the descriptive considerations in conjunction with assumptions about selective processes are only suggestive: they suggest the judgments and inferences in question *may* have a normative justification—that this is a possibility worth investigating. Showing that there *is* such a normative justification then requires an independent (normative) argument of some kind. It is this weaker conclusion that I will defend in subsequent chapters, but my point here is that this illustrates one possible way in which the descriptive and normative might be connected. Of course one might think that in some particular case or cases, this contention about selective processes operating on judgments and inferences in such a way as to preserve correctness is mistaken, but whether this is so is (I will argue) a broadly empirical matter. If this contention is correct in some cases, one can't argue that, as a matter of principle, descriptive claims about how people judge can have no bearing on how they ought to judge.

A similar point holds in the other direction. One general assumption that might be used to connect claims about normatively appropriate causal reasoning to empirical claims about how people in fact reason is the assumption that many people are "rational," in the sense that they make normatively appropriate causal inferences and judgments much of the time. To the extent that this assumption is correct, it will be reasonable to begin with normative models of causal reasoning and then investigate the extent to which, as an empirical matter, people conform to these models. In fact, as we shall see, a methodology of this kind is widely employed in psychological investigations of causal cognition, often with fruitful results. Again, I will not try to defend this methodology here (that is done in Chapter 3); my point is rather that the assumption of rationality is one obvious way in which the results of normative theorizing might be brought to bear on or shown to be relevant to descriptive investigation.

In fact, there are a number of different ways (in addition to what I have just described) in which descriptive and normative claims might be relevant to or brought to bear on one another. In my experience, philosophers tend to frame questions about the relevance of the descriptive to the normative as questions about whether descriptive claims can be "evidence" for normative claims (or vice versa) or whether there are nontrivial entailments between the two. But there are many other possibilities. A normative claim might provide *motivation* for engaging in a descriptive inquiry (to see whether people conform to the normative claim). The empirical fact that some feature is present in people's cognition might motivate an investigation into whether there is some normative rationale for the feature. (Plenty of examples of both possibilities will be discussed subsequently.) In neither case do we need to think of the relation between the normative and descriptive as a matter of one providing evidence for the other or as one entailing the other. Trying to force the descriptive/normative connections into such frameworks will make them seem more mysterious than in fact they are.

I.3.2. Metaphysics and Epistemology

I.3.2.1. Metaphysics

As a non-practitioner, I approach this subject with considerable trepidation. At a very general level I take it that "metaphysics" has to do, at least in part, with what the world is like, with what is "out there" in nature. In the case of causation, this will have to do with what true causal claims are answerable to in the world, and with the structures in nature that support the application of causal reasoning (what the world has to be like, very broadly speaking, if we are able to engage in successful causal reasoning about it).[5] Different conceptions of causation will give different answers to such questions. For example, if you are an interventionist like me, taking true causal claims to describe what would happen if certain interventions were to be performed, then (at a minimum) you will think that there are facts about what would happen if those interventions were to be performed—facts about what would happens if rocks with certain properties were to strike windows, facts about what happens to headaches when aspirin is ingested, and so on. In this sense there are intervention-supporting relationships "in" the world.[6] (I will say a bit more about this momentarily.)

If issues of this "what is out there" sort are all that is meant by metaphysics, I am fully on board regarding its importance. Indeed, I think it is impossible to talk coherently about successful causal learning and judgment without some conception of what causal relationships in the world involve, since "success" implies discovering such relationships However, as a naturalist, I take the relevant features and relations that are "out there" to be just those countenanced in science and in other kinds of empirical inquiry—it is ordinary empirical inquiry that tells us which are the intervention-supporting relationships. Moreover, if we ask *why* these particular intervention-supporting relationships obtain, the explanations will be ordinary scientific explanations appealing to, for example, the chemical structure of aspirin and the way in which this interferes with the

[5] For more on this theme of structures in the world that support causal reasoning, see Woodward, forthcoming c and Weinberger et al. forthcoming.

[6] In saying this, all that I mean is that claims about such intervention-supporting relationships are truth-apt and that whether they are true or false has to do with how matters stand in the world. In my view, saying this does not require an elaborate "ontology" of such relationships. To employ an analogy that I have used elsewhere: it is a factual matter whether a given roulette wheel has such-and-such probabilities of producing red or black outcomes when spun by a macroscopic agent. Moreover, there are physical explanations for why roulette wheels behave in this way in terms of the method of arbitrary functions and similar results. Acknowledging this does not require that one subscribe to any particular "ontology" of chance.

enzymes that synthesize prostaglandins. We don't require in addition some variety of "metaphysical" explanation. This view amounts to what I will call "minimalist" metaphysics or "minimal realism."[7]

Although the contrast may not be a sharp or clear one, I would distinguish the minimalism just described from more ambitious or expansive forms of metaphysical theorizing, which (it seems to me) go beyond what science tells us about nature or involve forms of explanation that are something other than scientific.[8] These have taken a variety of different forms in recent discussion concerning causation and related topics. Some approaches postulate special entities or relations—powers, relations of necessitation between universals, and so on—to serve as "truth-makers" for causal claims. In other cases criteria for the acceptability of some proposed metaphysical theory are imposed that (in my opinion) have no obvious scientific motivation, as in the demand that the theory be reductive and/or that it only make use of primitive notions ("perfectly natural properties," etc.) that are metaphysically acceptable.

I said earlier that it is an important fact that we have procedures for learning about and reasoning with causal relations that are sometimes successful and that to characterize such success we need some conception of what is out there in nature that such procedures discover. However, my view is that all the resources that we need to characterize success and to explain how it is possible will be of minimalist sort previously described. This is one reason why I do not think that the topics discussed in this book require more ambitious metaphysics.

Some will hold that even if there are facts about what would happen if rocks strike windows or aspirin is ingested, these cannot be part of the "fundamental ontology" of the world. I won't dispute this since I don't mean to be making claims about fundamental ontology. What matters for my purposes is that whatever may be the constituents of nature at more fundamental levels (and however "fundamental" is understood), these are somehow arranged in such a way that certain claims about what would happen if certain intervention were to occur turn out to be true and others turn out to be false. Suppose, for example, that fundamentally all that exists is a Humean mosaic of perfectly natural properties at spatiotemporal points of the sort characterized by Lewis (1986). My version of minimal realism assumes that under this supposition it will still turn out to be the case that there are also what-would- happen-if facts of the sort previously mentioned.[9] It is

[7] If this strikes you as just science, rather than metaphysics, I'm fully on board. I use the phrase "minimal metaphysics" to accommodate the many philosophers who think that any claims about what exists amount to metaphysics.

[8] I don't have a hard-and-fast criterion that distinguishes minimal from more ambitious metaphysics. Perhaps one diagnostic for the latter is whether some distinctively metaphysical (as opposed to "scientific") form of explanation is invoked, as in Loewer 2012.

[9] If this sort of minimal realism seems weak and unambitious, I have intentionally designed it to be so. It does rule out some views that have been entertained by prominent philosophers—for example, the claim that there are no worldly facts about what would happen if such and such were

these that we represent and reason about when we engage in causal inference and judgment. My picture is thus one in which we can legitimately talk about causal claims and associated what-would-happen-if counterfactuals at many different "levels," including "upper levels" involving social, psychological, and biological variables and relations, as well as facts disclosed by ordinary manipulative activities like throwing a rock at a window. Some of these intervention-supporting relationships will be more local (less invariant—see Chapter 5) than others but all are part of reality.

I.3.2.2. Epistemology

Metaphysics, whether taken in a minimalist or more expansive sense, is obviously a different enterprise from epistemology. I take the latter (and here we can also include methodology) to have to do with what we can know and how we know it and, more generally, with strategies for learning about various aspects of nature. Obviously what the world is like is a different matter from what we know or even can know about it. However, it does not follow from this truism that investigations of the epistemology and metaphysics (including, especially, minimal metaphysics) of causation should be pursued completely independently of each other. Instead, as I see it, these enterprises should mutually constrain one another. Thus we should resist arguments along the lines of "I'm interested in the metaphysics of causation. Your arguments, even if correct, have to do only with the epistemology of causation (or methodologies of causal inference), which is a completely different subject."

Why should we resist such arguments? To begin with, as emphasized previously, from a normative perspective, one wants claims about how we ought to learn about and judge causal relationships to fit with or track which such relationships hold in nature—a matter of metaphysics, even on a minimalist conception of that subject. Different conceptions of the conditions a relationship should meet to qualify as causal (and different views about whether there are any relations in nature that satisfy these conditions) will fit with different conceptions of the evidence and inferential strategies that are appropriate for discovering such relationships. We thus have the following desideratum on an account of causation: there is a strong prima facie case that it should be understandable, given our conception of what causal relationships involve, how the kinds of evidence and strategies we employ to discover such relationships sometimes actually successfully lead to their discovery. Again consider interventionism about causation: true causal claims describe what the results of hypothetical experiments would be were we to perform them. If this view is correct

the case, with modal claims of this sort rather being "grounded" in, or reflections of, our projective activities.

we should be able to see how normatively appropriate methods for discovering causal relationships provide reliable information about the outcomes of such hypothetical experiments, thus showing that the methods are normatively appropriate *because* they provide such information. Of course well-designed actual experiments will trivially meet this condition, but so will various inference strategies involving non-experimental data. For example, these may include the use of instrumental variables and regression discontinuity designs[10]—these techniques work because they provide information about what would happen if certain experiments were to be performed, without performing these experiments.[11] It is less clear why such techniques are particularly appropriate for discovering causal relationships on certain other conceptions of causation (or so I claim).

As a concrete illustration of how one would like the epistemological and the (minimal) metaphysical to be connected, consider a randomized experiment assessing the role of a certain pedagogical technique *P* in improving educational performance *E* among a group of students. If the experiment is properly performed, it is generally thought that it can provide evidence that *P* causes *E*, among these students, where this means something like the average causal effect (see Chapter 5) of *P* on *E* is positive. Now consider the common metaphysical suggestion that true causal claims are just those claims that relate events that "instantiate" laws of nature. Putting aside any other misgivings one may have about this suggestion (including, for example, the unclarity of the notion of "instantiation"), one question we should ask is how this metaphysical claim relates to the evidence provided in the experiment. Why are we entitled to think of the results of the experiment as evidence for the instantiation of a law of nature? After all, "*P* causes *E*" is not itself a law of nature on any reasonable conception of what a law is, and there is unlikely to be any other explicit invocation of laws in the experiment. Moreover, why does it matter that the experiment involves randomization, given that "randomization" is not a notion that plays any role in our notion of a law of nature? Of course if we are willing to *assume* that true causal claims must instantiate laws, then we can presumably conclude, when the experiment is properly carried out, that what is discovered is the fact that *P* and *E* instantiate some (unknown) law, but this gives us no insight into *why* it is normatively appropriate to interpret the experiment as discovering this.[12] Such a connection between the method employed in the experiment and the interpretation of the causal claim it establishes is one thing that we are looking for when we ask for a connection between the epistemology/methodology and metaphysics of causation.

[10] See, for example, Angrist and Pischke 2009.

[11] For more on this see Woodward 2015a.

[12] In other words, it is *consistent* to assume that (i) randomized experiments are good ways of discovering causal relationships and that (ii) what is discovered is instantiation of a law. But what one would like is a closer connection between (i) and (ii) than mere consistency.

To enlarge on this theme, consider the many strategies and procedures, experimental and non-experimental, that are commonly taken to be appropriate for learning about casual relations and judging which causal claims are true. Now consider the possibility that these fail to fit at all with the correct metaphysical account (either minimalist or expansive) of causation in the following sense: given the metaphysical account, the procedures and criteria we now employ are all completely inappropriate for discovering causal relationships. Going further, consider the possibility that there are *no* procedures or criteria for recognizing true causal claims (as described by some correct metaphysical theory) that human beings can actually carry out or implement.

Whatever we may think about these as logical possibilities, the idea that they might be the correct account of the relationship between the epistemology and metaphysics of causation ought to strike us as deeply puzzling, if one holds, as I do, that the notion of causation is a functionally useful one and human causal reasoning is to some extent "rational." For starters, why would we have developed conceptions of what it is for a causal claim to be true and conceptions of how to tell whether such claims are true that are so completely out of sync in this way? If we have done this, why wouldn't we notice this problem and take steps to correct it, either by adjusting our understanding of causation or its epistemology/methodology or both? Given such a failure of synchronization, both our methodology and our conception of what it is for causal claims to be true would be completely useless—the former because unreliable, the latter because, given this unreliability, we have no way of telling which causal claims are true.

I don't mean to suggest that this sort of systematic mismatch can never happen in any area of investigation. I take it to be very plausible, for example, that given a literal reading of what is required for various theological claims to be true, the procedures and arguments for determining which such claims are true that believers standardly employ are completely inadequate for this purpose. But I also take it to be a plausible starting point that causal claims and causal reasoning are not like theology in this respect. When it comes to causation, my working assumption will be that *how* we find out about causal relations can tell us something about *what* those relationships involve. Conversely, a satisfactory account of what causal relations involve can help us to understand how our procedures for finding out about them succeed to the extent that they do.

Let me add, by way of elaboration, that the arguments of the preceding paragraphs depend on (or at least will make most sense in the light of) a certain conception of what a defensible epistemology/methodology of causal reasoning involves. There is a tendency among philosophers to be rather abstract and nonspecific when it comes to epistemology/methodology: it may be thought sufficient, for example, to say that causal claims are established by "our usual inductive practices" or "inference to the best explanation," or by "making the appropriate adjustments in our web of belief," without further specifying what these involve. If one operates at this level of generality, it may be hard to see how epistemological/methodological considerations could impose any tight constraints on an account of causation, since the epistemology seems to amount to little more than "We somehow figure out which claims are true." However, as subsequent chapters and discussions elsewhere illustrate, actual strategies for reliable causal inference are far more specific and discriminating. It is these strategies that can provide useful restrictions on accounts of causation.

I.4. More on Epistemic Limitations and the Metaphysics and Epistemology of Causation

I have been defending the idea that the ways in which we think about causation are shaped by what we can know or find out, among other factors.[13] This is likely to prompt the reaction that even if this is true of how we think about causation, causation as it exists in the world is a very different matter. I have already agreed that there is an obvious sense in which this distinction between how we think and what the world is like is uncontroversial. However, we need to be careful about the conclusions we may be tempted to draw from this distinction. When we talk about causation as it exists in the world, we of course are still making use of conceptions and ways of thinking that we humans have developed. There is very strong empirical evidence that these conceptions are influenced or shaped by features of our psychology and situation in the world—including our epistemic limitations, goals, and interests. (This is true, I claim, of both ordinary lay thinking about causation and causal notions deployed in the various sciences.) Denying these empirical facts is not a viable strategy. An alternative possibility is to concede the empirical facts but to claim, as a normative matter, that in talking about causation as it exists in the world (or what causation "is") we should be aiming at an account of causation that completely abstracts away from such human influences—an account of causal relations as God would describe them or some such. But I see no reason to suppose that we have access to such a godlike conception of causation. Nor, to the extent that we attempt to

[13] What I understand by "shaped by" will become clearer subsequently, but very briefly I mean "causally influenced by in something like the way in the way that the holding of a set of ideas can be influenced by one's material circumstances, interests, and goals." I do *not* mean that built into the *semantics* of causal claims are references to epistemic limitations, goals, and the like, that "causation itself" is in some way interest- or goal-relative or anything similar. Various factors can influence how we think about causation without being part of the semantics of causal claims.

entirely abstract away from our goals and limitations, is it clear what the use (for us) of the resulting account would be—why, normatively speaking, it would be a good thing to have. My contrary assumption is that in talking about causation, we cannot avoid employing categories and modes of thinking that reflect characteristically human concerns.[14] Part of understanding causation as it exists in the world involves understanding these categories and modes of thought and what work they do for us. Moreover, a normatively useful notion of causation must connect with these human concerns.

I emphasize again that this does *not* imply some kind of subjectivism or projectivism about causation. Acknowledging that how we think about causal relations is shaped by human concerns does not mean that there is nothing in the world outside of us that is tracked by such relations or that these relations are "mind dependent" in some problematic way.[15] Given a specified way of thinking about causal relations such as interventionism, it is an objective matter whether the world contains any such relations and, if so, which ones. However, it is also a mistake to think that acknowledgment of the objectivity and mind independence of causal relationships requires that we adopt the causation-as-God-would-understand-it idea.

I.5. The Role of Goals and Interests

Alongside the tendency to look for accounts of causation (and other notions of scientific interest) that abstract entirely from our epistemic and calculational limitations (as well as our procedures for finding out about these), there is a parallel tendency to think that a suitably objective account of such notions should abstract away from any considerations having to do with goals and interests or what we want to *do* with these notions.[16] I view this tendency as misguided. As

[14] Some may think that fundamental physics gives us just such a God's view conception. But physical theorizing is just as much a human activity as any other. The idea that physics describes the universe from God's perspective is theology, not science.

[15] I acknowledge glossing over some complex issues here, not the least of which is what is meant by "mind dependent." It is clear that the acceptability of some kinds of causal judgments is more influenced than that of others by, for example, social norms. This appears to be true, for example, for "actual cause" judgments involving norm violations—see, for example, Hitchcock and Knobe 2009. In other cases, causal judgments are sensitive to which possibilities are regarded as serious, as opposed to far-fetched, although such assessments in turn are heavily influenced by "objective" facts—see Chapter 6. But even in these cases, there is a more objective core of dependency relations that underlies the judgments in question. Moreover, in many other paradigmatic cases, factors such as norms play little or no role.

[16] This is reflected in talk of "perfectly natural properties," "cutting nature at the joints," and the like, where at least in philosophy the assumption seems to be that these can be characterized without any reference to the purposes for which one may want to use such descriptions and distinctions—the descriptions and distinctions instead being supplied by nature alone. (If nature could talk, this is how she would describe herself.)

an interventionist, I see causal cognition as shaped by human goals and interests having to do with manipulation, control, and prediction. Again, I claim that this does not by itself introduce an objectionable kind of subjectivity. Connecting causal relationships to relationships that are exploitable for manipulation and control does not imply that whether or not it is true that manipulating *C* is a way of changing *E* is somehow produced or created by our goals or interests. My picture is this: our goals lead us to care about certain sorts of relationships—those that support manipulation—but which relationships in nature have these features is determined by what nature is like. The idea that acknowledging a role for goals and features of our epistemic situation in shaping how we think about causation undermines the "objectivity" of that notion seems to me to reflect a confusion about what a reasonable sort of objectivity requires. Again, as I will understand it, "objectivity" for causal relationships holds to the extent that causal claims are the sorts of things that are true or false, and which they are turns on how matters stand in nature and not on such factors as our projective activities. Relatedly, an important mark of the objectivity of causal claims is that there are procedures involving objective information about the world (statistics, results from experiments, etc.) that can sometimes be used to determine whether they are true or false.[17]

I.6. Philosophical Strategy and an Apology

Next a remark about philosophical strategy (or at least the strategy followed in this book) and standards of assessment. The standard I will adopt is *fruitfulness*. This standard evaluates in terms of criteria like the following: whether the approach and positions advocated lead to new discoveries or insights, either conceptual or empirical; solve previously unsolved problems; provide connections to work in other disciplines; generate new research problems; and provide resources for dealing with them. When one evaluates in terms of this standard, one assumes that virtually any philosophical program or set of positions will encounter problems and difficulties at some points, but the idea is that rather than focusing entirely on these, evaluation should focus more on the program's positive achievements and successes, at least as long as these continue. We should thus resist the temptation to think that the successes are not genuine as long as there are also unsolved problems, counterintuitive

[17] This is another illustration of the way in which the metaphysics and epistemology of causation are intertwined. Strongly subjective or projectivist views about the "nature" of causation are hard to reconcile with the apparent objectivity of the procedures we have for determining whether causal relationships hold.

implications, and so on. I take this to be the standard often employed in the evaluation of scientific research programs, and I think it is an appropriate one to employ in philosophy as well.

Finally, an apology. I'm aware that there is a fair amount of repetition in what follows. To some extent this is deliberate: I've re-iterated certain points in an effort to avoid being misunderstood, at the cost perhaps of trying the reader's patience.

1
The Normative and the Descriptive

1.1. Introduction

As noted in the introduction to this book, it has several interconnected themes. One has to do with the interrelations between, on the one hand, what I will call *normative/theoretical* work on causal representation and inference, of a sort conducted in statistics, computer science, and philosophy, and, on the other hand, *descriptive empirical* work on causal cognition of the sort conducted by psychologists, among others. I explore some interactions between these enterprises and how each can beneficially influence the other.[1]

A second thread, taken up in later chapters (5–8), has to do with what I call *distinctions within causation*. Suppose we adopt a "minimalist" account of causation, which provides a basis for distinguishing between causation and mere (non-causal) correlation but does not go beyond this.[2] This minimalist account might be provided by the interventionist treatment of causation I prefer (which is described in Chapter 2; see also Woodward 2003), but it also might be provided by a number of other treatments as well. Then among those relationships counted as causal by such a minimalist account, there are a number of other distinctions we might wish to draw—these are "distinctions within causation." For example, relationships that satisfy minimal requirements for causation can vary in the extent to which they are *invariant*, and, I shall argue, this in turn influences how we think (and ought to think) about such relationships and use them in causal reasoning and explanation. Invariance has to do, roughly, with the extent to which a causal relationship continues to hold as various changes occur—this is discussed in more detail in Chapter 5. Another distinction within causation has to do with the extent to which causal relationships satisfy a requirement of *proportionality*, in something like the sense in which this notion is used by Yablo (1992). Proportionality concerns the extent to which there is

[1] A remark about my use of phrases like "causal cognition" and "causal inference": I use these as handy umbrella terms for the full range of more specific topics I will considering; these include but are not necessarily limited to different forms of causal learning, causal reasoning strategies, and the use of more particular causal concepts and distinctions in causal judgment—e.g., the distinction between "direct" and "indirect" causes, or between "actual cause" and other sorts of causal judgments.

[2] This sort of minimalism has no direct connection with what I call minimal realism in the introduction.

Causation with a Human Face. James Woodward, Oxford University Press. © Oxford University Press 2021.
DOI: 10.1093/oso/9780197585412.003.0002

a match between the variation in the cause and the variation in the candidate effect—more precisely, the extent to which the causal claim is formulated in terms of variables that capture the full range of dependency relations that hold in the situation of interest without falsely implying dependency relations that do not hold. This is discussed in Chapter 8.

The relationship between this second theme (distinctions within causation) and the first (the interaction of the normative and the empirical) comes about, in part, because a substantial amount of recent empirical research on causal cognition is directed at distinctions within causation—that is, whether in fact people reason in accord with such distinctions, where the distinctions are commonly understood as having some normative foundation. In empirical studies of causal judgment it is common to ask subjects not just to make binary judgments about whether a causal relationship is present or not (given some scenario or body of evidence) but also, in cases in which a causal relationship is judged to be present, to make graded judgments of "causal strength" (along some nonbinary scale) about that relationship. The notion of causal strength is by no means unproblematic, but at least to some degree strength judgments seem to track distinctions within causation of the sort mentioned earlier—more invariant causal relationships are judged to be stronger and so on. In other words, people make discriminations in strength *among* relationships that they judge to be causal rather than non-causal, and this raises the question of the bases (both as an empirical and a normative matter) for such strength judgments.[3] Thus the use of strength measures and the normative idea that there are distinctions to be made among relationships that satisfy some minimal criterion for causality fit together naturally.

In addition, this book also has several goals that are methodological in nature and which may be of interest to those whose principal concern is not with causation or causal reasoning. I use the example of causal reasoning to explore these more general issues. First, there is the general question of how, if at all, empirical investigations into how people reason and behave might be relevant to the more "theoretical" or "conceptual" or "normative" investigations carried out by philosophers and others with similar interests. For example, how (if at all) do results about how we in fact reason bear on issues about how we ought to reason? Such questions about the relation between "is" and "ought" arise throughout philosophy. I use empirical work on causal cognition to illustrate some possible answers to this question, some of which may also apply elsewhere.[4]

[3] See Chapter 7.

[4] Since this is already an overly long book, I don't make any proposals about the extent to which my proposals about the relation between the descriptive and normative considerations in causal cognition generalize, leaving this for another occasion.

Second, there are much-discussed methodological issues concerning the role of appeals to "intuitions" (or, if you like, "judgments about cases") in philosophical argument and, along with this, the status of various programs advocated in experimental philosophy (understood as the programs described in, e.g., Sytsma and Buckwalter 2016). The philosophical literature on causation contains many appeals to intuitive judgments about hypothetical examples—examples in which certain relationships are described (e.g., between a failure to water flowers and their deaths) and a judgment is then made about whether the relationship is causal. I see such appeals as having a number of different roles, but sometimes (I will argue) they are best understood as empirical claims about how the intuiter and (typically) others in addition to the intuiter judge, often accompanied by the suggestion that such judgments are reasonable in some way.[5] I contrast this understanding of the significance of such claims with the common philosophical idea that intuitive judgments provide us with some sort of access into the "nature" of causation or the "nature of our causal concept."

When appeals to intuitive judgment are understood along the lines I recommend, they raise questions about how such empirical claims relate to normative/theoretical accounts of causation. Moreover, research both in experimental philosophy and in psychology also often collects reports of subjects' causal judgments, and claims about these judgments are obviously empirical in nature. This raises questions about how such empirical information might be relevant to philosophical and other sorts of theories about causation. I will explore some of these issues in subsequent chapters (in particular, Chapter 3). There I will describe some important differences between, on the one hand, uses of intuition in more traditional philosophical theorizing and in experimental philosophy and, on the other, empirical research of the sort conducted by psychologists. At the same time, I will suggest that when it comes to causation, the armchair philosopher who appeals to intuitions, the experimental philosopher, and the empirical psychologist are in the same business, at least in many respects. In particular, all three groups should be understood as advancing empirical claims about patterns of causal judgment. The intuitions of the armchair philosopher have, in this respect, no special authority. Again, I hope that some of the morals I will draw will apply to other topics besides causation.

[5] "Best understood" here means that this is the most defensible way of understanding the significance of such judgments, not that this is necessarily what is intended by philosophers who appeal to such judgments.

1.2. Functional Accounts of Causation and Causal Reasoning

My approach in this book involves a different way of thinking about causation and causal reasoning than is standard in a lot of philosophical discussion. It involves thinking about causal notions[6] in *functional* terms,[7] that is, focusing on what we want to *do* with such notions, with the function or goals or purposes causal thinking serves for agents like us (or that we would like it to serve). I take this focus on function to be importantly different from more traditional philosophical projects that focus instead on what causation "is" or on the analysis of "our concept" of causation.[8] A functional focus leads us to ask different questions (and to accept different answers) than these alternative approaches do. A focus on function is fairly common in the psychological literature on causal cognition and in disciplines like statistics and machine learning concerned with causal reasoning; I propose a similar stance with respect to such standard philosophical topics as the sorts of causal judgments that are appropriate about different kinds of cases, distinctions we may wish to draw among different causal concepts, and how causal claims connect to other sorts of claims involving counterfactuals and probability.

This emphasis on function provides one natural motivation for grouping together the various topics figuring in this book. For one thing, it brings together the descriptive and the normative. If we are to think about causal cognition in functional terms, then (I will argue) we need accurate empirical information about how adult humans (and I will suggest, other subjects such as small

[6] I speak of causal notions in the plural because I think there are a number of distinguishable causal concepts, all involved in causal cognition, although I shall take these to share a common interventionist core.

[7] To forestall a possible confusion, I should note that this way of thinking about causation in "functional" terms is very different from the kind of "functionalism" about causation and other concepts associated with the so-called Canberra plan. The latter involves assembling all of our folk truisms about causation and conjoining these to provide a kind of Ramsey-sentence-style (implicit) definition or analysis of causation. Among other differences, as I understand it, the Canberra plan assigns no role to the function of causal talk in the sense described here. The constraints are provided by the truisms, not by the goals of causal thinking. In addition, unlike the Canberra plan, I assign no special status to folk truisms about causation—indeed I think that a number of the candidates for such truisms offered by philosophers (such as the claim that causation is an "intrinsic" relation, as this is commonly understood) are wrong, both as claims about what the folk believe and as claims about what a normatively acceptable treatment of causation will look like. Such supposed truisms are of course different from the particular judgments about causal relationships (the impact of Suzy's rock caused the window to shatter) that the folk make—I'm not advocating general skepticism about the latter.

[8] I don't mean that an account of what causation is or of our concept of causation could not be organized around such functional questions but rather that for the most part this does not happen in the philosophical literature. The contrast between thinking of an account of causation as having the aim of capturing "our concept" of causation and the functional approach that I advocate is developed in more detail in Chapter 3.

children) judge, learn, and reason causally—*what* they do. We also need information about *why* they judge, etc., as they do—what factors influence judgment and reasoning—and about how these make for success or failure in their reasoning processes. We need such information to assess claims about what the goals or function of causal thinking might be (since there should be some connection between such goals and observed practices of reasoning) and to assess the extent to which various forms of reasoning and inference that subjects actually exhibit conduce to those goals.

As already intimated, this emphasis on function thus foregrounds *normative* concerns—given a claim about the goal or purposes served by some feature of causal thinking (whether this is a distinction among causal concepts,[9] a

[9] Since here and elsewhere I refer to causal "concepts" while at the same time expressing (in later chapters) skepticism that either psychological research or armchair judgments about cases are best interpreted as providing information about concepts, some explanation/clarification is in order. I would distinguish at least three uses of concept talk. The first is (I claim) relatively innocent and noncommittal: talk of "the concept of X" is just shorthand for something like (i) "way of thinking or reasoning about X" (or "framework for thinking about X" or "account of X") with nothing more specific intended. Unless I signal otherwise, this is what I have in mind when, for example, I speak of Hume's concept of causation. This use of "concept" should be distinguished from two other possibilities—which I will label (ii) $concept_p$ and (iii) $concept_n$. Interpreted as $concept_p$ (p for psychological), "concept of X" is a notion that figures in psychological explanation. When someone has a $concept_p$ of X, this is taken to imply that they have a mental representation of some kind of X and that their possession of this representation *explains* various aspects of their behavior—for example, their categorization judgments. I take this to be the notion of concept that is in play when Goldman (2007), whose views are discussed in Chapter 3, argues that intuitions or judgments about cases provide information about concepts. This way of thinking about concepts is common among philosophers and something like it is the target of the various "theories" of concepts (concepts as prototypes, etc.) discussed by psychologists. One way in which the $concept_p$ notion differs from the first notion is that the first notion can be understood in a purely descriptive vein, while the second makes a claim about explanation and thus involves us in questions about the details of the representation that does the explaining. My misgivings about interpreting either the results of judgments about cases or psychological research in terms of the $concept_p$ notion are several. First, it seems to require that we have some basis for distinguishing what belongs to the concept of X from other claims about X that are true (and may be of considerable interest) but are not part of the concept of X. Second, and relatedly, even if we are very liberal about what belongs to the concept of X, there will be much of interest to both philosophical and psychological investigation that is not plausibly regarded as part of that concept. In the case of causation, for example, there are important facts about learning strategies, default assumptions (see Chapters 2, 4, and 5), and much more that are uncovered in normative and empirical investigation but that are not plausibly viewed as "part" of "our concept" of causation. Taking the results of philosophical and psychological investigation to be confined to claims about the $concept_p$ of cause is to adopt an unduly restrictive view of the subject.

Finally, this second psychological notion of concept should be distinguished from a third notion. This is a notion of concept that lives within a formal, often normative theory and that is often a target of explicit definition. For example, Pearl 2001 distinguishes two concepts of indirect effect—natural and controlled—and describes how each may be estimated from statistical information. Pearl provides illustrations of this distinction, and he clearly assumes (or hopes) that the distinction will be recognizable or seem sensible to his readers. Nonetheless it is clear that he is not primarily putting his distinction forward as part of a psychological claim about how people think or their mental representations—he is not introducing a $concept_p$, but rather something we might describe as a $concept_n$ (n for normative). In other words, the rationale for his distinction has primarily to do with its normative usefulness, not its role in psychological description or explanation, although, as I have argued, once the distinction is formulated, we can go on to raise the question of whether, as a descriptive matter, it is reflected in people's behavior. $Concepts_n$ are common in mathematics and the

distinction between cause and correlation, a reasoning or learning strategy concerning causal relationships, or whatever), we can ask whether that feature is well adapted to or an effective means to that purpose or not. So what I call *methodology*[10] (which is concerned with how we ought to learn, reason, and judge, both in connection with causal thinking and more generally) will occupy a prominent place in my discussion. The descriptive and the normative also come together in my related advocacy of "rational" models of causal cognition—that is, empirical models of causal cognition that are informed by normative theory and that attempt to explain why human causal cognition is often relatively successful when evaluated in terms of the goals that it attempts to achieve.

The remainder of this chapter is organized as follows. Section 1.3 explains and illustrates what I mean by "normative" models of causation and causal reasoning and suggests that, contrary to what is sometimes supposed, most philosophical "theories" of causation have an important normative component. Section 1.4 discusses the notion of a descriptive model of causation, Section 1.5 expands on the notion of a functional account of causation, and Sections 1.6–1.7 discuss "rational" models of cognition. Section 1.8 sketches my views about the role of intuitions or judgments about cases in accounts of causation. Sections 1.9–1.11 discuss the role of cognitive goals in a functional account of causation and the role of "pragmatics." An appendix describes some of the commitments of a functional account of causation in more detail.

1.3. The Normative

In addition to philosophical work on causation, there is a burgeoning literature coming out of computer science, machine learning, statistics, and econometrics that is more computational in nature, and that focuses on problems of causal inference and learning, typically with substantial use of some formal apparatus such as Bayes nets and graphical models, structural equations (e.g., Pearl 2000; Spirtes et al. 1993/2000), or hierarchical Bayesian models (e.g., Griffiths and Tenenbaum 2009). Other examples of such work include the potential response

mathematical sciences and often usefully formulated in terms of necessary and sufficient conditions; by contrast accounts of concepts$_p$ in terms of necessary and sufficient conditions are in most cases very implausible.

The various causal concepts discussed in Woodward 2003 and in parts of Chapter 2 should be understood as proposals about concepts$_n$. Unlike the use of concepts$_p$ to interpret empirical psychological research, I see nothing problematic in principle about the use of concepts$_n$ in formulating normative proposals about causal reasoning. When I speak of conceptual engineering in what follows, it is of course concepts$_n$ that I have in mind.

[10] See Woodward 2015a.

framework developed by Rubin and others (e.g., Rubin 1974) and now considerably elaborated and widely used in econometrics (Angrist and Pischke 2009), as well as proposals about inferences about causal direction based on machine-learning ideas, such as those due to Bernhard Schollkopf and his collaborators (e.g., Janzing et al. 2012). Much of this work from outside of philosophy is explicitly "normative" in the sense that it attempts to describe how we *ought* to learn, reason, and judge about causal relationships. For example, it makes proposals about which inferences to causal conclusions are *justified* or involve *correct* reasoning—that is, about which inferences lead reliably to the achievement of some epistemic goal associated with causal thinking, including identification of which causal claims are true (given some account of what truth for such claims consists in). It also includes proposals about distinctions among causal relationships that we ought to make or recognize, as in Pearl's (2001) previously mentioned distinctions among various notions of indirect effect.

It is perhaps less common to think of the various accounts of causation and causal reasoning found in the philosophical literature as also normative in aspiration, but in fact it seems to me that many are (or can be fruitfully viewed in this way), although they may have other goals as well. In particular, many if not most such philosophical accounts may be viewed as recommendations for how we should think about causation or which causal concepts we should employ and, since these recommendations in turn entail particular causal judgments, also as normative proposals about the causal judgments we ought to make.[11]

As an illustration, consider the status of "double prevention" relations. These are cases in which *c* prevents the occurrence of *d* which, if it had occurred, would have prevented the occurrence of *e*, with the result that *e* occurs. The occurrence of *e* depends on the occurrence of *c*, in the sense that if *c* occurs, *e* does and if *c* had not occurred, *e* would not have occurred. For example (Hall 2004), Billy shoots down an enemy plane (*c*) that, had it not been shot down, would have shot down Suzy's plane (*d*), preventing Suzy from carrying out a bombing (*e*). Some accounts of causation require a "connecting process" between cause and effect (e.g., Dowe 2000). Since no such process is present in standard cases of double prevention, including the Billy and Suzy case, these accounts judge that the relation between *c* and *e* above is not causal. Such accounts can be viewed as normative proposals about how we should conceptualize the notion of causation and which causal judgments are correct or warranted when double-prevention relations are present. We can then ask what the normative justifications for these proposals might be and whether they are compelling. In other words we ask

[11] This may seem to raise an obvious worry about triviality—the worry is that every concept is trivially normative in the sense that it is correctly applied to some instances and not others. My reason for thinking that "cause" is different from, say, "dog" is that use of the former is connected to nontrivial issues about function that will emerge subsequently.

whether we *should*, given our goals and purposes in making causal judgments, distinguish double-prevention relations from causal relationships (or, perhaps alternatively, make distinctions among causal relationships, according to which double-prevention relations differ from other kinds of such relationships). Implicit in this way of viewing matters is the idea that we have choices about which causal concepts to adopt (and about what commitments should be carried by such concepts) and that these choices can be evaluated in terms of how well they serve our goals and purposes.[12] We may contrast this view of the matter with the common philosophical practice of thinking of the issue as simply one of whether double-prevention relations are "really" causal or not, where this is to be settled by, say, examining "our concept" of causation to see whether it applies to double-prevention relations.

As another illustration, David Lewis's well-known counterfactual theory of causation (Lewis 1973a) and the similarity measure characterizing closeness of possible worlds on which it relies may be regarded as (among other things) a normative proposal according to which causal claims should be judged as true or false depending on whether they are related to true counterfactuals in the way that Lewis's theory prescribes. According to this theory, moreover, the truth or falsity of counterfactuals should be determined by reference to the particular criteria for judging similarity among possible worlds that Lewis endorses.

Finally, the "interventionist" account of causation that I favor is also normative, both in the sense that it makes recommendations about which causal claims should be judged true or false, and in other respects as well—for example, as I explain both in what follows and elsewhere (e.g., Woodward 2015a, 2015b), the interventionist account imposes restrictions on the sorts of variables that can figure in well-posed causal claims, embodies ideas about which variables it is appropriate to "control for" in assessing "multi-level" causal claims (it implies that you should not control for supervenience bases—see Woodward 2015b), and describes what sort of evidence is relevant to assessing such claims. The distinctions among causal claims (with respect to features like invariance and proportionality) that I have explored in more recent work (e.g., Woodward 2010) and which are discussed in more detail in Chapters 5–8 are also intended to be taken normatively—I claim these are distinctions that it is appropriate or rational to make, given goals associated with causal cognition.

[12] Some readers may object to the idea that we have choices about which causal concepts to employ on the grounds that any such concepts that are different from the causal concepts that we currently employ will for that reason not qualify as causal concepts. If you are tempted by this line of thought, substitute the following: we face a choice between employing current ways of thinking about causation and an alternative set of concepts and distinctions—call these "smasual" concepts and stipulate that they are not causal concepts. We still face the question of whether we would be better off talking about smausality. Similarly if you don't like the idea that causal concepts can change—think of this as causal concepts being replaced with smasual concepts.

In my view, many philosophers working on causation tend to efface the normative dimensions of what they are doing when they describe their projects as providing an account of what causation "is" or as providing a "metaphysics" of causation or, alternatively, as describing "our concept" of causation. These characterizations make it sound as if their projects are merely reportorial or descriptive (of "causal reality" or the metaphysics of causation or of a causal concept we possess) and have no normative content (beyond such description). But of course there are many different ways of correctly describing the world. (There is nothing incorrect, e.g., about just reporting correlations). We thus face the question of why we should think causally at all and why, in doing so, we should employ some particular way of thinking about causation rather than any one of a number of possible alternatives. In my view, answers to these questions will inevitably have a normative dimension and will require reference to our goals and purposes.

Describing these accounts as normative of course raises the question of where their normativity comes from—what is the source of the "oughts" they embody? My answer has been foreshadowed in my earlier remarks (and I will say more about it in Sections 1.6–1.8 and Chapter 3), but in order to forestall possible puzzlement and confusion, the short answer is that for the purposes of this book I will treat this as a matter of means/ends justification, with the normative imperatives being hypothetical rather than categorical.[13] Inquirers have certain ends or goals, and various proposals are assessed normatively in terms of how well they contribute to these goals. When the methodological proposal in question concerns causal reasoning, we assess the proposal in terms of goals associated with the function of causal thinking. At a high level of abstraction these goals might include, for example, successful prediction, including predicting the results of interventions, but there are many other possibilities, some of which are described subsequently.

1.4. The Descriptive

Alongside the normative literature referenced in the previous section, there is also a very rich and rapidly growing body of research that is more descriptive in aspiration: it purports to describe, as a matter of empirical fact, how various populations (adult humans, human children of various ages, nonhuman animals) learn and (where appropriate) reason about and judge with respect to causal relationships. A great deal of this work is conducted by psychologists,

[13] I emphasize that I do *not* claim that there are no principled grounds for choosing among ends or goals—it is just that I do not discuss these in this book.

but important portions can also be found in the literature on animal learning, comparative animal cognition, and anthropology. Moreover, although many philosophers may prefer not to think about what they are doing in this way, the philosophical literature on causation is also full of what look, at least on first inspection, like empirical descriptive claims. This is perhaps most obvious when the philosopher advances claims about what "people" (a population that may not be very clearly specified) "would say" or how they would think or judge regarding the truth of various causal claims. A similar point holds for philosophers who report that they find some claims and not others "intuitive," where there is usually some accompanying assumption or expectation that these judgments are fairly widely shared, which again involves empirical assumptions.

As a more specific illustration of an apparently empirical claim (connected to the earlier remarks about double-prevention relations), one finds some philosophers contending that ordinary folk do not judge relationships between two candidate events to be causal (at least strictly and properly speaking) when there is counterfactual dependence but no "connecting process" linking the two events. By contrast, other philosophers claim that people judge at least some such scenarios to be causal (compare Dowe 2000 with Schaffer 2000). Similarly, Hall (2004) contrasts certain cases involving double prevention with other cases involving "productive" causes (these include but perhaps are not limited to cases involving transference of energy and momentum, as when a rock shatters a bottle), and claims (or at least strongly implies) that people have different reactions to these two kinds of cases qua causal claims—they will view the latter as possessing features (that are relevant to how we think about causation) that the former lack and perhaps as more paradigmatically causal. In fact, as we shall see, something in the neighborhood of Hall's empirical claim seems correct, in the sense that some double-prevention cases receive lower "causal strength" ratings than corresponding cases of productive or generative causation. However, it is also the case (see Chapters 6 and 7) that people distinguish *among* double-prevention claims in interesting ways—an observation that does not figure in Hall's discussion but which (I will claim) is relevant to our understanding of the bases on which people make such judgments.

In other cases, philosophers make what look like empirical claims about people's uncertainty regarding certain causal judgments or about the *absence* of agreement or consensus regarding such judgments. Consider cases of symmetric overdetermination involving individual events (both c_1 and c_2 are present; e occurs; if c_1 (c_2) had not been present, c_2 (c_1) would have caused e). Lewis (1986) claims that in such cases people are uncertain about which of the following three alternative judgments is correct: (i) c_1 causes e and c_2 causes e, (ii) neither c_1 nor c_2 cause e, (iii) c_1 and c_2 together cause e but not individually, as claimed in (i). Lewis uses this empirical claim about absence of agreement to help to motivate his own treatment of the case, which favors (iii) as the correct judgment.[14] In fact, however, a number of empirical studies show that the majority of subjects do not report uncertainty about which of (i)–(iii) is correct and that their judgments favor alternative (i). Of course this empirical observation does not by itself show that (i) is the normatively correct judgment. However, as we shall see (Chapter 5), there is a plausible normative theory that does yield this result.

Philosophers also make what appear to be empirical claims about the *basis* on which people make causal judgments, that is, about the factors that influence such judgments (and not just about the judgments themselves).[15] For example, in the course of a discussion of the direction of causation and non-backtracking interpretations of counterfactuals, Lewis (1979) advances various claims about the information on which people rely in assessing counterfactuals and the direction of causation—in particular he claims that "everyone believes" that "everything that happens leaves many and varied traces" (1979, 66), so that it would take a "big miracle or many and varied small miracles . . . to eradicate those traces." Moreover he contends that people evaluate counterfactuals and causal claims on the basis of this belief, that as an empirical matter the belief is true, and that reliance on this belief leads people to make *correct* judgments about causal direction.[16] Again these are testable empirical claims.

In addition to the examples just described, there are many empirical claims that, although perhaps not explicitly made by philosophical theories of causation, are connected to them through looser considerations of motivation and plausibility. Although it is not logically inconsistent to advance an account of what causation "is" or how one ought to reason causally and to deny that this

[14] Lewis argues that because of this uncertainty it is not problematic that his account yields the result that (iii) is the correct judgment—he claims that his account is not inconsistent with ordinary judgment but instead resolves people's uncertainty in a particular way. Of course, as commentators (e.g., Paul and Hall 2013) have noted, if (contrary to actual fact) it was the case that ordinary judgment is uncertain among (i)–(iii), then, in favoring (iii), Lewis's account would be inconsistent with ordinary judgment.

[15] For more on this distinction see Chapter 3. Obviously it is one thing (i) to claim that people make certain judgments and not others and a very different matter (ii) to advance claims about the factors that influence such judgments. In my view, even if appeals to intuition can sometimes be a useful source of information about (i), they are often not a reliable source of information about (ii).

[16] To the best of my knowledge there has been little empirical investigation of the information people rely on in making judgments of causal direction. Lewis's claim about the existence of an asymmetry of traces seems false as an unqualified empirical claim, for reasons described by Elga (2001), so that, to the extent that people succeed in making correct judgments about causal direction, they apparently don't do so on the basis described by Lewis. Recent work in machine learning (e.g., Janzing et al. 2012, discussed in Woodward 2019 and Woodward, forthcoming b) describes information and strategies/heuristics that provide what is arguably a normatively defensible basis for making judgments of causal direction. Such strategies have been shown to yield correct judgments at rates well above chance in cases in which causal direction is independently known. I am unaware of any empirical investigations of whether people commonly employ such strategies, but given the general framework advocated in this book about the relation between the normative and the empirical, it would not be surprising if they do. (Hint to psychologists.)

has any implications at all for how people in fact reason, learn, and judge about causation, most philosophical accounts of causation do not take this form and would strike many of us as arbitrary and unmotivated if they did. For example, it seems clear that Lewis thinks that his theory has something to do with how people in fact think about causation and that in particular that he supposes that in their thinking about causation, people connect causation and counterfactuals in a fairly close way. It is (let us suppose) logically possible that (i) Lewis's theory might be correct when understood as an account of what causation is or how we ought to think about it, even though, (ii) as an empirical matter, people rarely use counterfactuals, tend to get confused when they use them, cannot agree on the truth-status of most counterfactuals, and fail to connect causal claims and counterfactuals in any way. Nonetheless this combination of (i) and (ii) would be a puzzling state of affairs and certainly does not seem to be what Lewis expected to be true. At the very least some explanation would need to be forthcoming of how (i) and (ii) could both be true at once and of why we should accept (i) even though (ii) is the case.[17] This is what I mean when I say that Lewis's theory would seem unmotivated if certain supporting empirical facts were not true, even if the denial of the theory is logically consistent with those facts not holding.

In fact some psychologists (see Section 2.9) have contended that even adult humans do not think about or represent causal claims by means of counterfactuals (at least usually and readily). It also has been argued that this is at least true for young children, who, it is claimed, use and understand causal notions but have great difficulty using counterfactuals (see, e.g., Beck et al. 2011). Relatedly, there are also philosophers (e.g., Bogen 2004) who reject counterfactual accounts of causation in favor of mechanistic or causal process accounts (see Chapter 2) and who also contend that is a matter of fact people do not associate causal claims with counterfactuals in the way that counterfactual theories of causation suggest. My own reading of the empirical literature is that these last contentions are mistaken and that, in fact, both adults and relatively young children very readily connect causal claims and counterfactuals (see Chapter 4) and reason with counterfactuals in assessing causal claims. Thus, broadly speaking, Lewis's expectations about this are correct. However, I also think, as suggested previously, that if instead the empirical claims about people's failures to connect causal claims and counterfactuals were correct, advocates of normative versions of counterfactual theories of causation would have considerable explaining to do.

Let me also emphasize, for future reference, that although empirical results about causal cognition sometimes seem to be *relevant* to normative theorizing in, for example, the ways I have described, it seems misleading (as suggested in the introduction) to describe those results as straightforwardly *evidence* (still less

[17] Recall the arguments in the introduction about the implausibility of this sort of mismatch.

conclusive evidence) for the correctness of some associated normative theory; instead the relationship between the empirical results and the normative theory is looser and more indirect than that. This is so in part simply because there is no direct inference, without additional assumptions, from "is" to "ought" (again, see the introduction). So I do not mean to claim, by the preceding remarks, that we can somehow straightforwardly infer from the observation that people connect causal claims and counterfactuals to the conclusion that some counterfactual theory of causation is normatively correct or even that such empirical results are "evidence" for the correctness of the normative theory. I will return to this issue of the bearing of empirical results on normative theory in subsequent chapters, but we should recognize that there are various ways that empirical results can be relevant to or bear on normative theory (and conversely) without being, at least in any straightforward way, being evidence for the latter or premises from which the latter can be inferred.

As another illustration of these points about possible connections between empirical claims and normative thinking, consider Hume's account of learning about causal relationships in terms of learning regularities, relationships of spatial and temporal contiguity, and expectations associated with these. Again one might think of this purely as an account of what causation is or how, as a normative matter, we ought to think about it. Nonetheless, I submit it would be surprising and puzzling if Hume's account was correct when construed as an account of what causation is but people did not learn (or had great difficulty learning about) causal relationships from regular associations or if their causal judgments did not track the information they received about regular associations or spatial and temporal cues relatively closely. Certainly this would be contrary to what Hume seemed to assume, as an empirical matter, about human causal learning and reasoning. In fact, the empirical evidence seems to show, on the one hand, that human causal judgment (and in the case of animals, behavioral analogues of such judgments) are strongly influenced by information about regularities and spatiotemporal relations but, on the other hand, that such causal judgments seem to have additional structure that goes beyond what is present in regularity information[18]—see Chapter 4.

Ranging a bit further afield, on an Humean account of (iii) it would also be natural to expect a considerable amount of continuity between human causal learning and judgment and what is described by "associative" models of learning

[18] I emphasize that I mean simply that as an empirical matter people represent causal relationships in ways that have additional structure beyond regularity information and that the regularity information that is accessible to them underdetermines the representations they learn. Moreover, reasoning in accord with this additional structure is often successful—see subsequent chapters. Whether at some fundamental level all that exists are particular matters of fact and regularities and spatiotemporal relations involving these is a different matter.

and conditioning in the animal-learning literature since, after all, nonhuman animals are certainly capable of learning associations and forming expectations on the basis of these, as well as making use of information about spatial and temporal contiguity. In fact, a prominent approach to human causal cognition (see Shanks and Dickinson 1988 and various others) is built around this idea, relying on a well-known model of associative learning (the Rescorla-Wagner model—see Chapter 4) and subsequent extensions. One can think of such associative models as one way of implementing the idea that causal learning and representation in both many animals and humans is "Humean." Again the predictions of such associative models about how humans and other subjects learn and judge about causal relationships are testable. If, as I will suggest is the case (again see Chapter 4), these models face some empirical problems, this can be an important source of clues about the respect in which human causal cognition may not be purely "Humean" (or, if it is Humean, this will need to be in some more subtle way).

1.5. Functional Approaches to Causation

I claimed in the previous section that normative (including philosophical accounts) of causation often have empirical presuppositions or at least seem unmotivated if those presuppositions do not hold.[19] Nonetheless there are many questions about how, if at all, these normative and descriptive/empirical considerations might be related to one another. Without attempting to offer a complete answer to this question (as I have already suggested, the possible connections are numerous and complex, with several being explored in more detail in Chapter 3), I want in this section to say more about the functional approach I favor.[20] This, as intimated earlier, provides one natural framework for thinking about these normative/empirical connections.

By a functional approach to causation, I mean an approach that takes as its point of departure the idea that causal cognition or thinking in causal terms is at least sometimes useful or functional in the sense of successfully serving various goals and purposes that we human beings (and perhaps other creatures) have.[21]

[19] In addition to those explicitly discussed in what follows, a number of philosophers and psychologists have adopted approaches to causation that are broadly functional in some respects. See, e.g., Hitchcock 2012. For a functional approach to explanation by a psychologist see Lombrozo 2011. Of course these writers don't necessarily endorse everything that I have in mind by a functional approach.

[20] For more on this sort of approach, see Woodward 2014.

[21] I sometimes put this in terms of the functions of causal concepts, inference strategies, etc., but if you don't think that concepts, etc., can have functions, just translate this into claims about the goals of users.

It then proceeds by trying to understand what these goals are, and attempts to relate various features of causal reasoning—the causal concepts we employ, the learning strategies we use, the different kinds of causal judgments we make, and so on—to the achievement of those goals. Doing this can be illuminating in a number of different ways. It can help to explain why we care about or value the discovery of causal relationships as distinct from other sorts of relationships—for example, non-causal correlations. It can provide a motivation or justification for distinguishing among different sorts of causal relationships or causal concepts, since these may conduce to or be rationally related to somewhat different goals. For example, within an interventionist framework, distinctions among different sorts of causal judgments are related to more specific goals that fall under the general rubric of manipulation and control: the distinctions (e.g., between type-level and actual cause judgments) provide information relevant to different manipulation-related questions. Moreover, to the extent that some form of causal reasoning or judgment is successful in allowing us to achieve goals associated with causal cognition, a functional analysis can help to explain this success—it can give us insight into why various practices around causal cognition work (to the extent that they do) in achieving our goals and perhaps also an understanding of when and why they fail to work in achieving our goals. In this case a functional analysis may also suggest alternative practices that are better suited to achieving our goals.

So far I have characterized a functional approach to causation as having to do with connecting various features of causal cognition to goals that we have and with assessing whether and how those features contribute to those goals. In some circumstances, it may be reasonable to take the additional step of concluding, not just that various features of causal cognition contribute to goals we have, but that they are there *because* they contribute to those goals. In my view, this additional step requires that it be true that some process is at work that selects, from among various possible forms of reasoning, those that best contribute to goals that we have. This selective process might involve, among other possibilities, natural selection and/or learning with feedback, with the latter involving both individual development and the acquisition of cultural knowledge over a more extended period of time. If it is possible to show this, then we will have a functional or selective *explanation* for the presence of these features in causal thinking, in the sense described by Wright (1976). However, in my view functional analyses can be useful and illuminating in virtue of providing the kind of information described in the previous paragraph even if they don't carry this sort of strong commitment about functional explanation. That is, we can hold that features F contribute to goal G and that the presence of F explains success in attaining G without also claiming that F is present because it contributes to G. Explaining success in this way is a familiar feature of theorizing in disciplines like psychology and neurobiology. For example, the success of the human visual system in extracting relatively reliable information about aspects of the external environment is explained in terms of various computations carried out by that system on input information and the reliability characteristics of those computations.

In general the functional approach involves thinking about different forms of causal reasoning, distinctions we may make among different sorts of causal claims or concepts, different strategies for causal learning, and so on, from the perspective of an epistemic or conceptual engineer or designer,[22] focusing on what various aspects of causal thinking are or might be used for, and employing this as our point of entry into the topic. Causal thinking is viewed as (in part) a tool or technology for achieving certain goals. This orientation thus puts the normative dimension of causal thinking and judgment at the center of our inquiry. As we shall see, it is also an orientation that connects very naturally to the descriptive work on causal cognition we will be discussing. Very briefly this is because, in addition to its making normative sense to evaluate features of causal cognition in terms of whether these conduce to various goals, it is also true to some considerable extent, as an empirical matter, that these features do in fact conduce to the goals in question.

As a simple illustration, consider the distinction between "mere correlation" and causation. Virtually all accounts of causation agree that there is such a contrast, although they provide different accounts of it. A functional approach to causation asks, among other questions, why we employ a notion (or, more accurately, notions) of causation that incorporate such a contrast (i.e., what work the contrast does for us) and what the contrast consists in. The (or at least an) answer provided by the interventionist framework is that we value the discovery of relationships that are potentially exploitable for purposes of manipulation and control; not all correlational relationships are so exploitable, so we find it worthwhile to distinguish those that are from those that are not, labeling the former as causal and the latter as merely correlational. I acknowledge that in this particular

[22] A number of recent books and papers discuss various versions of conceptual engineering. Examples include Burgess and Plunkett (2013) on "conceptual ethics," Haslanger (2002), and Cappelen (2012). These philosophers stress the normative character of many philosophical concepts and (in some cases) the possibility of revising them. However, several of them have views of how to go about conceptual engineering that are rather different from the approach I advocate. For example, Cappelen denies that thinking about conceptual revision in terms of the functions of various concepts is appropriate, while I think that this is crucial. Haslanger's view on this matter is closer to my own. I should also add, though (what may be obvious), that although I think it important to understand various features of causal cognition in functional terms, in this book, I don't propose any wholesale engineering or re-engineering that discards present ways of thinking about causation in favor of some completely different alternative, so in this sense I am not doing revisionary conceptual engineering of the sort that some philosophers have in mind. Also the functional perspective on causation that I favor extends to far more than just causal concepts. It covers many other aspects of causal cognition, and to that extent involves more than just an engineering perspective on concepts.

case the functional question and the interventionist answer may not seem very deep; but in other cases this sort of approach can be quite illuminating, as I hope to illustrate in what follows.

One way of appreciating what is distinctive about the functional approach is to consider what it contrasts with. We noted previously that a standard approach to many philosophical problems, prominent in analytical philosophy, is to present examples and scenarios involving some target concept X (e.g., causation, justice, knowledge) and to then form (and encourage readers to form) intuitions or judgments about these. The philosopher then attempts to construct a theory of X (or perhaps a theory of our concept of X) that systematizes these judgments, perhaps also assuming other constraints (e.g., of a more metaphysical nature). Many philosophical discussions of causation adopt some form of this methodology. In my view, there are many reasons to be dissatisfied with this approach, but one basic problem is that, as usually practiced, it does not provide answers to functional questions and neglects the role that functional considerations can play in constraining theories of X. That is, even if an otherwise satisfactory systemization of our intuitive judgments about causation could be produced via the procedure described (something I consider unlikely, for reasons that will emerge later), this would not by itself tell us what causal concepts and our practices of causal cognition are *for*. Among other limitations, the reconstruction of intuitions described earlier would not tell us why we have this particular causal concept or concepts and not others or whether the concepts that we do have are well suited to our purposes. Moreover, by neglecting functional considerations, important constraints, in addition to those that might be provided by intuitive judgment, are ignored. The functional focus brings these back into the picture. In this, and, in a number of other respects, it moves us away from projects that just focus on the reconstruction of intuitions. I will add that in addition to the questions just described, there are many other important questions about causation and causal reasoning that cannot possibly be answered by an intuition-driven methodology even if (contrary to what I think is the case) it provided us with reliable information about the nature of causation (or the nature of our concept of cause).

As another illustration of what is distinctive about the functional approach, consider the contrast mentioned earlier between causal claims involving production and those involving double-prevention relations. As we noted, it is plausible that many people recognize some such distinction in their intuitive judgments. But rather than just recording this fact, a functional approach asks additional questions. To the extent that we make such a distinction, what function might it serve? That is, is there some basis or rationale for the distinction, given goals that we associate with causal inquiry? As noted earlier, it turns out, as an empirical fact, that people not only distinguish between some productive and some double-prevention relations but that they also make distinctions *among* double-prevention relations, regarding some as better or more paradigmatic cases of causation (or as having greater "causal strength") than others. A functional approach will again ask whether there is some rational basis for such distinctions, where (to repeat) "rational" means "conducive to goals we have." I will suggest in Chapter 7 that the empirical evidence supports the conclusion that the distinctions people make in these cases track a distinction between those causal relationships that are more and those that are less invariant. That is (roughly), productive causal relationships are often (but not always) more invariant (more stable under various sorts of changes) than double-prevention relationships and, among double-prevention relationships, those that are more invariant tend to be viewed as more paradigmatically causal. Moreover, this set of distinctions makes sense, given that people have good reason to value more invariant relationships in terms of goals associated with causal thinking.

Note that this treatment of double prevention makes use of information about people's patterns of judgment but assigns this information a rather different role than in more traditional philosophical analyses. Rather than thinking of people's judgments as sources of information that reflect some sort of direct intuitive grasping of the nature of causation (or as a source of information about "our concept" of causation), we instead think of such judgments (as well as information about the factors that influence these) as a possible source of information about shared practices of causal judgment. We then try to understand these practices in terms of their functions and how they relate to various goals that we have. Thus rather than trying to use intuitions to answer the question of whether double-prevention relations are "really" causal, we instead replace this question with the more functional questions described previously.[23] Again, because our analysis is functional, we do *not* assume that people's practices of causal judgment and inference are necessarily normatively appropriate in terms of goals associated with causal thinking. Instead further analysis is required to establish whether this is the case—that is, independent arguments that such practices are or are not effective in achieving our goals. As a consequence, we do not automatically take reports of intuitions as normatively justified or correct when understood as claims about how we ought to think or judge—taken by themselves, such reports are relevant (at best) to claims about how people in fact judge. I will say more about all this in Chapter 3.

[23] More precisely, we replace inquiries into the nature of causation or our concept of it with (i) descriptive accounts of how people and other subjects reason causally, and explanations (computational and causal) of such reasoning and (ii) normative accounts of how we ought to reason causally. Let me emphasize again, though, that this does not mean that there is no fact of the matter regarding which causal claims are true; the point is rather that we need to specify, via some normative account, what it is for a causal claim to be true or correct before we can assess the truth value of particular causal claims.

There are a number of other features that can be distinctively associated with a functional approach to causation, but rather than breaking up the flow of the discussion in this chapter I have relegated them to an appendix. The reader is encouraged to look at this to get more of a sense of what I take a functional approach to involve.

Turning now to some comparisons, other philosophers, including Price (e.g., 2011) and Ismael (2015), have emphasized the importance of a functional perspective on causation. However, my view differs from theirs in an important respect: these writers contrast a functional perspective with a more straightforwardly realistic one and think of adoption of a functional perspective as pushing us in the direction of non-realism about some features of causal claims. Roughly speaking, their idea is that focusing on the function of causal thinking should lead us to recognize that some of the content of causal claims—in particular their distinctively modal commitments—does not have to do with the representation of (or does not reflect information about) what is in the world. For example, according to Ismael (2015), what is in the world is just patterns or regularities involving occurrent events, presumably something like Lewis's "Humean mosaic" (Lewis 1986). The modal commitments of causal claims go beyond the representation of such patterns and involve an "inductive projection of those patterns onto the unknown and the purely hypothetical" (2015, 119). This seems to suggest that the difference between (i) *X* causes *Y* and (ii) "*X* is correlated with *Y* but does not cause it" amounts to or derives from a difference in our willingness to project certain patterns and not others, but where these don't track independently existing features in the world.

On my view, a functional perspective on causation need not (and should not) commit us to projectivism or anti-realism about causation. My alternative picture consists in what I earlier called a kind of minimal realism about causation (see the introduction). This holds that the difference between those relations that are merely correlational and those that are causal has its source "out there" in the world and is not the upshot of our projecting our inductive commitments onto the world. Modulo some important qualifications, causal claims (once their content is made explicit) are thus truth apt in the sense that they are straightforwardly true or false, depending on how matters stand in the world. Thus rather than appealing to our inductive practices to explain the difference between (i) causal and (ii) correlational claims, it is the "objective" differences between what (i) and (ii) claim about the world, along with goals associated with causal thinking, that should be used to explain why our inductive practices take the form they do in distinguishing between them. According to this picture, the world contains many different relationships. Some of these are merely correlational, but others are not—some relationships in the world are such that they are exploitable for manipulation and control in the way that interventionists describe, while other correlational relationships are not. The intervention-supporting relationships in turn differ along other dimensions—for example, some are more invariant than others. Of course, it is a fact about us and our goals and interests that we value the discovery of relationships that support manipulation and control, and thus find it worthwhile to distinguish relationships that have this feature from those that do not, but this is not to say that our goals (or epistemic attitudes) determine which are the intervention-supporting relationships. Instead this is determined by what nature is like. A similar point holds for the other features that causal relationships may possess, such as relative invariance. Our goals explain why we value the discovery of invariant relationships, but this does not mean that which invariant relationships the world contains or which relationships are more rather than less invariant (or whether the world contains any invariant relationships at all) is somehow *due* to us.

More generally, as emphasized in the introduction, successful causal learning and reasoning requires that certain relations be present in nature—I think it is hard to tell a coherent story about what *learning* and *success* consist in if this is not true. There seems, after all, to be a fact of the matter about whether an intervention that involves taking a certain drug will increase the chances of subjects in some population of recovering from cancer. We do experiments to *discover* such facts—our experiments and the inferences we draw from them do not create such facts, nor do our expectations about the outcomes of such experiments. This dependence on what the world is like is also reflected in the observation that, on the version of functionalism about causation I favor, it is entirely possible (cf. appendix to this chapter) that in some worldly circumstances, the empirical preconditions for *any* application of causal thinking and representation may not be satisfied—there may be features of nature that are not (and cannot be understood as) causal.

Although I will not discuss this in what follows, I will add that this "functional" approach can be (and in some cases has already been) fruitfully applied to other topics on which philosophers focus. Something like this approach has been adopted by some epistemologists (e.g., Craig 1990) in connection with claims about knowledge ascription: they propose to approach this topic by asking what the point or goal of knowledge ascriptions is or might be, what role these play in our thinking and interactions with one another, or what distinction(s) we mean to mark when we describe someone as knowing or not.[24] Having identified

[24] Cf. Craig: "We take some prima facie plausible hypothesis about what the concept of knowledge does for us, what its role in our life might be, and then ask what a concept having that role would be like, what conditions would govern its application . . . then see to what extent it matches our everyday practice with the concept of knowledge as actually found" (1990, 2). Craig's discussion in his opening pages is very effective in addressing the kind of skepticism that holds, prior to looking at the details, that such functional projects cannot possibly be illuminating. Let me add that one can agree with Craig about the fruitfulness of functional approaches without endorsing his specific proposal

some candidate goal or point, one can then ask about the extent to which our actual practices fit or contribute to this goal. This contrasts with the traditional approach in analytical epistemology of constructing various scenarios (Gettier cases and so on) and then consulting our intuitions about whether these are genuinely cases of knowledge. Similar functional projects might be pursued for "free choice," "morally responsible," and so on.[25] I see what follows as a sort of testing ground, in connection with one particular set of concepts, of the merits of this approach.[26]

1.6. Rational Models of Causal Cognition and the Importance of Explaining Success

I turn now to an empirical claim that I have been presupposing in much of the discussion in this chapter and its implications for the functional project.[27] This is that as an empirical matter, human causal cognition is comparatively successful in enabling us to get around in and cope with the world. It is of course true that people have many false causal beliefs and make many unjustified causal

about the function of knowledge claims, which in his view has to do with the identification of good informants.

[25] With respect to free will, something like this approach can be found in portions of Dennett 1984 and in Vargas 2013.

[26] Let me acknowledge the obvious point that is possible to trivialize functional approaches by adopting "unreasonable" or unilluminating claims about functions or goals. For example, if one holds a theory of causation according to which causes must resemble their effects, one might adopt an account according to which the function or goal of causal reasoning is to provide such resemblance information, thereby providing "support" for one's theory of causation. More generally, if your theory says causal claims must possess feature R, you might try to argue for your theory using a functional strategy by claiming that the function of causal thinking is to provide information about R. My response: (i) We are looking for claims about the function of causal reasoning that allow us to capture and evaluate important features of that reasoning; "resemblance" clearly fails this requirement, in the sense that even the most enthusiastic advocate of a resemblance theory will need to concede that there are many features of causal reasoning that her theory does not capture. (ii) Relatedly, we want to identify a function that has nontrivial, non-obvious implications regarding (i). I claim that interventionism identifies such a function, "resemblance" does not. (iii) The burden should not be on the defender of a functionalist approach to explain why one should not adopt an interpretation of that approach that trivializes it; rather the burden falls on the trivializer to justify adopting her trivializing proposal. This is why, for example, we should not adopt proposals like "The function of causal thinking is to describe the causal facts."

[27] The claim that normative theorizing about causal cognition is best approached from a functional perspective is, as nearly as I can see, independent of any claims about whether this or that specific feature of causal cognition is in fact successful in the sense of conducing to our goals. However, it would make little sense to view the normative aspects of a functional account as complementing descriptive investigations in the way that I envision if as, as a matter of fact, human causal cognition was not successful in many respects and was instead systematically unsuccessful and dysfunctional. In that case the role of normative theory would primarily be to highlight the errors to which we are subject as a descriptive matter and to motivate adoption of a theory that is very different from how we currently think.

inferences. Nonetheless looking at the matter in comparative perspective, one sees (cf. Chapter 4) that humans are much more successful at causal learning and reasoning than other animals, including other primates. (What such "success" involves is discussed and illustrated in more detail subsequently, but for present purposes we need not interpret it along purely interventionist lines. Think of "success" in this context as involving discovering relationships that would be regarded as causal on virtually any theory of causation.) Indeed, even very young human children are better in some important respects at causal cognition than other primates. To the extent this is so, it becomes an important project to explain how we are able to learn and reason so successfully about causal relationships. It seems plausible that this project of explaining success is something that can *only* be addressed by a combination of normative theorizing and descriptive results, giving us another point of interaction between these two enterprises. Moreover, as we shall also see, once this is recognized as an explanatory goal, it shapes or constrains both enterprises in important ways.

In this connection an analogy with "rational" or "ideal observer" theories of visual perception is highly suggestive.[28] The human visual system contains mechanisms and carries out computations that produce visual judgments and other outputs (e.g., appropriate nonverbal behavior) in response to input from the environment. But to think of the visual system *only* in this way is to miss one of its most important features—that although subject to errors of various sorts, it produces outputs that are reliable enough to enable us to get around in the world successfully and that are informative about portions of our local surroundings.[29] In other words, thinking of the visual system only in terms of computations that produce certain outputs neglects its function—what it used for. Any adequate descriptive theory of the visual system needs to explain these facts having to do with success and functionality. That is, it needs to explain how, from the very limited information that impinges on the retina, the brain is able to reach conclusions that are veridical enough for many practical purposes about the three-dimensional world of medium-sized objects that lies around us. Explaining this is a matter of devising theories that specify representations and computations that show how it is possible to, for example, derive results about object segmentation from information available to the visual system concerning shading, edges, and so on, where these results, moreover, are in some important respects accurate enough for the organism's purposes. To the extent that

[28] I first heard about this analogy from Alison Gopnik.

[29] This claim is not meant to commit to any very specific conception of how to understand the character of this output. For example, it is consistent with the idea that the goal of the visual system is not to produce a representation that has the character of highly accurate photograph—this may not be what the whole organism needs or can make use of. My point is simply that however the output may be understood in detail, it is reliably informative about ecologically relevant aspects of the visual environment.

the visual system can be shown empirically to operate in accord with principles that we know are reliable in the sense of often issuing in an accurate-enough reconstruction of the visual scene before us, we have at least a partial explanation of why (or how it is possible that) the visual system "succeeds." Obviously, to carry out this project we need, among other things, to think of the visual system and the principles by which it operates in broadly functional terms—the (or a) function or "goal" of the system's operation is to produce outputs that contain accurate-enough information about aspects of the surrounding environment to guide successful behavior.

In the case of the visual system a standard way of approaching the problem (of explaining why its operation is successful to the extent that it is) is in terms of an "ideal observer" or an "optimality" (or "rational") analysis. For our purposes we may think of this as involving specifying (i) a "task" carried out by the visual system or a goal that it has, (ii) the inputs or stimuli from the environment that are available and relevant for accomplishing this task/goal, and (iii) other constraints to which the system is subject. One then proceeds by deriving (in the case of visual processing this is usually or mainly a mathematical exercise) what an "optimal" solution to this task would look like, given (i)–(iii) along with a set of computations that will achieve this solution or something close to it.

Proceeding in this way can be useful (illuminating) for a number of different reasons that are worth distinguishing. (This is not to say the possibilities about to be described *are* clearly distinguished, particularly in discussions of cognitive processing—the distinctions that follow are mine.) First, if the behavior is close to optimal and there is evidence that the envisioned computations are actually being performed, this can explain how the organism is able to achieve successful performance or something close to it—thus answering the question about success previously posed. Second, the analysis can also be very useful in forcing one to (i) clarify what the task or goal is (of course this raises the question of whether the organism actually has the specified goal rather than some other goal) and to (ii) clarify what information in the environment/stimulus is required or relevant to achieving that goal (and whether that information is actually accessible to the organism). As we shall see, (i) can be particularly important in the interpretation of causal inference tasks in which the output is verbal judgments, since subjects may interpret the tasks they are given in ways that are different from what the experimenter intends—that is, they may be successfully solving a different task, associated with a different goal, from the one the experimenter has in mind. Moreover, on a descriptive level, even when the output is well short of optimal, the "ideal observer" can provide a "benchmark against which to compare the performance of . . . real systems" (Geisler 2011, 772). For example, if the organism is using the same environmental stimuli as the ideal observer and some of the same computations but making use of these in a suboptimal way, one might expect that its performance should "parallel" that of the ideal observer so that, for example, stimuli that make the task harder for the ideal observer should also be harder for the real system. Relatedly, if the system is employing some heuristic that imperfectly tracks the optimal solution, such an analysis may give us some insight into why the system works as well as it does. Moreover, given a clear understanding of what success consists in, one can also diagnose failures and mistakes. In particular, one's theory of how success is achieved may also afford some insight into why and when such mistakes occur, which can be important (among other reasons) if one wants to improve performance.

In recent years, beginning with the pioneering work of Anderson (1990), a large number of rational analyses of cognitive (and not just perceptual) processes have been developed that are similar in spirit to what has been described here.[30] Such analyses have been applied to phenomena such as categorization, the organization and behavior of memory, and many sorts of reasoning tasks (such as reasoning with conditionals). Oaksford and Chater, two prominent advocates of this approach, write: "This research has shown that a great many empirical generalizations about cognition can be viewed as arising from the rational adaptation of cognitive system to the problems and constraints that it faces" (2000, 32). This "rational model" approach has also increasingly been adopted in empirical research by psychologists on causal cognition—indeed most, although not all, of the psychological research described in this book adopts some version of this approach.[31] In some cases the rational model employed is spelled out in considerable mathematical detail and is accompanied by claims that human performance is optimal or nearly optimal relative to this model or at least that the model comes closer to fitting performance than alternative (especially nonrational) models, where a nonrational model is one that lacks any relevant normative rationale. Examples discussed (or at least mentioned) subsequently include Cheng's power PC model of causal strength judgment, Tenenbaum and Griffiths's causal support model (and more generally the use of Bayesian models of causal learning), and the use of constraint-based theories of causal inference of the sort described in Spirtes et al. (1993/2000) to model human causal learning, as in Gopnik et al. (2004). In other cases, the rational model (and the associated normative theory) is less formal—examples include results about the connections between subjects' causal judgments and

[30] In addition to the considerations just described, Anderson also emphasizes the way in which a preference for rational models can cut down sharply on the search space of models considered and thus help with "underdetermination" problems. Note, however, that this consideration requires that the correct model be among the rational models.

[31] As paper titles such as "Causal Learning and Inference as a Rational Process: The New Synthesis" (Holyoak and Cheng 2011) indicate. Other examples of this sort of work include Gopnik et al. 2001 and Sobel et al. 2004.

the interventionist counterfactuals they endorse (they tend to endorse the normatively correct ones)—and may be much more loosely specified, as in, for example, Lombrozo's results about the relationship between causal strength judgments and invariance, discussed in Chapter 7.

The models described previously are optimality models that pay little attention to the fact that subjects operate under resource constraints of various sorts—they have limited computational power, memory, and so on.[32] Recently "resource-constrained optimality models" (Lieder and Griffiths 2018) have become popular, in connection with both causal cognition and many other kinds of psychological activities. These model subjects as "rational" and as attempting to optimize in some way, but as doing so under various sorts of constraints, so that they adopt strategies and heuristics that perform well or even optimally given the constraints. These models have the advantage that they can sometimes show apparently "irrational" judgment and behavior to be rational and normatively appropriate, given the constraints people face.

Finally, although there seem to be few examples of this in the recent literature on causal cognition, there is also the possibility of models built around some notion of satisficing, in the sense of Simon (1957). Here one might begin with some notion of "good enough" (even if not optimal) performance, given the organism's goals and then show that the organism is following strategies that meet this good-enough standard and are "rational" in this sense.

I will understand "rational model" in what follows to cover all of the possibilities described previously so that this category includes models with an explicit formal structure and an accompanying optimality analysis, models that involve resource constrained rationality, and non-formal models that nonetheless see causal cognition as well adapted to various goals.

As remarked, rational analyses have been carried out in connection with tasks involving learning causal relationships from various sorts of data and in connection with judgment tasks involving ratings of causal strength, among others. I will be suggesting a similar sort of analysis for other aspects of causal cognition—for example, for tasks involving distinctions among causal concepts, and tasks that involve sensitivity to more specific features of causal relationships like proportionality and invariance. In other words, I will argue that people often connect causal claims and invariance judgments and causal claims and proportionality and make distinctions among causal claims that can be understood as rational.

[32] Of course, in another, more indirect sense these models are "resource-constrained" since they are built around the assumption that subjects have access only to certain kinds of input and not others.

1.7. The "But People Are Stupid" Objection

My emphasis on the rationality or adaptiveness of much human causal thinking may strike some readers as fundamentally misguided. Haven't psychologists like Kahneman and Tversky (and many others) shown that people make all sorts of elementary errors in reasoning? Doesn't ordinary observation show that people believe many mistaken and unfounded causal claims?

There are a number of complex issues here that would require a very detailed discussion to address satisfactorily. Here I will only respond in a rather cursory way. First, and most fundamentally, whether a given population performs well or badly on a given causal reasoning task is a question that can only be settled by detailed empirical investigation, not by seat-of-the-pants hunches or anecdotes about Trump supporters. As we shall see, available empirical evidence, some of which is described in what follows, does support the claim that many people are successful at a range of causal reasoning tasks. That they (or some) are unsuccessful at others does not undercut the evidence for success when it occurs, although it does of course pose the question of why they succeed at some tasks and not others and why some but not others succeed at those tasks. Second, at least in some cases, experimental results that are sometimes taken to provide evidence for inferential mistakes may show nothing of the kind. This is so for a variety of reasons, including the possibility that the subjects interpret the tasks differently from the way in which experimenters expect, with subjects solving *their* interpretations of the experimental task in a rational or successful way. Some have argued that this is the case for such well-known examples of alleged inferential errors as the case of Linda, the feminist bank teller case, and others.[33] Examples of this sort point to a general problem, also arising in the psychological literature on causal cognition and discussed subsequently, which has to do with ambiguities in the way in which subjects interpret verbally described scenarios and instructions, with the result that those subjects may be solving different problems than those that researchers think that they are solving.

Another possibility is that the normative theory that is used to support the claim that the subjects are behaving irrationally is wrong; a more adequate normative theory would show that the behavior is rational. Examples of this will be discussed in subsequent chapters. A related possibility, noted previously, is that subjects are behaving rationally but are subject to various sorts of resource constraints. As a consequence they employ strategies and heuristics that are optimal given these constraints, even if they sometimes produce errors that would not be made by unconstrained reasoners. In such cases, rational models can still be illuminating, for the reasons described previously.

[33] See, e.g., Hertwig and Gigerenzer 1999.

Another relevant consideration, illustrated in Chapter 7, is that performance on reasoning tasks, causal and otherwise, is often sensitive to various features of the format in which those tasks are presented. In particular, a fairly generic although by no means universal result is this: people often do considerably worse on verbally described tasks calling for explicit reasoning that are described in a relatively abstract way—for example, by reference to some formal theory—than they do on analogous tasks that are more concrete and involve nonverbal stimuli and tests for success that are nonverbal. For example, when subjects are given verbally described information about probabilities and asked to solve some problem that requires application of Bayes' theorem, they may perform poorly. At the same time, there is considerable evidence that not just human beings but members of other species can reliably track information about the frequency of various events in their environment and reliably update this information in a normatively appropriate manner, including updating in accordance with Bayes' theorem (e.g., Biernaskie 2009). In other words, in some situations self-conscious deliberative explicit reasoning can perform less well than implicit processing as revealed in nonverbal behavior—the two can dissociate. This has interesting implications, which are briefly discussed in Chapter 3, both for experimental paradigms that explore causal reasoning by focusing on verbal responses to verbally described scenarios and for appeals to intuition in philosophy (which also typically involve verbal responses to verbally described scenarios). It may be that these mispresent human competences in important respects.

Yet another point is that the fact that some significant number of subjects may perform badly on some causal reasoning task is by no means the only observation of interest if we want to understand causal cognition. That some considerable number also perform well (and how they manage to perform well) is also of great theoretical interest. It may be that the behavior of the latter can potentially be understood in terms of rationality analyses, even if the former cannot. This is another reason why it is myopic to dismiss rationality or optimality analyses on the grounds that people often make mistakes.[34]

[34] Another relevant point is that many if not most of the causal beliefs that humans acquire are acquired from others, either through explicit instruction or in other ways—there is really no alternative to this. Many people (think of religious figures and politicians) have a strong interest in inducing others to acquire causal beliefs that further their (i.e., the inducers) own interests, whether these beliefs are true or false. This may be particularly true of causal beliefs about social, economic, and political relationships. So many mistaken causal beliefs may not reflect human irrationality but rather the fact that we have to rely on others for many of our beliefs and many of those others have self-interested (or other) reasons for promulgating false beliefs. Another possibility is that existing networks and institutions involved in the spread of beliefs may have suboptimal properties so that even rational agents who are forced to rely on information provided by others for many of their beliefs end up with many false beliefs. These are all ways in which people can end up with false beliefs without being stupid or irrational and without following normatively defective reasoning strategies. The normatively appropriate behavior that is often seen in experimental contexts may reflect in part the fact that in those contexts subjects are not acquiring causal beliefs from others but rather on the

Finally, let me comment briefly on a closely related issue: to what extent is there continuity or similarity between learning and reasoning strategies employed by ordinary people and by scientists or other experts? One possible view is that there is very little continuity—in virtue of their training and education (or their superior epistemic virtue and intelligence), scientists, statisticians, and perhaps even philosophers have acquired patterns of reasoning, standards for the evaluation of evidence, and so on, that are very different from those of ordinary people. This explains why ordinary people have so many false beliefs and science is so successful. An alternative view is that ordinary people, even children, employ reasoning strategies that exhibit considerable similarity to those employed in science, although the latter may be more explicit, rigorous, formally developed, and so on. This is a "layperson as scientist" or "scientist in the crib" (Gopnik and Meltzoff 1999) view of ordinary human subjects. Of course there is a gradation of possible positions between these two alternatives, but, as should be obvious, I am more sympathetic to some version of the second alternative. Again I would ask that those who think that the first alternative is obviously correct to look at the empirical evidence, both as described in what follows and in other sources.

1.8. Intuitive Judgment

My views about the role of appeals to intuition or judgments about cases (hereafter just "intuitions") are described in more detail in Chapter 3. However, since these views are nonstandard, a bit more foreshadowing may be helpful for parts of what follows in this and the following chapter. Philosophers have of course held many different views about the role and significance of intuitions. It is not clear to me that these share a common, generally accepted core. In any case what follows is not an attempt to describe such a core. Rather it is my reconstruction of what I regard as a defensible element in such appeals, regardless of what those who appeal to intuition may think of themselves as doing. I see appeals to intuition (or judgments about cases) in the causation literature as claims about practices of causal judgment or inference—the intuiter uses her own reactions/judgments about cases to support or motivate the claim that these judgments reflect more widely shared judgments and distinctions. Sometimes such appeals will also include claims about the bases on which such judgments are made or what causes people to judge one way rather than another, although as we shall see, such claims are often less reliable than straightforward descriptive claims about

basis of data presented to them; when beliefs are acquired on the basis of social processes, matters are different.

how people judge. Understood in this way, as claims about patterns of judgment, intuitions report empirical claims that are either true or false and are typically empirically testable. Moreover (I contend), although claims by philosophers about shared practices of judgment can certainly be mistaken, as we see from previous examples, they turn out to be correct in many cases involving causal cognition, as will be illustrated subsequently. That is, in cases of causal judgment philosophers are fairly good, although far from infallible, at detecting how others (including non-philosophers) will judge. This suggests one way in which appeals to intuition can be genuinely informative, although of course it is always desirable to check claims about shared judgments empirically.

I will add, however, that this "vindication" of intuition is very limited. It does not establish that intuition can be a source of information about the "nature" of causation (assuming that this goes beyond a description of practices of causal judgment) or about which views of causation are normatively justified. Moreover, if we think of intuition simply as a possible source of information about how the intuiter and others judge, then it becomes clear that there are many other sources of such information—for example, empirical studies of verbally expressed causal judgments of the sort conducted by psychologists as well as studies that report various forms of nonverbal behavior. (See subsequent chapters for numerous examples.) It may be that some or many intuitive judgments by philosophers are, in addition to being accurate reports of how other people judge, normatively defensible (or in some way relevant to normatively defensible general accounts of causation), but if so, this is not because of some special status these judgments have just in virtue of involving intuitions or being produced by philosophers.

I have emphasized that the fact that people conform to some pattern of judgment or reasoning does not by itself show that they are normatively correct to do so. To the extent that the metaphysics of causation has to do with finding a correct or veridical view of causation, this is one reason why intuition does not directly validate such metaphysical claims.[35] Instead, as argued earlier, one supports the claim that some pattern of judgment or reasoning is normatively correct by appealing to means/ends arguments connecting such patterns to goals associated with causal cognition.[36] Nonetheless, if one accepts the idea that human causal cognition is often rational or adaptive, this suggests that the practices of judgment and reasoning reflected in appeals to intuition may sometimes, perhaps

[35] Critics of experimental philosophy complain that in some versions this enterprise unwarrantedly attempts to draw philosophically substantial conclusions from mere surveys of how people judge. To the extent that "philosophically substantial conclusions" are those that are normatively warranted, I agree that they don't follow from descriptions of how people judge. However, this is also true for philosopher's reports of their intuitions and claims, even if accurate, of how the folk judge.

[36] So at the risk of repetition, the idea is *not* that an intuition is shown to be normatively justifiable just by constructing some normative theory according to which it is justified. The normative theory must itself be justifiable via the kind of means/ends analysis described earlier.

frequently, be normatively justified. Thus we can use intuition/judgments about cases to generate hypotheses about shared practices of reasoning and judgment and then check to see whether there is some normative justification for those practices. Again, I stress that this is a necessary *independent* step—normative justification does not follow just from the intuition itself (or from the fact that it presents itself as normatively justified) or from the fact that is shared. Nonetheless the fact that this independent step is necessary does not show that intuitions and judgments about cases can only play a trivial role in the process of producing normatively justifiable conclusions. For one thing, we may not recognize that some possible pattern of judgment or reasoning is normatively justifiable until we see it in our practices and think to ask whether it may have some normative justification—examples will be given subsequently.

As an additional illustration of this conception of the relation between judgment about cases and normative justification, consider the debates among philosophers of science in the 1950s and 1960s over the role of the directional or asymmetric features in causation and causal explanation. Critics of the deductive-nomological (DN) model such as Scriven argued that the model was defective because it did not reflect these directional features, mistakenly allowing the length of the shadow cast by a flagpole to explain the length of the pole, as well as correctly allowing the length of the pole to explain the length of the shadow. In a recent discussion, Stich and Tobia 2016 treat such criticisms as amounting to the claim that the DN model should be rejected because it is contrary to our "intuitions" about flagpole cases, which deny that shadows explain flagpole lengths. I agree with Stich and Tobia (and with Hempel, in his earlier discussion of this example) that it is not justifiable to reject the DN model merely because it is "contrary to intuition." If Hempel's conception of the goal of explanation (that it involves law-based grounds for prediction) is correct, we should regard the preceding intuition as normatively mistaken. On the other hand, the intuition we have about the flagpole example can motivate us to ask whether there is some alternative normative basis for distinguishing between flagpole-to-shadow explanations and the reverse. In fact, as I discuss elsewhere (Woodward, forthcoming b), interventionism is one of several normative theories that provide such a rationale: according to interventionism the flagpole-length-to-shadow explanation, or causal claim, is correct because intervening on the former will change the latter; the derivation or causal claim in the reverse direction is non-explanatory because intervening on the shadow will not change the flagpole length. This asymmetry in behavior under interventions is in turn related to statistical asymmetries in the relationships among the variables representing pole height, shadow length, and angle of the sun—asymmetries that track additional plausible ideas about explanation. I will not argue in support of this diagnosis here; my intent is simply to illustrate the significance I think we should assign to appeals to intuition. Although an intuition/case judgment can suggest shared practices of judgment, claims about the normative justifiability of the intuition require an independent appeal to normative theory justified in the means/ends way already described.

This picture of the significance of intuition or judgment about cases as a possible source of information about practices that, to the extent that these are rational or success conducive, can also have normative significance may strike some as far-fetched or at least unduly indirect. It contrasts with views according to which intuition by itself provides straightforward access to the truth about items of philosophical interest such as causation as it exists in the world. I suggest, however, that a picture rather like mine can be found at number of (perhaps surprising) places in the philosophical literature. Consider Austin's essay, "A Plea for Excuses," (1956–1957) often thought to be a paradigm of "ordinary language philosophy." Responding to the question of why we should care about various distinctions in ordinary language among forms of excusing, Austin writes:

> Our common stock of words embodies all the distinctions men have found worth drawing, and the connexions they have found worth marking, in the lifetimes of many generations: these surely are likely to be more numerous, more sound, since they have stood up to the long test of the survival of the fittest, and more subtle, at least in all ordinary and reasonably practical matters, than any that you or I are likely to think up in our armchairs of an afternoon—the most favoured alternative method.

He goes on to say, though, that "ordinary language has no claim to be the last word, if there is such a thing." Instead, according to Austin, ordinary language is merely a good starting point. In connection with the study of excuses, he discusses at some length two additional sources of information—the law and empirical psychology—and makes it clear that he does not privilege ordinary language over these. He also adds that ordinary language can also be, in some cases, an inadequate instrument—some of the distinctions it involves may be the product of superstition, confusion, and so on. Thus the distinctions it embodies need to be evaluated functionally, in terms of how well they work.

Austin's views about ordinary language are similar to those I recommend about intuitive judgment. To the extent it is plausible that our ways of talking reflect the operation of selective processes (Austin's "survival of the fittest"), we have a rationale for paying attention to them.[37] But we are not entitled to assume,

[37] Suppose that no such selective process is operative. Then I take Austin to be suggesting that there is no obvious other rationale for focusing on how we talk. Let me add that I am very aware that some readers will think that talk of processes that sometimes select for good or correct judgments is pollyannaish or a just-so story, with no empirical backing. Suppose this is so. Assume also that an

just because we talk in certain ways, that it is, normatively speaking, correct or good practice to do so. Moreover, ordinary language is not the only or a privileged source of information for philosophizing, and our task is not just to describe its deliverances. Substitute "intuitive judgment" for "ordinary language" and the resulting view is very close to my own.

As a second illustration of this role for intuitive judgment consider a common practice in philosophy of science. This consists in selecting some episode of theory choice or hypothesis evaluation by scientists (revealed in verbal "judgments" and/or other behavior such as decisions to purse certain research programs) and then trying to reconstruct it as "rational"—that is, as conforming to some defensible methodology. What should we make of this procedure? If it is to be convincing, it is not enough to merely describe the judgments reached during the episode—it does not follow that they are correct merely from the fact that the scientists endorse them.[38] Moreover, the goal should not be merely to systematize the judgments without any reference to independently defensible normative standards. However, to the extent that there is reason to believe that the science around the episode is "successful," this can prompt us to look for a normative rationale that helps to explain this success. Here scientists' choices and judgments play the role I assign to intuition and judgments about cases in non-scientific contexts. They reflect empirical facts about scientific practice. Whether that practice is normatively justified requires additional argument.

1.9. Goals, Realism, and Objectivity

I spoke earlier of goals or purposes associated with causal thinking. I assume in what follows that good candidates for these goals will be generic and common or shared among many different subjects—the goals that we are interested in should be those that many people think it valuable to achieve, even if they may weigh individual goals differently. The goals that are relevant for this sort of analysis should also have an epistemic dimension, even if they are not "purely epistemic," and may also include what some philosophers would regard as a "pragmatic"

alternative account according to which intuitive judgments give us rational insight into the nature of things is unsupportable. Then I think that the correct conclusion is that there is no good reason to pay attention to such judgments, at least in theorizing that has a normative component. In other words, if you want a role for intuitive judgment, I think you need something like the selection story. Of course, there is no reason to think that such a story is available for all intuitive judgments. For example, folk judgments of mereology may lack any such basis.

[38] There is some tendency in the philosophy of science literature to argue that a normative theory can be "confirmed" by exhibiting episodes in the history of science in which scientists judge in conformity with the theory. For reasons analogous to those in connection with appeals to intuition, I think this is wrongheaded.

component. (E.g., some philosophers regard whether a theory can be used by us to calculate experimental results as a "pragmatic" matter. Whether or not this is accepted, I regard this feature as epistemically valuable—see the discussion of "pragmatic" in Section 1.11 of this chapter.) I will not attempt to provide a general account of what makes a goal at least in part epistemic, but will rely on what I hope are uncontroversial judgments about this—discovering relationships that allow us to reliably predict is a goal with an epistemic dimension; putting forward ideas that you think will please your boss is not.

Epistemic goals can be described at varying levels of generality and specificity. Speaking abstractly and not necessarily confining ourselves to goals that are distinctively associated with causal cognition, we can point to goals like the discovery of relationships that are useful for prediction or for manipulation, or finding investigative procedures that lead to avoidance of certain kinds of errors or that will produce large gains of certain kinds of information. Such goals can in turn be made more precise and specific in different ways—for example, with a specification of a more particular target and more precise criteria for what counts as success or failure in achieving that target. Rather than just "prediction," the goal might be further specified as finding a predictor of some specific target quantity that is optimal by some criterion—for example, predicting on the basis of observations of the past behavior of some random variable what the mean of new observations of the value of that variable will be on the assumption that the new values are i.i.d. (independent and identically distributed) draws from the same stationary distribution as the past values. Given such a specification, methodological investigation and assessment ("normative theory") will then have to do with finding effective means for achieving the goals or fulfilling the criteria. For example, a criterion like the Akaike Information Criterion is one proposal about the optimal predictor for the problem described here—it can be normatively justified within a frequentist framework as an asymptotically unbiased estimator of a quantity that (roughly) minimizes the "distance"[39] between the predicted values of the target and previously observed values.

What might be good candidates for general goals associated specifically with causal thinking? It may be tempting to propose as a goal the discovery of true causal claims, but of course different accounts of causation may disagree about just what this involves, so that this goal seems too undiscriminating and uninformative to be helpful.[40] Interventionists think that the identification of

[39] More specifically, the Kulbeck-Leibler distance.

[40] At a workshop at which I discussed some of these ideas, one philosopher responded that the function of the causal concept is to "describe causal reality." Of course I don't think this is false, but if one wants to pursue the functional project I have outlined, it is unilluminating. "Describing causal reality" is not a function or goal that can provide any independent purchase on distinctions among causal concepts or strategies of causal learning and reasoning.

relationships that are exploitable for purposes of manipulation and control is a central goal of causal thinking, but in principle a "functional" approach might associate causal thinking with other goals instead (or as well)—possibilities include the compact and unified representation of certain regularities, the codification of our commitments to various inductive strategies (as in Spohn 2012), or perhaps the achievement of certain information-theoretic goals (implicit, arguably, in the developments in machine-learning approaches to causal inference such as Janzing et al. 2012, mentioned earlier). Following a functional approach, we can ask whether and what respects, various features of, or proposals about causal judgment, inference, and reasoning contribute to (or fail to contribute to) these goals.

1.10. Considerations Relevant to Choosing among Goals

As noted earlier, it is a fact about us that we have certain goals, although some of these may be shared by any "rational" agent that is bounded in certain ways. However, once such goals are specified, it can be a perfectly "objective" matter whether some means (whether this is the use of certain distinctions, inferential strategies, patterns of judgment, or whatever) does or does not effectively get us to the goal. Goals may "come from us" in the sense of reflecting generic concerns that humans and perhaps other agents have, but, given the goals, it is not up to us whether some means or strategy contributes to their achievement. Classical statistics is one source of illustrations of this general theme. Suppose that I want to arrive at a "good" or "accurate" estimate of the value of some quantity m on the basis of noisy measurements resulting in a set of observations d_i that reflect the influence both of m and some randomly distributed error u_i that varies from measurement to measurement: $d_i = m + u_i$. Assume that the estimator will be a random variable m^* that is some function of the d_i. We then proceed by adopting criteria for what it is for an estimator to be "good." One widely used criterion is that the estimator should be *unbiased* in the sense that its expectation value $E(m^*)$ is equal to m—that is, to the true value of the quantity being estimated. There many such unbiased estimators of m. We thus appeal to a second criterion: choose the estimator that, among the unbiased estimators, has the smallest variance. Given these criteria and certain other assumptions, such as the assumption that the estimator must satisfy a linearity requirement, one can then prove mathematically that the best linear unbiased estimator (BLUE) for m must take a certain specific form.

What I want to emphasize is not the details of this idea but rather the general form that the justification takes. One first specifies a general goal: finding a good estimator for m, and then subsidiary goals or criteria that specify what counts as a good estimator in this context—unbiasedness, minimum variance, and so on. One then shows that a certain choice of an estimating function is an effective (indeed optimal) means to this goal, thereby providing a means/end justification for the procedure. Of course one might not have the goal of estimating *m* accurately at all (according to any standard of accuracy). Or one might deny that unbiasedness is among the criteria for a good estimator. (Maybe one thinks that a slightly biased estimator with smaller variance would be preferable.) So there is an element of "subjectivity" here, in the sense that specifications of goals reflect choices we make. However, given that one has the goal of estimation and accepts the preceding criteria for what makes an estimating procedure good, it is an objective matter whether some proposed means is a good one for achieving this goal.

I suggest that we should think about the normative dimension of causal reasoning and judgment in a similar way. Supposing that one does have the goal of discovering relationships that are exploitable for purposes of manipulation and control (and perhaps given some further specification or specifications of what this involves), it is an objective matter whether a particular procedure, applied to evidence in some context, leads to the discovery of such relationships or not. Which such procedures are effective at such discoveries is one of the topics of methodology as I conceive it. Given some normatively justifiable procedure, we can then ask about the extent that, as an empirical matter, it is found in people's behavior. Similarly and consistently with a broadly interventionist viewpoint, one might distinguish among relationships that can be exploited for manipulation, regarding some as potentially more useful than others in the service of this goal, because, for example, they hold over a wider range of circumstances (are "more invariant"). One can then ask which inference procedures and associated ways of thinking about causation will facilitate achievement of this goal. In addition, as noted later, a similar approach can be adopted for other facets of causal reasoning, such as the distinctions we make among different causal concepts or choices of which variables are best for causal analysis.

Here are some briefly described examples of this sort of approach, some but not all of which will receive more attention subsequently.

Suppose that X and Y are observed to be correlated (where the context involves passive observation rather than experimental manipulation) with $Y = bX + U$. Here U is an unobserved error term that may be correlated with X, and b is interpreted as the causal effect (if any) of X on Y. It can be shown that we can reliably estimate b if we can find an instrumental variable Z (also passively observed) for X with respect to Y. Such an instrument Z is a variable that (i) is associated with X, (ii) is independent of U, and (iii) is independent of Y given X and U. Under these circumstances $b^* = cov\,(Y,Z)\ /\ cov\,(X,Z)$ is an estimator for b. Intuitively, this works because the instrument Z functions (in terms of the information it provides) like an intervention on X with respect to Y, telling us what would happen if an intervention on *X* were to be performed—it provides information about this goal and is justified as a measure of causal effect accordingly.[41]

Suppose one wants a measure of the strength of a causal relationship that generalizes well across certain different conditions—for example, a measure that assigns the same strength to the causal relation from *C* to *E* across variations in the frequency with which *C* occurs, where this relation is intervention-supporting in the way described earlier. (We look for such a measure because we assume that the relation between *C* and *E* can remain the same when the frequency of *C* changes—see Chapter 5). Then, given certain additional assumptions, a measure of causal strength derived by Cheng (1997) can be shown, as a matter of mathematics, to better satisfy this goal than alternative measures (see Chapter 7).

Suppose that we adopt the goal of representing the full range of causal conditions on which effect *E* depends, where "depends" is understood along interventionist lines. Then in certain situations we should employ variables characterizing those causal conditions that satisfy a condition related to what Yablo (1992) calls "proportionality"—this will satisfy this goal better than alternative choices of variables. The proportionality condition is justified in terms of this goal.

At an abstract level, I will understand the notion of means/end justification as covering a variety of different possibilities. Sometimes such justifications will be mathematical, as in some of the examples already mentioned. Mathematical justifications themselves can take a number of different forms. For example, they may be "frequentist," showing that some procedure for forming causal judgments has good error characteristics. Or they may justify some patterns of causal reasoning within a Bayesian framework, as in work by Tenenbaum and colleagues (e.g., Tenenbaum and Griffiths 2001). Alternatively, the justification may be "empirical" and proceed by showing that when an inference procedure is applied to data for which the causal relationships are independently known, it generates, as an empirical matter, correct results. For example, Janzing et al. (2012) appeal to empirical facts about the reliability of their procedures for inferring causal direction in this way, along with purely mathematical arguments. Again the justification is that the procedures "work" in getting us to certain goals, in this case the reliable identification of causal direction.

Let me also emphasize that adopting this general means/ends approach to normative justification is not tantamount to arguing that we ought as a normative matter to have this or that general goal, such as discovering relationships that support manipulation or that are more rather than less informative about what happens under manipulations.[42] My claim is rather that *given* that we have

[41] See, e.g., Angrist and Pischke 2009.

[42] Since I think of justification in terms of means and ends, some ends (those that are not means to some further goal) cannot be justified in this way.

such goals, we can deploy means/ends analysis to argue for various distinctions, inferential procedures, choices of variables, and so on the grounds that they are effective means to such goals. Of course my view is that thinking of causal reasoning in terms of the goal of discovering relationships that support manipulation affords important insights, but the case for this claim consists in the detailed working out of these insights.[43]

The notion of means/ends normative assessment makes (I hope) obvious sense when applied to the assessment of causal discovery and reasoning strategies. However, I wish to apply such assessments more broadly to other aspects of causal cognition and judgment—for example, to the evaluation of distinctions we might make among causal relations. Although explicit use of means/ends assessment is not so common in the latter cases, I think it is entirely appropriate. For example, the distinction between causal contribution of *C* to an effect along a particular route as opposed to the total effect of *C* (see Chapter 2) can be regarded as a reasonable and warranted distinction within a broadly interventionist framework because these two causal notions correspond to different possibilities for manipulation and control and different kinds of information are required to estimate them—they correspond to different manipulation-related questions one may be interested in answering. Someone with an interest in manipulation will want to make use of both of these causal notions and will distinguish between them. Similarly also for the distinction between actual cause judgment and other sorts of causal judgments—these also correspond to different questions about the results of manipulations (see Chapter 2). Thus although there is, according to interventionism, a general connection between causation and our interest in manipulation and control, it is also the case that distinctions among causal notions correspond to more nuanced distinctions among different sorts of manipulations that we may be interested in.

At a more abstract level, overall ways of thinking about causation have undergone considerable historical evolution. As a number of scholars have noted (e.g., Clatterbaugh 1999), medieval and some early modern treatments of causation show what from a modern perspective looks like tendency to run together logical and causal relationships—the accounts seem to assume that when *C* causes *E* there will be logical or conceptual connection of some kind between *C* and *E*.

[43] Some readers may think that is "circular" or question-begging to hold that prediction and control are goals associated with causal thinking since these are goals closely associated with interventionism, the account that I favor. In response I would claim that, first, (i) it is uncontroversial that many people do have the goal of identifying relationships that can be used for manipulation and control—that is just a fact, not a question-begging assumption. Second, (ii) as I attempt to show in subsequent chapters, a framework that associates causal reasoning with this goal makes good sense of our practices of causal reasoning. Those who think that causal thinking should be understood, partially or entirely, in terms of other goals are, as I have said, welcome to formulate such alternative accounts, but my prediction is that they will fare less well than interventionism.

Over the course of the early modern period and culminating with Hume, these two kinds of relations—the empirical relations we find in efficient causation and logical relationships—become conceptualized in a way that more clearly separates them one another. Again I see this as a change that is functionally desirable—it facilitates causal reasoning, at least insofar as this is connected to the goal of manipulation and control.

1.11. Pragmatism and Pragmatics

In its emphasis on means/ends justification and goals associated with causal cognition there is an obvious sense in which the approach I favor has a "pragmatic" component. However, "pragmatism" as a philosophical position can take many different forms. Similarly "pragmatic" is used in many different ways in discussions of causation (and allied topics like explanation). These need to be distinguished—my approach is "pragmatic" (or reflects themes associated with pragmatism) in some respects but not others. In general, loose talk of "pragmatic" and failure to distinguish among different senses of this notion has been a source of considerable confusion in recent philosophy.[44]

First, some philosophers who think of themselves (or are identified by others) as "pragmatists" hold that what is *true* is just a matter of what some community of inquirers can or does agree on or perhaps just a matter of what "works" in some sense that is not further specified—views that I emphatically reject.[45] Second, many philosophers who discuss the "pragmatics" of causal judgment or the role of "pragmatic factors" in such judgments (or in causal explanation) often have in mind factors that they think of as idiosyncratic to particular contexts, and/or to the particular interests and background knowledge of the speaker and her audience.[46] They view the influence of these factors as unsystematic, and (often) as lacking in any principled justification—thus arbitrary and subjective. For example, "causal selection" (the practice of selecting just a few factors as "the

[44] A helpful standard would be this: philosophers should not label (or dismiss) a consideration as "merely pragmatic" unless they accompany this with an explanation of what they mean by "pragmatic" and why this understanding justifies dismissal. Similarly, someone who claims to provide a "non-pragmatic" account of X (explanation, confirmation, etc.) should explain what is meant by "non-pragmatic" and why such an account is desirable.

[45] For further discussion of this and other issues having to do with what a reasonable pragmatism should involve, see Woodward, forthcoming c. I emphasize that the version of pragmatism that I favor, with its emphasis on means/ends analysis, requires the availability of a notion of truth that does not reduce to what people agree to or to what works. We need to be able to make sense of claims that it is true that certain means conduce to certain ends, and such truths can't just be a matter of what people agree on.

[46] My guess is that one important source of this understanding of "pragmatics" in philosophy comes via linguistics, where syntax and semantics are contrasted with "pragmatics," but I know of no historical investigation into this.

causes" from a much larger set of factors that are causally relevant to some outcome) is often taken to be a matter of "pragmatics" in just this sense. Labeling the factors that influence causal selection as "pragmatic" in this sense thus suggests that there is no interesting general account that relates them to widely shared epistemic ends or goals. Often the conclusion is that they are unworthy of further philosophical attention.[47] "Pragmatic" theories of explanation often conceptualize the factors influencing explanatory judgments along similar lines—these are taken to be highly contextual and tied to particular interests a speaker or her audience may have. The upshot is that the test for whether an explanation is good becomes largely a matter of whether the audience finds it satisfying or acceptable.

To distinguish this conception of the "pragmatic," let us say it involves pragmatic_2 considerations and let us say that the way in which means/ends justification is "pragmatic" involves pragmatic_1 considerations. Obviously when it is claimed that some set of judgments reflects the role of pragmatic_2 factors, this carries with it a very different set of implications from the implications carried by the claim that means/ends justification is pragmatic_1; the latter claim does not imply that the considerations under discussion are arbitrary, unprincipled, dependent on idiosyncratic interests, or unrelated to more general epistemic goals. On the contrary, means/ends analysis is one way of providing principled justifications related to such goals. The justification for the use of a best linear unbiased estimator described previously is pragmatic_1, but not pragmatic_2: the justification is intended to apply across a wide variety of contexts, once the relevant goals are specified. Moreover, although the goals themselves reflect our "interests" in the sense that if we had different interests, we might have had different goals, the goals themselves (e.g., choose an estimator that is unbiased or, more broadly, reliably estimates some quantity of interest) are specified in an interest-independent way. Given these goals and related specification of criteria for goodness, it is not in any further way a subjective or audience-dependent matter whether an estimating procedure is good or not.[48]

[47] In a frequently cited passage, Lewis (1973a, 558–559) describes various proposals about the basis for causal selection as "principles of invidious discrimination" prior to declaring he will not discuss them further. This is illustrative of the way in labeling a consideration "pragmatic" is often treated as a license for dismissing it.

[48] Some readers may be tempted to respond that as long as a consideration reflects the influence of goals or interests at all, it should be labeled "pragmatic," so that there is no real difference between pragmatic_1 and pragmatic_2. But on my view there is a huge difference between simply saying that inquiry is guided by what "interests" us where this is not further specified, which is what pragmatics_2 involves, and specifying a specific goal that interests us in objective, non-psychological terms that allows for the application of illuminating means/ends analysis. Note also that if "pragmatic" considerations are contrasted with those that are truth-related, pragmatic_1 considerations are truth-related: it is true or false that some choice of means effectively conduce to some end, and in the sorts of cases we are considering, the conclusions reached on the basis of means/ends analysis will also be true or false.

Yet another consideration that philosophers sometimes have in mind when they use the label "pragmatic" (call this pragmatic$_3$) has to do with features of cognition or reasoning (whether everyday or scientific) that reflect human limitations (or possibly limitations of any rational agent remotely like us): computational limitations, limitations in what we can know, limitations in what we can manipulate, and so on. These are contrasted with what allegedly would be possible in the absence of such limitations. Thus it may be claimed that it is "possible in principle" to calculate the future behavior of the stock market from the physics of the standard model—but that "pragmatic" limitations (often said to be "*merely* pragmatic") stand in the way of actually doing so. Often this is followed by the suggestion that the limitations in question are philosophically inconsequential or uninteresting (because they are "pragmatic") and that they ought to be abstracted away from in philosophical discussion. Again I observe that there is a big difference between pragmatic$_2$ considerations that vary idiosyncratically from person to person and context to context and defy systemization and the pragmatic$_3$ considerations just described (which are arguably features faced by any human or bounded rational agent). I will also add, echoing remarks in the introduction, that it seems crazy to seek to understand human cognition or scientific practices by abstracting away from all pragmatic$_3$ considerations—of course such considerations will influence the way that we think and theorize, including the way in which we think about causation and related matters. In any case, a distaste (which I certainly share) for accounts that focus entirely on pragmatic$_2$ considerations in trying to understand causation and causal cognition should not lead us to reject pragmatic$_1$ treatments or to insist on treatments that deny the significance of pragmatic$_3$ considerations.

1.12. More on Minimal Realism and the Epistemology/ Metaphysics Connection

My discussion in this chapter and in the introduction has invoked the notion of a minimal realism about (or a minimal metaphysics of) causation and along with this the idea that causal cognition is often successful in, for example, discovering and reasoning about such relations. This fact of success is what establishes a bridge or connection between, on the one hand, how we think about causation and the procedures we use for discovering and reasoning about them and, on the other hand, what is "out there"—that is, between the epistemology and minimal metaphysics of causation. I emphasize, however, that I am not simply assuming that if we think about causal relations as having certain features or employ procedures for detecting causal relations that assume that they have certain features, it follows automatically that causal relations as they exist in the world have such features. It is crucial in addition that the patterns of thinking that we have *work*—that they are successful in tracking worldly behavior. In other words, we have evidence that the world, to some considerable degree, cooperates with or supports how we think and infer in connection with causal relations.[49] In my view, this is a good reason for resisting the idea that there is a complete mismatch between how we think about causal relations and what is out there (a kind of radical error theory). Or more cautiously: at the very least someone who favors the mismatch view owes us an account of why our practices of causal judgment and inference work as well as they do. It is not an explanation of this to simply say that the practices are "useful heuristics"—what is needed is an account with details of why the practices work.

To illustrate with an example from Chapter 4: as an empirical matter, human causal representation seems to go beyond the representation of regularities or associations to include representations of counterfactuals: representations of what would happen if certain interventions or other possibilities were realized. This by itself does not establish that there is anything in the world corresponding to or supporting this additional counterfactual content, although from a functional or rational model perspective it does raise the question of why we have such representations if they are completely erroneous. But in fact these representations often allow us to reason successfully: a representation that has the counterfactual content "If I were to do *X*, then *Y*" can be learned both by performing manipulations and (if certain additional assumptions are satisfied) from observations not involving manipulations. And such a representation can be (and often is) correct or successful in the straightforward sense that if in fact I do *X*, *Y* follows. Moreover, the information (statistical information and empirical background assumptions) from which we can learn such representations straightforwardly has to do with how matters stand in the world. This makes it hard to see how what is learned or represented in such cases completely fails to track "what is out there."

Appendix: More on a Functional Approach to Causation

In order to further flesh out some of the distinctive aspects of a functional approach, here are some additional examples of questions/problems/ideas/strategies it suggests. In some cases, more detail will be provided in subsequent chapters; in other cases the reader is directed to separate papers.

[49] As indicated in the appendix to this chapter, I don't claim that such support is always present. There may be systems and accompanying theories that do not lend themselves to causal analysis.

A.1. Nonfunctional Aspects of Causal Thinking

I have suggested (and will argue in more detail in subsequent chapters) that many features of causal thinking are reasonably well adapted to our goals. But from a functional perspective we should not assume that this is always true, and we should certainly not assume that all of the various proposals that have been made regarding causation and good causal thinking over the entire history of human thinking are necessarily well suited to our goals. In principle we might discover that some common ways of thinking about causation or some patterns of causal reasoning and judgment are not functional at all. This might happen in a number of different ways: the reasoning might, for example, rest on mistaken empirical presuppositions, or it might turn out that, contrary to what many people think, some candidate causal concept, characterized in a certain way, has no or very few real-world applications. Or perhaps some recommended way of thinking, when applied, blocks or undermines various goals we are trying to achieve rather than furthering them. A more extreme possibility is that certain ways of thinking about causation turn out on examination to be logically or conceptually incoherent. Thus implicit in a functional approach to causation is the thought that there may be other practices of causal judgment and inference that would better serve our goals.

A.2. Scope and Limits of Causal Thinking

What are the scope and limits of causal thinking? Under what circumstances and conditions is thinking causally useful or fruitful? Under what conditions, if any, is it not likely to be illuminating? Are there certain empirical conditions that a system must satisfy before it is profitable to try to analyze its behavior in causal terms? Once one begins thinking about the extent to which various elements of causal cognition are well adapted or not to our goals, one is struck by the possibility that these features may be functional in some contexts or with respect to certain kinds of problems—for example, when certain empirical presuppositions are satisfied[50]—and yet not be functional with respect to other contexts or problems. After all, a tool or technology can be effective in connection with some problems or in some circumstances but not others.

[50] Woodward, forthcoming b, describes some of the empirical assumptions that underlie successful reasoning about causal direction.

A.3. Causal Thinking Is Adapted to Empirical Features of Our World and Those Broadly Similar, but Not to All Possible Worlds

Although features of causal thinking are relatively well adapted to our world, there is no reason to suppose that causal thinking will be well adapted to worlds that are very different from our own. Moreover, to the extent that people are willing to make judgments about which causal relations hold in such worlds, there is no reason to think that such judgments are a reliable source of information since there is no selective story to be told about them. Thus inquiries into the causal relationships that would obtain in fictions in which magicians turn humans into frogs are unlikely to be fruitful. Similarly we should not expect a functional account of causation to apply to worlds in which there is systematic violation of various generic empirical patterns that typically obtain in our world.[51] Perhaps we can imagine a world in which anti-thermodynamic behavior is rampant with the causally independent fragments of shattered vases regularly reassembling into intact vases and so on, but, needless to say, this doesn't happen regularly in our world. Or one might imagine an "occasionalist" world in which, unlike ours, whenever we think we have intervened on *C* with respect to *E*, we are mistaken (although this fact is hidden from us) with *E* in fact being caused by other events *X* (e.g., Divine Volitions) that are correlated with our interventions. Causal thinking works by assuming that things like this don't regularly happen.

A.4. Different Areas of Inquiry May Require Somewhat Different Forms of Causal Thinking

I've suggested that there is a broad continuity between common-sense forms of causal thinking in ordinary life and in the various sciences, with the continuity having to do with concerns organized around manipulation and control. However, the details of how this works will depend on various more local considerations, including more specific goals, facts about different subject matters and about what we can know (among other considerations)—again a conclusion that follows naturally from a focus on functionality. In some common-sense contexts, so-called actual cause judgments play an important role, often connected to the assignment of responsibility. (See the appendix to Chapter 2.) I see these as connected to and motivated by a concern with manipulation, but the details of this concern are importantly different from the concerns motivating causal analysis in, say, physics—contrary to what some philosophers

[51] Again, for more on this see Woodward, forthcoming b.

suppose, actual cause judgments are not very central to physics. More generally, different causal notions are differentially important in different sciences. The control-related notion of causal specificity (a notion not discussed in this book, but see Woodward 2010) is important in biology but arguably not in physics, in part because of contingent facts about biological organisms and perhaps other designed structures. So in stressing the continuity among various forms of causal thinking I do not mean to imply that, say, every feature of common-sense causal thinking is reproduced throughout science.

A.5. Eliminativism

Some philosophers are commonly interpreted as holding that thinking about the world in terms of cause and effect is fundamentally misguided or confused—perhaps because such thinking is inconsistent with what "fundamental physics" tells us about the world or perhaps for some other reason.[52] These writers favor elimination of the notion of cause from our thinking or at least from important parts of it (such as scientific inquiry) and its replacement with something else—often claims about correlations or regularities interpreted along "Humean" lines. As explained earlier, advocates of a functional perspective can agree that the application of causal notions has empirical preconditions and that it is possible that these will not be satisfied in certain domains of inquiry. But, as argued, functionalists will also insist that just because these preconditions are not satisfied in certain domains, it does not follow that they are unsatisfied in all others and that causal thinking is inappropriate everywhere. Moreover, functionalists will (or should) ask about the proposed replacement for causal thinking (if any) and about the extent to which it can fulfill the functions previously associated with causal thinking. Suppose, for example, that the replacement does not provide the resources to distinguish between those relationships that are exploitable for manipulation and control and those that are not. Advocates of a functional approach (or at least advocates who think that the interventionist framework captures aspects of the function of causal thinking) will see this as highly problematic since there is an important function of causal thinking that is not supplied by the replacement. (Note that the replacement is rejected on that basis and not just because it is contrary to "intuition.")

52 One candidate for such a view is Russell 1912.

A.6. Causation as a Natural Kind

A common view about natural kinds such as gold is that these have a unified nature or essence that may be sharply distinct from how most people recognize or conceptualize them. Most people who think about gold presumably do so in terms of a standard stereotype of a precious, yellowish, ductile metal but (on this view) its underlying nature is described by something more like "element with atomic number 79." This underlying nature may be unknown to most ordinary people, but it is still what people are referring to when they use the word "gold." Could something similar be true for "cause," with its underlying nature (perhaps given by physics in terms of energy/momentum transmission) being sharply distinct from how we think about it? If you think of cause as a notion that should be partly understood in terms of its function, another possibility presents itself: whatever unity is possessed by the various relationships we describe as causal may have more to do with their sharing a common function than with a common underlying nature. It might be the case, for example, that some relationships we ordinarily describe as causal involve transmission of energy/momentum, while many others (such as relations of double prevention) do not. If one expects "causation" to be behave as a natural kind term, it may be tempting to regard causal claims falling into this second category as failing to track the underlying nature of causation, assuming this has to do with energy/momentum transmission. On the other hand, from a functional perspective, classifying relations involving energy/momentum transmission and at least some of those involving double prevention together as causal will make perfectly good sense to the extent that this classification picks out a common function such as identifying relationships that are exploitable for manipulation and control. In this respect "cause" may be more like "screwdriver" than "gold."

A.7. The Problem of Variable Choice

The results of causal analysis (judgments, learning strategies, and so on) are extremely sensitive to the variables employed (Woodward 2016a). Transformations that "mix" uncorrelated variables can result in variables that are correlated and conversely, so that analyses that are guided by correlational evidence may reach different causal conclusions depending on the variables employed. When one is trying to discover "upper level" causal relations (e.g., relations among psychological variables) and has more fine-grained information about "lower level" variables (e.g., fMRI data) most ways of aggregating the lower-level variables into upper-level variables will produce badly behaved or uninformative results, and other aggregations (sometimes only one) will produce much better-behaved results. "Better" here means that the variables allow for the formulation of relationships that better capture difference-making relationships, are more invariant, require the postulation of fewer causal connections to capture observed correlations, and have other desirable features (see, e.g., Glymour 2007). However, most standard causal inference procedures have little to say about variable choice—to a large extent they assume that the investigator already has available an acceptable set of variables used to describe the evidence and then purport to tell us what causal conclusions can be drawn on this basis. The issue of "where the variables come from" is not addressed. Thus, to the extent that it is possible, it would be normatively desirable to formulate criteria or heuristics that might guide variable choice. Again, I see this as part of the functional project.

At the same time, the issue of variable choice also has an empirical dimension, which so far has been largely unexplored in psychological studies of causal cognition.[53] Many experiments provide subjects with a pre-specified set of variables and ask them to make judgments about the causal relations among these, rather than asking subjects to formulate appropriate variables. How do ordinary adults and others come up with the variables they use in causal analysis and judgment, assuming that in real-life tasks those variables are not always supplied ready-made? When and how do people learn new variables for causal analysis when the old ones are less than satisfactory? When people are given a choice among different (but not necessarily inconsistent) causal claims that are formulated in terms of different variables, what guides their choices? I will explore some of these issues in Chapters 7 and 8.

[53] An important exception is Goddu and Gopnik 2020, as well as some of the research discussed in Chapters 7 and 8.

2
Theories of Causation

This chapter reprises some standard philosophical theories of causation. I have not attempted to provide a detailed survey—there are many books and papers that do that. Rather my goal is provide an overview that will make subsequent chapters more intelligible for readers who may not be familiar with the philosophical theories and to provide some common background for subsequent discussion. I devote more attention to interventionist treatments of causation than to alternatives, but here too I have not tried to be comprehensive. Readers who are familiar with this material may wish to skip this chapter.[1]

2.1. Type- versus Token-Causal Claims

As a point of departure, we should note that there are a number of different causal concepts that philosophical theories have attempted to explicate. One important distinction (but far from the only one) has to do with the contrast between what are called *type*-causal claims and *token*-causal claims.[2] A type-causal claim—"generic" might be a better description—asserts that some type of event or variable or factor (or whatever one thinks the relata of causal relationships are) causes some other type of effect, as in

Smoking causes lung cancer. (2.1)

In this case, the types "smoking" and "lung cancer" are binary, or "Boolean"—they are either present or not. In other cases, type-causal claims relate variables[3] or factors that can take many different values, as in "An impressed force of magnitude F causes a mass m to accelerate with acceleration $a = F/m$."

[1] Although it does illustrate some of the ideas from Chapter 1.

[2] The "type" label is misleading in a number of ways, for reasons described in Woodward 2018. (See also the appendix to this chapter.) I continue to employ it because it is such entrenched usage. Similarly for "token causation"—as noted in the appendix, "actual causation" is in many respects preferable.

[3] In contexts like this I will sometimes use "variable" to refer both to quantities or properties like mass and to the words or symbols we use to describe these. The resulting conflation of use and mention is harmless and simplifies the exposition.

Causation with a Human Face. James Woodward, Oxford University Press. © Oxford University Press 2021.
DOI: 10.1093/oso/9780197585412.003.0003

A token-causal claim (sometimes called a singular causal claim or a claim of actual causation) is a claim about the causation of some particular event by another particular event or about the causation of the presence of a feature in a particular individual or unit (which might be represented as a variable taking a certain value for that unit), as in

Jones's smoking caused his lung cancer. (2.2)

When we consider that (2.1) might be true and that it might also be true that Jones smoked and that he developed lung cancer and that, consistently with this, (2.2) might be false (his lung cancer was instead caused by exposure to asbestos), we see that the relationship between (2.1) and (2.2) is far from straightforward.

In this book I will focus mainly but not exclusively on type-causal claims, the main exception being Chapter 6, in which there is discussion of both type and token / actual cause claims. I discuss token-causal claims briefly in an appendix to this chapter.[4] There are several motivations for this restriction. First, type-causal claims are the primary targets of many philosophical/normative theories, including interventionism. This is in part because in many areas of science, the focus is on discovering type-causal relationships rather token-causal relationships. Second, much of the psychological research I will discuss also focuses on learning and judgment involving type-causal relations (or at least on cause to effect inferences in which the goal is to determine the kind of cause that will produce an effect—see appendix). Third, in my assessment, normative theories of type-causal relationships are at present better developed than normative theories of token cause judgments. In my view, this is closely related to the fact we lack a generally accepted account of the function of the latter judgments. In the absence of such an account, it is not obvious how to employ the methodology described in this book. Fourth, token-causal judgments have distinctive features (other than the obvious ones) that are not shared by type-causal claims. This is partly because token judgments are addressed to an inferential problem that differs in important ways from the problems that motivate type-causal judgments and partly because considerations having to do with causal selection play a more important role in token-causal judgment than in type-causal judgment—again, see the appendix to this chapter.

A common (but by no means universal) view, perhaps particularly among metaphysicians, is that token-causal relations are primary or more fundamental, with type-causal relations being derivative from them. This may suggest that we

[4] For a more detailed discussion of actual cause claims, see Woodward, forthcoming d. This provides additional support for a number of claims made in the appendix.

first have to figure out the correct account of token causation and then use this to construct an account of type causation. My contrary view is that discussion of the various type-causal notions can proceed at least largely independently of an account of token-causal claims—in fact this is the standard practice in the literature concerned with causal inference in statistics and computer science.[5] A related point is that, as discussed in the appendix, token-cause claims are in many respects rather different from the varieties of type-causal claims and have features not shared by the latter. For this reason, my view is that it is a mistake to focus, as some writers do, on token causal claims and assume that their features will transfer smoothly to type causal claims. (Token causation ≠ causation.) Let me add, though, that I don't intend by these claims to suggest that investigations into actual cause judgments are uninteresting—on the contrary. This topic has recently been the subject of both important empirical investigations and computational theorizing.[6]

In what follows I will adopt the convention of using lowercase letters (*c*, *e*) to refer to the relata of token-causal claims and uppercase letters (*C*, *E*) to refer to the relata of type-causal claims.

2.2. Difference-Making Theories: Regularity Versions

It is fruitful to divide philosophical accounts into theories that conceptualize causes as "difference-makers" (where this in turn might be explicated in a variety of different ways) and those that do not. The basic idea of difference-making (DM) accounts is that causes "make a difference" to their effects. Let us focus on type-causal claims like "*C* causes *E*," where *C* and *E* are binary and where the situation is one in which no other potential causes of *E* are present—no overdetermination, no additional backup factors that will cause *E* if *C* does not, and so on.[7] According to DM accounts, such claims are understood as having to do with

[5] One reason for thinking that this ("can proceed independently") claim is correct is that, as a number of examples including (2.1)/(2.2) show, it appears that one can have full information—call this *I*—about the relevant type-causal relations and about which token events occur, and yet for it to be unclear which token or actual cause relations obtain (or even what evidence would settle this question). In other words, it appears that we can know what is going on at the type level, as specified by *I*, and reason successfully about this, without knowing what actual cause relations obtain (or even whether there is a determinate answer to questions about this). This suggests that even if is true that actual cause relations are primary, metaphysically speaking, we don't have to appeal to an account of these in order to provide an account of the various more type-level notions.

[6] See, for example, Halpern 2016; Hitchcock 2017; and Icard et al. 2017, among others. See the appendix to this chapter for additional discussion.

[7] Suppose the situation is one in which causal overdetermination or a backup cause is present. Then *C* will not make a difference for *E* in the sense that the absence of *C* will be followed by the absence of *E*. Advocates of difference-making theories (in particular theories that understand difference-making in terms of counterfactuals—see subsequent discussion) have explored a variety of strategies for dealing with such problem cases, mainly but not exclusively in cases involving token

a contrast between (i) a situation (or situations) in which *E* occurs and (ii) some alternative situation (or situations) in which *E* does not, with the presence of the cause *C* being what makes the difference between (i) and (ii).[8] Thus causation has to do with the existence of a relationship of *dependency* (or *contingency*, as psychologists often say) of some kind between *C* and *E*. There are a number of ways of unpacking this idea of difference-making, which vary along a number of different dimensions, including whether the alternative (ii) is taken to be some actually occurring situation or merely possible or hypothetical.

An example of a theory with the former structure is Mackie's (1974) INUS condition account[9]: *C* causes *E* if and only if *C* is a *nonredundant* part (where *C* is typically but not always by itself *insufficient* for *E*) of a *sufficient* (but typically not necessary) condition for *E*. (Again, *C*, *E*, and the other factors that enter into the account are taken to be binary.) The relevant notions of sufficiency, necessity, and nonredundancy are explicated in terms of regularities: short circuits *S* cause fires *F*, because *S* is a nonredundant part or conjunct in a complex of conditions *X* (which might also include, for example, the presence of oxygen *O*) that are sufficient for *F* in the sense that $X = S.O$ is regularly followed by *F*. *S* is nonredundant in the sense that if one were to remove *S* from the conjunct *S.O*, *F* would not regularly follow, even though *S* is not strictly necessary for *F* since *F* may be caused in some other way—for example, through the occurrence of a lighted match *L* and *O*. Thus the difference-making role of *S* is captured by the contrast between the presence of *S* in the conjunct *X* (in which case *X* is sufficient for *F*) and its absence from the conjunct, in which case *F* would not follow from the presence of *O* alone in *X*.

In the version just described, Mackie's account is an example of a *reductive* (sometimes called "Humean" or regularity) theory of causation in the sense that it purports to reduce causal claims to claims involving regularities, just understood as patterns of co-occurrence that (it is claimed) do not make use of causal or modal language. A number of philosophers hold that reductive accounts of

causation—see, e.g., Lewis 1986; Woodward 2003; Halpern 2016. I will not discuss these strategies here; the important point is that there are ways of dealing with overdetermination and so on within a broadly difference-making framework, roughly by appropriately refining the notion of difference-making, so that it amounts to difference-making in some appropriately specified context. (This can be spelled out formally.) It thus would be a great mistake to conclude from the possibility of overdetermination, backup causes, and so on that the difference-making picture of causation is misguided.

[8] When the factors or variables figuring in a causal claim can take many different values, the corresponding idea is that at least some variations in the value of the cause variable will make a difference for the value of the effect variable, according to whatever relationship (in the deterministic case this can be represented by a function) connects the two, as in $a = F/m$. In other words, at least two different values of the cause variable must be mapped into different values of the effect variable.

[9] INUS is an acronym for *insufficient* but non-redundant part of a condition which is itself *unnecessary* but *sufficient* for the result.

causation are desirable or perhaps even required—a viewpoint that does not seem to be widely shared outside of philosophy.[10]

As described, Mackie's account assumes that the regularities associated with causal claims are deterministic. It is possible to construct theories that are similar in spirit to Mackie's, but which assume that causes act in accordance with probabilistic regularities. Theories of this sort, commonly called *probabilistic* theories of causation (e.g., Suppes 1970; Ells 1991), are usually formulated in terms of the idea that C causes E if and only if the probability of E conditional on C in some suitably chosen background condition(s) K is higher than the probability of E in K when C is absent. Following the standard convention in presenting such theories of taking C and E to represent the presence of these factors and $-C$ to represent the absence of C, we may express this as

$$C \text{ causes } E \text{ iff } Pr(E/C.K) > Pr(E/-C.K) \tag{2.3}$$

for some appropriate K. (It is far from obvious how to characterize the appropriate K, particularly in non-reductive terms,[11] but I put this consideration aside in what follows.) Provided that the notion of probability is itself understood non-modally—for example, in terms of relative frequencies—(2.3) is a probabilistic version of a regularity theory. What (2.3) attempts to capture is the notion of a *positive* or *promoting* type cause as opposed to a more general notion of C's being causally relevant, either positively or negatively, to E. Spraying weeds with weed killer is causally relevant to their survival in the sense that it makes a difference to whether they survive, but it does not promote survival; by contrast, moderate watering likely does. To anticipate later discussion, the familiar measure of contingency $\Delta p = Pr(E/C) - Pr(E/-C)$ used in the psychological literature as one measure of causal strength can be thought of as related to a version of (2.3) in which one conditionalizes only on C and $-C$ and takes the difference between the probability of E under these two conditions as indicating C's strength as a positive causal factor.

Suppose that we take the notion of "regularity" literally in the theories (deterministic and probabilistic) just described and require that regularities must

[10] For example, this assumption about the desirability of a reduction is not made in many areas of statistics that deal with causation or in machine learning or econometrics. That the notion of causation that we in fact operate with is one that can be reduced to claims about regularities is also not assumed in much of the discussion of causal cognition in psychology.

[11] For discussion see Cartwright 1983 and Woodward 2003. The general problem is that specification of K seems to require not just information about other causes of E besides C but also how those other causes are connected (i.e., via which causal routes or paths) to C and to one another—"connected" in the sense described by the directed graph or structural equations representations discussed in Section 2.4. Needless to say, specifying all of this in non-causal terms is not straightforward.

have "instances" that actually occur—that for there to be a regularity involving smoking and lung cancer, there must be actually occurring cases of smoking followed by lung cancer and so on. Then the INUS account will not judge that smoking causes lung cancer if no smoking ever occurs and similarly if there are no instances of nonsmoking. Relatedly, in the probabilistic case if $Pr(C)$ or $Pr(-C) = 0$, (2.3) will be undefined.[12]

One question we might ask is whether, as an empirical matter, people think about causal relationships in this way—that is, whether, if there are no cases in which *C* (or not *C*) occurs, people regard the claim that *C* causes *E* as undefined or lacking in truth value. Presumably this would mean that (i) they judge that there is no fact of the matter about whether *C* causes *E* or something similar. This should be distinguished from their thinking that (ii) "*C* causes *E*" has a truth value but that they lack sufficient information to determine what that truth value is. Both casual observation and more careful empirical investigation suggest, unsurprisingly, that many people opt for the second alternative (ii) in cases in which *C* always (or never) occurs—see, for example, the discussion of Cheng's power PC model in Chapter 7.

A standard philosophical criticism of regularity theories like Mackie's is that they lack the resources to distinguish causal from non-causal correlational relationships. For example, in a case in which *C* acts as a common cause of two joint effects *X* and *Y*, with no direct causal connection between *X* and *Y*, *X* may be an INUS condition for *Y* and *Y* may be an INUS condition for *X*, even though, by hypothesis, neither causes the other. Similarly, when *C* causes *E*, *E* may be an INUS condition for *C*, as well as *C* being an INUS condition for *E*, so that the theory appears to fail to distinguish cases in which *C* causes *E* from cases in which *E* causes *C*. Parallel problems arise for probabilistic theories of causation insofar as these are pure regularity theories, although I will postpone more detailed discussion until Section 2.7.[13]

[12] In the case of probabilistic theories, it is usually assumed that "*C* causes *E*" should be understood with reference to some underlying population; thus the non-occurrence of *C* or -*C* means non-occurrence within this population.

[13] Very briefly: even on the assumption of strong connecting principles (like the Causal Markov and Faithfulness assumptions described in Section 2.7) linking probabilistic information and causal structure, probabilistic information in the form of (conditional) dependence and independence relations may fail to uniquely determine causal structure. In fact this is generally the case—see Spirtes, Glymour, and Scheines 1993/2000 for the extent of this underdetermination. This is compatible, however, with the possibility that information about the full joint probability distribution (i.e., information in addition to independence and dependence information) governing the candidate effect and cause variables uniquely fixes causal structure, at least given certain generic restrictions on the functional forms linking cause and effect. As best I am aware, there are no formal results that establish such a unique fixing of causal structure for arbitrary probability distributions, although there are results of a more restricted (but still relatively general) nature concerning inferences assuming particular functional forms. I focus in what follows on the role of independence and independence information in fixing causal structure since here there are formal results and probabilistic theories of causation make use of this sort of information. Information about the temporal order of variables can

These (apparent) "counterexamples" point to an accompanying methodological issue: causal claims are (at least apparently) often underdetermined by evidence having to do just with correlations or regularities (or at least regularity evidence of a sort we can obtain). There may be a number of different, incompatible causal claims that are not only consistent with but even imply the same body of correlational evidence. For example, in a probabilistic context, the correlational evidence may be consistent both with a chain structure $X \rightarrow C \rightarrow Y$ and a common-cause structure $X \leftarrow C \rightarrow Y$, since both imply the same conditional independence or screening-off relations—see Section 2.7.

It is widely recognized by scientists that such underdetermination is common. Typically it is addressed by making use of additional plausible assumptions that go beyond the immediately available pure regularity or correlational information. These assumptions are then used in conjunction with the regularity information to restrict the class of possible causal structures. Indeed identifying such assumptions is a primary methodological focus of many of the accounts of causal inference and learning in the non-philosophical literature. The assumptions themselves can be either relatively domain general (as with the Markov and Faithfulness assumptions discussed in Section 2.7) or much more specific. For example, they may involve the assumption that certain variables are causally or probabilistically independent of others,[14] or that the causal relations of interest conform to particular functional forms. However, in none of these cases are the assumptions reducible in any obvious way to claims about regularities, thus apparently undermining the reductive aspirations of regularity theorists.

It is of course an empirical question whether ordinary subjects (and not just scientists) are also aware of this underdetermination problem, but as we shall see, the evidence seems to support a positive answer. Assuming that subjects sometimes "solve" this underdetermination problem (or at least infer to one or a limited number of causal hypotheses from all those that are consistent with the observed regularities), they must be making additional assumptions of some kind that lead them to a more restricted set of causal hypotheses (or they must be doing something equivalent to this, such as only considering a more limited set of hypotheses). It thus becomes an interesting question in empirical psychology which such additional assumptions subjects tend to make and whether those assumptions enable correct or reliable inferences, according to some standard of correctness. Both of these questions will be addressed (and answered in the affirmative) in subsequent chapters.[15]

also play an important role in addressing underdetermination claims—see, e.g., Stern, forthcoming. However, as noted later in the chapter and in Chapter 4, such information is not always available.

[14] As with so-called identifying restrictions in econometrics.

[15] Two additional observations. First, note that this is another case in which normative inquiry (revealing the presence of an underdetermination problem) suggests an empirical psychological

One possible response that regularity theorists may make to these observations about underdetermination is to appeal to temporal and perhaps spatial information in addition to information about regularities. For example, if causes must temporally precede their effects, this might be used to distinguish (i) *C* causes *E* from (ii) *E* causes *C* when both are INUS conditions for each other.[16] Similarly it may be possible in some cases to use temporal information to distinguish a chain from a common-cause structure. Both temporal and spatial information is generally regarded as acceptable for use in a Humean, reductive program. Perhaps then underdetermination problems may be resolved or at least made less severe, both as an empirical and as a normative matter, in this way.

I will say more about this suggestion about the use of temporal information in a later chapter, and here will confine myself to a few remarks. First, there is plenty of empirical evidence that humans are sometimes able to use temporal information to facilitate causal learning and to distinguish among different possible causal structures, so that there is certainly some cogency to the suggestion.[17] Of course temporal information is particularly useful in inferring causal direction when presented with two correlated variables. On the other hand, there are also many cases in which the temporal information that would be required to distinguish among alternative causal structures is unavailable, sometimes even in principle.[18] Moreover, as a matter of methodology, there are strategies and procedures that do not rely on temporal information and that can sometimes be used to distinguish among different causal structures (see, e.g., Hausman 1998; Janzing et al. 2012; Spirtes et al. 1993/2001; and Woodward, forthcoming b). To the best of my knowledge, there is little or no empirical work on the extent to which ordinary people employ such strategies, but it would not be surprising if they sometimes do. It seems clear that scientists sometimes employ such strategies. The conclusion that I draw from this is that, as a matter of both methodology

hypothesis: that people make additional assumptions that enable them to solve or mitigate that problem. Second, another standard way of putting these points about underdetermination is that subjects engaged in causal inference face an "inverse" problem, just as the visual system does. (See, e.g., Gopnik et al. 2004.) The visual system faces the problem of inferring "inversely" to recover information about the three-dimensional world of objects and their spatial relationships from perceptual information falling on the retina (where the latter is in part generated by the former). A number of different hypotheses about the 3D structure of the environment are consistent with the information available to the retina. Similarly subjects involved in causal inference face the problem of recovering information about causal relationships from correlational and other sorts of information. In both cases, these problems are insoluble without additional assumptions.

[16] Note that this solution will not be available if people are able to distinguish cause and effect when the two are simultaneous, as the evidence suggests that they are sometimes able to do.

[17] See, e.g., Rottman 2017.

[18] It may be, for example, that the variables involved in some candidate causal relationship are defined on time scales such that there is no well-defined sense in which one of them temporally precedes the other. See Woodward 2019.

and the empirical psychology of causal inference, (i) we cannot always rely on temporal information in conjunction with regularity information to distinguish among alternative causal structures and (ii) there are other non-temporal considerations that are relevant to distinguishing among such structures.[19]

So far I have neglected a more radical argument apparently available to the regularity theorist: to the extent that we face an underdetermination problem in which apparently different causal structures cannot be distinguished on the basis of regularity (and perhaps temporal) information, why not conclude that there is no fact of the matter about which of these structures obtain or that there is no factual difference between the structures? Suppose, for example, *C* is an INUS condition for *X*, *C* is an INUS condition for *Y*, and *X* and *Y* are INUS conditions for each other. Rather than worrying that this information underdetermines whether, say, (i) *X* causes *C* which causes *Y* or alternatively, (ii) *C* is a common cause of *X* and *Y*, why not conclude instead that there is no fact of the matter about which of (i) or (ii) (or some other alternative) is the correct structure? Why not say instead that all that "really" exists are the facts about regularities described by the INUS condition claims? We humans may think that some additional structure is present corresponding to either (i) or (ii) (maybe because for some reason we "project" certain inductive expectations we have formed and not others) but we may be mistaken to think this. Indeed, one might think that this is a very natural strategy for a regularity theorist to adopt—if all that really exists is regularities, why try to find a basis for additional distinctions among these, taking some regularities to reflect causal relations and others not to do so?[20]

At this point, the functional perspective advocated in Chapter 1 can be helpful. This leads us to ask: what, if any, function might be served by the additional structure that appears to go beyond regularity (or correlational) information—what work might it do (in addition to recording information about which regularities obtain)? One of the strengths of the interventionist account of causation is that it attempts to provide an answer to this question: the additional structure should be thought of as encoding information about what would happen if various interventions were to occur. Such information is important to us because, for example, of its role in predicting the outcomes of manipulations and in planning. Because this is information has to do with whether certain counterfactuals are true (counterfactuals having to do with the outcomes of interventions), it is underdetermined by purely regularity or correlational information, but it is nonetheless information we care about obtaining and representing. This is what the additional structure beyond regularity information does.

[19] Note that this suggests that although it may be true that effects are never temporally prior to their causes, there is more to the notion of causal priority or direction than temporal priority. This something more is what non-temporal strategies for inferring causal direction pick up on.

[20] And even if we can find such a basis, why should we care about distinctions based on it?

I will postpone more detailed discussion of this until Section 2.4 (and Chapter 4), but note the following points: (a) To the extent that we can support the claim that this additional information (about what would happen under interventions) reflects fact about what the world is like (it is "objective" in the minimal realist sense described in the introduction) and to the extent that such information is sometimes discoverable and can be assessed for reliability or correctness, we have an answer to the question posed in the previous paragraph about why we should go beyond representing actually obtaining regularities.[21] Assuming this and that the additional structure represented by (i) (*X* causes *C* which causes *Y*) versus (ii) (*C* is a common cause of *X* and *Y*) outruns what is present in regularity information, something important would be lost if we just represented the regularity information. (b) Note that this is *not* an argument that (just) appeals to the "intuition" that there is a difference between (i) and (ii) or to the observation that our concept of causation recognizes such a difference, taking this by itself to establish the inadequacy of the regularity account. Instead the argument is that we have good reasons (in terms of our goals) to distinguish (i) from (ii)—that it is rational to have a concept of causation that does so and that an account of causation that failed to make this distinction would fail to fulfill a function that our actual concept does fulfill. Arguably it is these functional considerations that underlie the "intuitions" we have about the difference between (i) and (ii). Of course, as already remarked, this argument also requires that there be some "objective" or "worldly" basis for the distinction between (i) and (ii). The claim of functional usefulness also requires that there be reliable methods for determining whether (i) or (ii) is correct. I contend that both of these requirements are met.

2.3. Counterfactual Theories

An alternative approach to causation, which also falls under the general rubric of difference-making theories, connects causation to counterfactuals, where the connection may or may not take the form of a reduction. Within philosophy a very influential version of this approach is Lewis 1973a. Lewis's theory is an

[21] To reiterate a point made earlier: I am *not* claiming that the additional structure that goes beyond regularity information is guaranteed to have an objective basis simply because we think about causation in terms of such additional structure. An independent argument needs to be supplied that the additional structure is objective in some relevant sense (supported by features of the world) and that there are ways of discovering it. Such an argument can be provided: we can sometimes check whether claims about additional structure are correct by performing the relevant interventions. It is hard to understand what this might involve if the claims about additional structure are not the sorts of things that can be objectively correct or incorrect. It is also the case that with the right evidence we are able to reliably predict the results of interventions not yet undertaken—again something hard to understand if the additional structure is not "out there."

account of token causation. He begins by formulating a notion of counterfactual dependence between individual events: *e* counterfactually depends on event *c* if and only if the following counterfactuals are true:

If *c* were to occur, *e* would occur (2.4)

and

If *c* were not to occur, *e* would not occur. (2.5)

Lewis then claims that *c* causes *e* iff there is a *causal chain* from *c* to *e*: a finite sequence of events *c*, *d*, *f*, . . . *e* such that *d* causally depends on *c*, *f* on *d*, . . . and *e* on *f*. Causation is thus understood as the *ancestral* or transitive closure of counterfactual dependence.[22] Lewis appeals to causal chains because (he claims) this allows him to deal with certain difficulties involving causal preemption that arise for simpler versions of a counterfactual theory.

The counterfactuals (2.4) and (2.5) are in turn understood in terms of Lewis's account of possible worlds: roughly, "If *c* were the case, *e* would be the case" is true if and only if the possible worlds in which *c* and *e* are the case are "closer" or more similar to the actual world than any possible world in which *c* is the case and *e* is not. Closeness of worlds is understood in terms of a complex similarity measure involving a number of different criteria that are ranked in order of importance. These are described by Lewis as follows:

(S1) It is of the first importance to avoid big, widespread, diverse violations of law.
(S2) It is of the second importance to maximize the spatiotemporal region throughout which perfect match of particular fact prevails.
(S3) It is of the third importance to avoid even small, localized simple violations of law.
(S4) It is of little or no importance to secure approximate similarity of particular facts, even in matters that concern us greatly. (1986, 47)

Lewis takes these criteria to imply that two worlds that exhibit a perfect match of matters of fact over most of their history and then diverge because of a "small

[22] As critics have noted, this makes causation transitive even though this does not seem correct from a normative perspective (see, e.g., Hitchcock 2001; Woodward 2003) and also does not seem to fit ordinary causal judgment as an empirical matter.

miracle" (a local violation of the laws of nature) are more similar than worlds that do not involve any such miracle but exhibit a less perfect match in matters of fact. Like regularity theorists, Lewis aspires to provide a reductive account. This requires that the similarity measure must not itself incorporate causal information, on pain of circularity—for example, the measure cannot assume as a primitive that two worlds are similar to the extent that the same causal facts obtain in both. Using this similarity measure, Lewis argues, for example, that the joint effects of a common cause are not, in the relevant sense, counterfactually dependent on one another and that while effects can be counterfactually dependent on their causes, the converse is not true. The similarity measure thus enforces what is sometimes called a *non-backtracking* interpretation of counterfactuals, according to which, for example, counterfactuals that claim that if an effect were not to occur, its cause would not occur are false. In this way, Lewis attempts to use counterfactuals to capture differences among causal structures that regularity theories have difficulty distinguishing. For example, in a common-cause structure in which c causes x and c causes y with no causal relation between x and y, and no other causes present ($x \leftarrow -c \rightarrow y$), it will be true (according to Lewis's interpretation of counterfactuals) that if c had not occurred, x would not have occurred, that if c had not occurred y would not have occurred and not true that if x had not occurred, y would not have occurred. An alternative causal structure like $x \rightarrow c \rightarrow y$ will be associated with a different set of counterfactuals.

Lewis's theory is, as I have said, an account of token causation. It is also natural to connect type-causal claims of various sorts to counterfactuals. For example, if

Smoking causes lung cancer, (2.1)

then one might expect it to be true that

Lung cancer is counterfactually dependent on smoking (2.6)

under some interpretation of (2.1)–(2.6). However, Lewis has little to say about an account like his might be extended to the type case. The interventionist account described in Section 2.4 provides one possible treatment of claims like (2.1) and (2.6).

Lewis's similarity measure attempts to provide an answer to a very general question that arises when counterfactuals are employed. Counterfactuals, as their name implies, encompass[23] situations in which their antecedents are "contrary to

[23] I have written "encompass" because I will follow the standard usage of philosophers of extending

fact." Envisioning a situation S in which a contrary-to-fact antecedent *C* holds requires not just that *C* be thought of as true in S but that some other features of S also depart from actuality—for example, effects that will follow if *C* is the case but which do not obtain in the actual world will hold in S. On the other hand, the use of counterfactuals requires that many features of the actual world be retained in S. We thus face the general question of what should be changed (from the actual situation) and what should not be changed (held fixed) in assessing the truth of a counterfactual. When I claim, in a situation in which I do not drop the wine glass, that if I were to drop it, it would fall to the ground, it seems natural to consider a situation S (a "possible world," according to Lewis) in which I release the glass, but in which much else remains just as it is in the actual world—gravity still operates; if there are no barriers between the glass and ground in the actual world, this is also the case in S; and so on. However, unlike what happens in the actual world, the glass falls to the ground in S. Lewis's similarity measure attempts to answer the question of what should be altered and what should be left unchanged when we consider a situation in which the antecedent of a counterfactual is true.

As is usual in philosophy, there are many purported counterexamples to Lewis's theory involving cases in which (it is claimed) the theory yields judgments that do not agree with what "we" find intuitive. From the perspective of the philosopher or methodologist of science, however, a more central difficulty is this: the various criteria that go into the similarity measure and the way in which these trade off with one another are very vague and unclear—so much so that a common reaction in the scientific literature is that the theory fails to provide useful guidance for the assessment of counterfactuals in many scientific contexts. For example, even if the notion of "miracles" is accepted as legitimate, it is left unclear how one is count the number of miracles or how to tell how diverse or "big" they are. Moreover, although the criteria S1–S4 are ordered in terms of "importance," it is left unclear whether, for example, a sufficiently big gain in perfect match of particular fact can outweigh the postulation of slightly "bigger" violation of law or whether avoiding the latter always has lexical priority over the former. As a consequence, although one occasionally sees references in passing to Lewis's theory in non-philosophical discussions of causal inference problems (usually when the researcher is attempting to legitimate appeals to counterfactuals), the theory is rarely if ever actually *used* in problems of causal analysis and inference in science.[24] With respect to ordinary people (as opposed

"counterfactual" to include some conditionals of form "if p were the case . . ." with true antecedents, so that not all counterfactuals have false antecedents.

[24] See, e.g., the assessment in Heckman 2005, discussed in Woodward 2016b. It is also worth underscoring that insofar as Lewis's theory is mainly or exclusively an account of token causation,

to research scientists), the theory does seem to do a reasonable, although far from perfect, job of reproducing lay judgment, but of course it is a further question whether ordinary subjects use anything like Lewis's apparatus in arriving at their judgments. I will return to this issue in Chapter 4.

Another problem with Lewis's theory that is particularly salient from a functional perspective is this: it is not accompanied by any story about the point or function of the concept Lewis describes. That is, Lewis provides no account of why we should have or make use of a notion according to which causation is understood as the ancestral of counterfactual dependence and counterfactual dependence in turn is understood in terms of the particular similarity measure Lewis describes. (The measure itself looks rather baroque, and its normative rationale is far from transparent). Instead, Lewis's justification for his similarity measure is (at least in large part) simply that it reproduces our ordinary judgments reasonably well. If we ask why we should not employ some different similarity measure (or measures) that would result in a causal concept that behaves somewhat differently and issues in different particular causal judgments, Lewis has (at least as I see it) no answer except that he is trying to explicate our concept and judgments and that his framework captures these. Ideally, from a functional perspective one would like more—some account of what job the resulting concept is doing and whether it is well designed for that job.

Given this and the unclarity of the criteria that go into Lewis's similarity measure, there are grounds for skepticism that ordinary subjects rely on or use this measure in arriving at their judgments, however well his theory may fit these judgments post hoc.[25] Nonetheless, many features of Lewis's theory are (as we shall see in Chapter 4) of considerable interest when construed as psychological theses. As an empirical matter, both adults and children do readily associate causal claims and counterfactuals in something roughly like the way that Lewis's theory suggests. Moreover, despite the limitations of Lewis's similarity measure, there is empirical evidence that often when one evaluates a counterfactual, one does something broadly similar to what Lewis's account suggests: one imagines a situation in which the antecedent of the counterfactual holds, much else is

it will fail to apply to many causal judgment problems in science and also many common-sense problems that are not concerned with token causation.

[25] That is, given the vagueness of the criteria, one might wonder to what extent they can usefully guide ordinary people in making causal judgements, even if, because of their flexibility, the criteria can be interpreted so that they fit many judgments post hoc. Moreover, given their vagueness, one might also think that people are likely to interpret and apply the criteria differently, in which case one would expect less agreement in causal judgment than actually exists. These are reasons to doubt that the *details* of Lewis's theory describe a "psychologically real" process for the assessment of counterfactuals. As I have remarked, this is consistent, however, with a number of the general claims associated with Lewis's account (such as the overall connection between causation and counterfactuals) being psychologically plausible.

left unchanged, and then one "runs" a mental simulation forward to see what happens in the imagined situation—this is suggested not just by introspection but by at least some empirical research (again see Chapter 4). Moreover, the empirical evidence shows that the counterfactuals people associate with type causal claims (and which they sometimes take to help *elucidate* such claims[26]) are indeed non-backtracking counterfactuals, even if these counterfactuals don't always behave as Lewis claims.[27] When presented with counterfactuals, people do seem (at least sometimes) able to make judgments regarding how similar their antecedents are to the actual world, and such judgments influence some aspects of causal judgment, including but not limited to judgments having to do with "causal selection."[28] These empirical observations fit with Lewis's willingness (noted in Chapter 1) to suggest or assume that his theory and his accompanying similarity measure capture at least some aspects of the way in which people assess counterfactuals and causal claims in ordinary contexts.

Somewhat ironically, the outsize influence of Lewis's theory of counterfactuals in portions of philosophy combined with its apparent vagueness and the paucity of serious scientific applications has helped to encourage skepticism about counterfactuals among philosophers of science and in other quarters. Caricaturing only slightly, the inference goes like this: the best theory we have for understanding counterfactuals is something like Lewis's theory. However, although the notions on which it relies (similarity relations among possible worlds and so on) may capture aspects of ordinary reasoning regarding counterfactuals, these are too unclear, epistemically inaccessible, and metaphysically extravagant for scientific use. Science (the argument goes) is concerned with this world, not with goings on in possible worlds distinct from but as real as our own, which is what Lewis claims his theory commits us to. If Lewis is right about counterfactuals, it is hard to see how they can play any legitimate role in science or in good causal reasoning more generally.

[26] By "elucidate" I have in mind the use of counterfactuals to clarify type causal claims—see Chapter 4. Let me add that I do *not* mean to deny that backtracking counterfactuals are used in some causal reasoning contexts—for example, in diagnostic reasoning. But when I infer backward from the occurrence of *e* to a possible cause *c* of *e*, I am not using a backward counterfactual to elucidate what it is for *c* to cause *e*. The counterfactual that elucidates this causal claim will a non-backtracking counterfactual.

[27] Call a counterfactual an "interventionist counterfactual" if it is a counterfactual the antecedent of which is made true by an intervention (see Section 2.4). As noted later, such counterfactuals behave broadly like Lewisian non-backtracking counterfactuals but differ from them in several important respects. As an empirical matter, the counterfactuals ordinary subjects employ in elucidating causal claims seem to behave like interventionist counterfactuals.

[28] In a well-known experiment, Kahneman and Tversky (1982) described a scenario in which two people arrive at an airport half an hour after the scheduled departure of their flights. C's flight has left on time, while T's flight was delayed and left just minutes before he arrived. Subjects judge that T will show more regret—a possible world in which T makes his flight seems "closer" to the actual world than a possible world in which C makes his flight, thus illustrating the role of such similarity judgments in people's reasoning.

I think this skepticism about counterfactuals is misguided. Science is full of counterfactual claims, and there is a great deal of useful theorizing in statistics and other disciplines that explicitly understands causation in counterfactual terms. What is mistaken in the preceding inference is the assumption that counterfactuals can only be understood in terms of a Lewisian semantics involving possible worlds. In other words, problematic features of that semantics are being used to cast doubt on the usefulness of counterfactuals generally. As has become clear over the past few decades, counterfactual claims in science as well as in disciplines like statistics and computer science can instead be explicated by means of devices like equations, directed graphs, and other formalisms with explicit rules governing the allowable manipulations of contrary-to-fact antecedents and what follows from these. Unlike the Lewisian framework, these can be made precise and applicable to real scientific problems.[29] Moreover, as we shall see, these representational devices can also fruitfully be used to model aspects of causal cognition in ordinary subjects.

2.4. Interventionist Theories

These can be thought of as one particular version of a non-reductive counterfactual theory (and thus also a version of a difference-making theory). The core idea is that what is distinctive about causal relationships (as opposed to relationships that are merely correlational) is that these are potentially exploitable for purposes of manipulation and control. Roughly speaking, if *C* causes *E*, then if we (or nature) were able to manipulate *C* in the right way, there would be some associated change in *E*. I have described one version of such a theory at length in Woodward 2003. Here I will provide only a sketch.[30]

[29] Another example in addition to those discussed subsequently of an important framework that relies heavily on counterfactuals to elucidate methodological problems associated with causal inference is the *potential outcomes* framework for understanding causal claims developed by Rubin (1974) and Holland (1986). This is now widely employed both in statistics and in econometrics. In a simple version, causation is conceptualized in terms of the responses to possible treatments imposed on different "units" u_i. The *causal effect* of treatment t with respect to an alternative treatment t' for u_i is defined as $Y_t(u_i) - Y_{t'}(u_i)$ where $Y_t(u_i)$ is the value Y would have assumed for u_i if it had been assigned treatment t and $Y_{t'}(u_i)$ is the value Y would have assumed had u_i instead been assigned treatment t'. (The definition of causal effect is thus explicitly given in terms of counterfactuals.) When dealing with a population of such units, and thinking of $Y_t(u)$ and $Y_{t'}(u)$ as random variables ranging over the units, the average or expected effect of the treatment can then be defined as $E[(Y_t(u)) - (Y_{t'}(u)])$. One can then use this framework and sophisticated extensions of it to characterize the additional assumptions that must be met for reliable causal inference to quantities like $Y_t(u_i) - Y_{t'}(u_i)$ and $E[(Y_t(u)) - (Y_{t'}(u)])$.

[30] Readers unfamiliar with this approach are encouraged to consult Woodward 2003, Hitchcock and Woodward 2003, or Woodward and Hitchcock 2003. I hesitate to describe the views of other writers as "interventionist" since they may differ in some respects from the view to which I attach that label, but an approach bearing a family resemblance (and which has greatly influenced my own views) can be found in Pearl 2000. Much of the work on causal inference described in Spirtes, Glymour, and

Although there are interventionist accounts of token cause claims, the central focus is on type-level notions of causal relevance. (As noted in Chapter 1, it turns out that there are a variety of such notions, and one of the attractions of the interventionist approach is that it can be used to distinguish among them by associating different kinds of causal claims with concerns with different kinds of manipulation and control.) Interventionists think of the relata of type-level causal relationships as variables. (By contrast, the relata of token-causal claims are most naturally thought of as values taken by variables for particular units or systems.) Variables must be capable of taking at least two values, although they may take more.[31] Examples of variables are mass, which may take any nonnegative real value measured in, for example, grams, and number of days worked in a given year, which may take any integer value from 0 to 365. Within this framework, we may think of values of variables as "possessed" by or as characterizing particular units or systems—as when a particular cannon ball weighs 5 kg and a particular person, Jones, smokes or not. (Here the variable is binary and takes two possible values—{smokes, does not smoke}.) When we are concerned with type-level causal claims, what we often have in mind is a claim that has to do with a relationship that holds in some actual or hypothetical population of such units—for example, in the case of (2.1) a population of people, who either smoke or not and who either have lung cancer or not.

To keep things as simple as possible, let us suppose first that we are dealing with a population that is homogeneous with respect to the causal relationship of interest in the sense that the relationship (if any and whatever it may be) is the same for all individuals in the population. What follows is a characterization of one possible type-level causal notion, which I will label TC (for total cause) since it corresponds to C's having a total or overall effect on *E:*

(TC) *C* (a variable representing the putative cause) causes *E* (a variable representing the effect) in background circumstances B iff (i) there is some possible intervention that changes the value of *C* such that (ii) if that intervention were to occur in B, there would be an associated change in the value of *E* or in the probability distribution *P*(*E*) of those values.

Scheines 2001 and subsequent papers can (I think) be naturally interpreted in terms of an interventionist conception of causation, although this work does not *require* such an interpretation.

[31] The properties or event-types that figure centrally in philosophical discussions of causation can be conceptualized as a special case—as binary variables that can take values corresponding to the presence or absence of some property. We should note, however, that this is not the way that philosophers usually think about properties—see Woodward, forthcoming e.

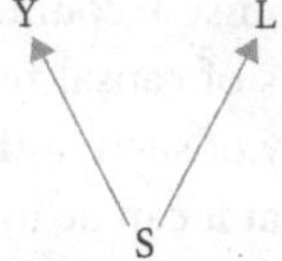

Figure 2.1

This requires some unpacking. First and most importantly, what is an intervention? I will first describe the idea informally, then illustrate it, and then provide a more precise characterization. Interventions *I* on some variable *C* are always relative to a second variable *E*, in the sense that *I* might be an intervention on *C* with respect to *E* and not with respect to some distinct variable *E**. Heuristically, an intervention on *C* is an experimental (or experiment-like) manipulation of the value of *C* that is unconfounded from the point of view of reliably determining whether *C* causes *E*.[32] Put differently, an intervention on *C* with respect to *E* produces a change in the value of *C* such that any change in the value of *E* (should it occur at all) occurs only via a route that goes through *C* and not in some other way. A randomized experiment to determine whether *C* causes *E* is one paradigmatic case of an intervention on *C* with respect to *E*.

To illustrate the idea, let us suppose that (the variable) *S* = *SMOKING* causes both *L* = *LUNG CANCER* and *Y* = *YELLOW FINGERS*—that is, *S* is a common cause of *L* and *Y* but that *Y* does not cause *L* and *L* does not cause *Y*. All variables are binary. Suppose also that these relationships are deterministic as well as common to everyone in the population—that is, when anyone in the population smokes, they develop both lung cancer and yellow fingers and when they do not smoke, they develop neither. In what I hope is a transparent diagrammatic representation (I will say more about such representations shortly) the causal structure is shown in Figure 2.1.

Suppose we manipulate *S*, where "manipulation" means that we change its value from not smoking to smoking or conversely but with no further restrictions imposed. In this way we also manipulate the value of *Y*. Assume that under such a manipulation of *Y*, the value of *L* will also change (since *S* causes *L*). So if our criterion for whether *Y* causes *L* was simply whether *L* changes when we manipulate *Y* in some way, we would mistakenly conclude that *Y* causes *L*. Intuitively a manipulation of *Y* (with respect to *L*) via changing *S* is confounded from the point of view of determining whether *Y* causes *L*—*S* affects *L* directly

[32] Unconfounded means just that—there *are* no confounding causes. It does not mean merely that there are no known confounded causes or that the model we are employing does not contain or represent such causes.

via a route that does not go through *Y*. The notion of an intervention is characterized in such a way as to exclude manipulations that involve this and other sorts of confounding.

Now contrast this with another possible experimental manipulation: we manipulate *Y* in a way that is independent of and uncorrelated with *S*. For example, we might randomly assign colors to subjects' fingers, independently of whether or not they are smokers, either dyeing them yellow or bleaching them. Under such manipulations of *Y*, the value of *L* will not change, assuming Figure 2.1 represents the correct structure. Such a manipulation is an intervention on *Y* with respect to *L*. The intuition behind (TC) is that *Y* causes *L* if and only if were such an intervention to occur, *L* would change. This also illustrates the idea behind the slogan that causal relationships are potentially exploitable for purposes of manipulation and control. If you were to try to exploit the correlation between *Y* and *L* in the common-cause structure in Figure 2.1 by intervening on *Y* in order to change *L* you would fail—under such an intervention on *Y* the correlation between *Y* and *L* would disappear. This corresponds to (or captures) the fact that *Y* does not cause *L*—you can't change whether someone does or does not develop lung cancer by intervening to change the color of their fingers. By contrast, you can change whether someone develops lung cancer by intervening to change whether or not they smoke—this corresponds to the fact that smoking causes lung cancer.

In my 2003, I characterized the notion of an intervention variable more precisely as follows (IV): *I* is an intervention variable for *X* with respect to *Y* iff

IV1. *I* causes *X*.
IV2. *I* acts as a switch for all the other variables that cause *X*. This is, certain values of *I* are such that when *I* attains those values, *X* ceases to depend on the values of other variables that cause *X* and instead depends on the value taken by *I*.
IV3. Any directed path from *I* to *Y* goes through *X*. That is, *I* does not directly cause *Y* and is not a cause of any causes of *Y* that are distinct from *X* except, of course, for those causes of *Y*, if any, that are built into the *I*-*X*-*Y* connection itself; that is, except for (a) any causes of *Y* that are effects of *X* (i.e., variables that are causally between X and *Y*) and (b) any causes of *Y* that are between *I* and X and have no effect on *Y* independently of *X*.
IV4. *I* is (statistically) independent of any variable *Z* that causes *Y* and that is on a directed path that does not go through *X*.

Condition IV2 is meant to capture the idea that the intervention *I* on *X* puts *X* entirely under the control of the intervention in the sense that the value of *X* is determined only by the intervention and nothing else. In the preceding example this corresponds to dyeing or bleaching fingers in a way that is completely independent of whether the subject smokes—finger color is determined for each subject only by dyeing/bleaching and nothing else. The other conditions IV3–IV4 are intended to rule out various possible forms of confounding. For example, the intervention *I* should not be a common cause of *X* and *Y*, should not itself be caused by some variable *Z* that also causes *Y* via a route that does not go through *X*, and so on.

(IV) characterizes what has come to be called a "hard" intervention (cf. Eberhardt and Scheines 2007). It is hard in the sense that, as explained previously, *I* removes the causal influence of all other variables on *X*. This contrasts with "soft" interventions, which do not entirely remove the influence of other variables on *X* but supply an additional exogenous source of variation in *X* in such a way that this variation affects *Y*, if at all, only through its relationship to *X* and is not correlated with other causes of *Y* that are not on the route from *I* to *X* to *Y*. As an illustration, consider an experiment designed to assess whether changes in income level affect frequency of visits to the doctor's office. Suppose the intervention consists in randomly providing a cash grant of $*n* to some subjects (the treatment group) and nothing to others (the control group) and the dependent variable is frequency of visits in both groups. In this case subjects' income reflects both their "endogenous" earnings that are independent of the grant and whatever they receive from the grant so that their (total) income level (the independent variable) is not entirely determined just by the intervention. Nonetheless, if the grants are distributed randomly so they are not correlated with subjects' endogenous incomes or other variables that may affect doctor's visits independently of income, the treatment will have an intervention-like character, and (oversimplifying somewhat) this fact can be exploited to estimate the effect of income on frequency of visits. The *instrumental variables* techniques mentioned previously and widely used in the social sciences and some areas of biomedicine to estimate causal effects can be thought of as procedures that make use of information about soft-intervention-like processes. In cases in which it may not be possible or ethical to perform hard interventions, soft interventions are possible alternatives. Moreover, as Eberhardt and Scheines (2007) show, it is possible to get information about causal relationships from soft interventions that is not available just from hard interventions. Nonetheless, soft interventions also have important limitations, one principal one being that it can be difficult to tell whether one has successfully executed such an intervention (see Eberhardt 2014).

The interventions characterized in (IV) are deterministic: the intervention *I* deterministically fixes the value of *X* for all of the units intervened on. It is possible to relax this requirement as well and consider "probabilistic" interventions that merely impose possible values on the individuals in a population in accord with some probability distribution. Information from such interventions can also be exploited for causal inference purposes.

Although both soft and non-deterministic interventions can be very useful, I will focus in what follows on the notion of a hard, deterministic intervention characterized by (IV). There are several justifications for this choice: first, the connection between hard interventions and causal claims has been the main focus of the psychological literature. Moreover, from a foundational point of view, soft and non-deterministic interventions do not seem to raise additional issues that are not already present in connection with the hard deterministic variety.

The notion of intervention characterized in (IV) makes no reference to human beings, human action, or for that matter the behavior of any animate creature. Instead (IV) is characterized in terms of causal and statistical notions. One consequence of this is that a purely natural process not involving human action at any point can qualify as an intervention if it has the right causal and statistical characteristics. This is attractive from a methodological point of view, since it yields non-anthropocentric notions of intervention and causation. On the other hand, as we shall see, it is important for the empirical psychology of causal cognition that it is fairly common, as an empirical matter, for human actions (and some animal behaviors) to have the characteristics of interventions, either of the hard or soft variety.

As noted above, (TC) connects causal claims and counterfactuals—counterfactuals concerning what would happen to *Y* if there were to be an intervention on *X*. Counterfactuals of this sort, the antecedents of which have to do with the occurrence of interventions, have come to be called *interventionist counterfactuals*. They have a number of features that distinguish them from counterfactuals interpreted in terms of a Lewisian possible worlds semantics.[33] I stress this point to reinforce the claim, made earlier, that it is not inevitable that we understand counterfactuals in the manner advocated by Lewis.

That a plausible version of an interventionist theory of causation needs to be formulated in terms of counterfactuals should be obvious. No one supposes that a causal relationship holds between *X* and *Y* only when there actually is an intervention on *X* (from a functional point of view this would deprive the notion of causation of much of its generality and usefulness); instead what matters is what would happen *if* there were an intervention on *X*.

[33] Although Woodward 2003 described examples that illustrated how Lewis's similarity measure on possible worlds led to different assessments of causal claims than the interventionist framework, I did not fully appreciate at the time how different the two frameworks are. Some additional discussion of these differences can be found in Woodward 2016c as well as in Briggs 2012. There is a lot more to be said on this topic, but not here.

Relatedly, it is *not* part of the interventionist account that one can only learn about causal relationships by observing the effects of experimental manipulations or interventions. Rather there are many different ways of learning about causal relationships, including learning from passive observational data that are not the results of experimental interventions. What the interventionist account claims is that when one learns about a causal relationship (for example, of the sort described by TC), what one learns about is what would happen if an intervention were to occur. In other words, the idea is that we should conceptualize causal learning from non-experimental observations in the following way: what one is trying to learn is what would happen if a certain experiment were performed (or an intervention were to occur) without actually performing the experiment/intervention. As suggested earlier, one systematic approach to answering such a question relies on information from observed correlations in conjunction with additional assumptions of various sorts that are not purely correlational in content, at least in their overt meaning. Although some philosophers (and perhaps psychologists) may wonder how it could be useful to think about causal relationships as claims about the outcomes of hypothetical experiments when these experiments may not actually be performed, in fact this basic approach has been widely adopted in the non-philosophical literature on causal inference: for example, as noted earlier, it underlies instrumental variables and regression discontinuity techniques in addition to other inference procedures. I will add that it is also a commonplace in this literature that one can often clarify how causal claims can be interpreted and what evidence is relevant to assessing them by connecting them to outcomes of possible experiments in this way[34] —a fact that is also of considerable psychological interest and which I discuss further in Chapter 4.

As already noted the characterization (IV) makes use of causal notions at several points—most obviously in IV1, which requires that an intervention *I* on *X* involves a causal relationship between *I* and *X*, but also in other conditions as well. This may seem to raise a worry about "circularity": In (TC) "*X* causes *Y*" is elucidated in terms of the notion of an intervention on *X* that is itself characterized using causal notions. This observation shows that (TC) is not a reductive account of causation: it does not explain "*X* causes *Y*" entirely in terms of notions that are themselves purely non-causal (like "regularity" or "counterfactual dependence" when understood along Lewisian lines). I have elsewhere offered several defenses of my contention that (TC) can be illuminating and not viciously circular even if not reductive. I will not repeat these here but will confine myself to the following observations. First, note that while the characterization of an intervention on *X* makes use of information about causal relationships, these

[34] For discussion and illustrations see, e.g., Angrist and Pischke 2009.

are not the same relationships as the relationship between *X* and *Y* that we are trying to elucidate. That is, the characterization of an intervention *I* on *X* refers to a causal relationship between *I* and *X* and to the relationships between *I* and certain other causes of *Y*, but not to the *X*-*Y* relationship itself. In using (TC) we are in effect using our grasp or understanding of some causal relationships (involving *I*) to elucidate another, distinct causal relationship. In other words, the picture is a familiar non-foundational one according to which we use some portions of our background knowledge to understand and learn about others. (There is much more to be said about how, as a developmental matter, we might break into this "circle" of causal relationships—I discuss this briefly in Chapter 4.)

Second, recall the functional perspective advocated in Chapter 1. Although interventionism does not provide a reductive analysis of causation, it can still provide insight into what distinguishes causal from merely correlation claims and why we value causal knowledge. As we shall see, the interventionist framework can also be used to distinguish among different varieties of causal claims. In addition, as observed earlier, researchers in many different disciplines contend that they can clarify the content of various causal claims, distinguish them from other causal claims with which they may be confused, and better understand how they can be tested by connecting them to hypothetical experiments and interventionist counterfactuals along the lines described by (TC). Similar observations seem to hold for causal cognition involving more ordinary subjects. If so, characterizations like (TC) can be useful and illuminating from a functional perspective even if not reductive and can provide insights into how people reason about causal relationships.[35]

I will add that any satisfactory story about how a connection between causal claims and interventionist counterfactuals might be illuminating will need to have a psychological dimension as well as a normative one. That is, we need an account of how this connection can function *psychologically* in a way that is informative, despite its circularity. One possibility is that thinking about causal claims in this way somehow allows one to access relevant information and patterns of inference that are not so salient or readily available when one thinks only about the original causal claim itself without explicitly considering associated counterfactuals and experiments. Roughly, the causal claim may not be initially psychologically represented in a way that makes its counterfactual implications obvious but drawing out these implications can help to clarify the

[35] Put slightly differently, if our goal is to elucidate causal notions and reasoning from a functional perspective, it may not matter if our account fails to be reductive. From a functional perspective, what we want in accounts of causation is that they elucidate use and function. The goal is not an account of what causation "is," where this is understood as independent of issues about function. So even if it is true that non-reductive account of causation cannot tell us what causation is, this need not be seen as damaging to functional projects.

causal claim. This would explain why, for example, when asked whether some causal claim is true, people often try to answer this question by generating and evaluating counterfactuals (as evidence shows that they do—see Chapter 4) even though the counterfactuals so employed seem to be interventionist and thus to presuppose causal notions of some sort.

I stress this point because a common philosophical reaction (and indeed the reaction of some psychologists as well—see Ober 2017) when presented with claims about the relationship between causation and counterfactuals like (TC) is to ask which "comes first" or is more basic or fundamental. Should (i) causation be understood in terms of counterfactuals or (ii) should counterfactuals be understood in terms of causation? (This is the so-called Euthyphro problem posed for interventionism by Menzies 2006.) Lewis opts for the first alternative, (i), but if we employ causal notions in the characterization of interventionist counterfactuals, the first alternative is excluded. On the other hand, if we opt for the second alternative, (ii) (as some philosophers do—see Edgington 2011), this may seem to preclude the use of counterfactuals to elucidate causal claims in the manner of (TC).

My way out of this supposed dilemma is to reject the assumption that (i) and (ii) are the only alternatives. I don't think that either causal notions or counterfactuals are more basic or fundamental than the other. Indeed I suspect categories like "basic" and "fundamental" are often inappropriate or inapplicable when the context has to do with methodology or psychology.[36] From the point of view of psychology, there is no reason why those features of the human mind that have to do with causal cognition and counterfactual reasoning should be organized in terms of a hierarchy of fundamentality, with one being more basic than the other.[37] Similarly, my own view, for what it is worth, is that the question of which is more fundamental seems unlikely to be fruitful in the methodology of causal reasoning.[38]

[36] Perhaps they are more appropriate in other contexts like metaphysics.

[37] Moreover, even if there is a notion of fundamentality that is appropriate in psychology, I see no reason why this should correspond to metaphysical fundamentality.

[38] In Woodward 2003 I described (TC) as well as characterizations of other causal notions as "definitions." In doing so I was following a standard practice in the non-philosophical literature, which is replete with characterizations of various causal concepts that are labeled as definitions even though these employ *definiens* making use of causal concepts—see, e.g., Pearl 2000. Nonetheless this created confusion among some philosophical readers, who interpreted these definitions as claims that the defining terms are "more fundamental" than the defined term or that the defined term "metaphysically depends" on the defining terms. I myself don't see why "definitions" need to be understood in these ways, but in any case I'm happy to drop the term "definition" and to use other words instead for the relationship described by (TC). For example, we might instead think of (TC) as a proposal about the interconnections (or mutual constraints) between a causal notion and interventionist counterfactuals with no accompanying claims about what is most fundamental.

2.4.1. Invariance

Another notion that will be important in our subsequent discussion is the notion of invariance. Suppose a relationship R between *X* and *Y* satisfies (TC). This implies that *Y* will change under some set of interventions on *X* and, when R is specified more precisely (e.g., by an equation as described later), that if R is a correct causal claim, it will correctly describe how *Y* changes under interventions on *X*. When R continues to hold in this way (i.e., correctly describes how *Y* changes under interventions on *X*), I will say that R is *invariant* under these interventions on *X*. Obviously, a relationship R can be invariant under some interventions on *X* and not under others, so that invariance in this sense is a matter of degree. Moreover, in addition to invariance under interventions, we can also characterize the extent to which R is invariant under other sorts of changes. For example, R might be invariant under some range of background conditions for *X* and *Y* and not others, where a background condition is (roughly) some factor that is distinct from *X* and *Y* and not causally between *X* and *Y*. (The color of the shirt an aspirin-taker wears is a background condition for the aspirin ingestion → headache relief relation.) As discussed in more detail in Chapter 5, other things being equal, we seem to prefer, as a descriptive matter, candidate causal relationships that are invariant over a larger range of interventions and other sorts of changes to relationships that are less invariant. Moreover, there seem to be good normative reasons for this preference. In any case, the notion of invariance is a natural supplement to an interventionist treatment of causation, for reasons discussed in Woodward 2003 and subsequent chapters of this book.

2.5. Directed Graphs and Equations as Devices for Representing Causal Relationships

So far I have introduced the notion of an intervention and used it to characterize one causal notion—(TC). However there are a number of other causal notions and relations that can be characterized in interventionist terms. In discussing these it will be helpful to first introduce several additional devices for the representation of causal relationships. These are, in the first instance, devices for public representation (as in research articles), but one may also consider whether they can be used in models of various subjects' mental representations of causal relationships—graphs, in particular, are extensively used for this purpose in the recent psychological literature.

2.5.1. Directed Graphs

I've already used these informally to describe various examples. One may think of a directed graph as an ordered pair V, E, where V—the set of vertices of the graph—contains the variables serving as causal relata[39] and E is the set of directed edges connecting the vertices. An arrow or directed edge ($X \rightarrow Y$) from X to Y means that X *directly causes* Y. The basic idea is that X is a direct cause of Y if X has a causal influence on Y that is not mediated by any of the other variables in V. We may characterize this more precisely as follows:

(DC) X is a direct cause of Y if and only if there exist values for all other variables Z_i ($Z_i \neq X, Y$) in V such that if those Z_i are held fixed at those values by independent interventions, there is a possible intervention on X that will change the value (or probability distribution) of Y.

A sequence of variables $\{V_{1 \ldots} V_n\}$ is a *directed path* or *route* from V_1 to V_n if and only if for all i, ($1 \leq i \leq n$) there is a directed edge from V_i to V_{i+1}. Y is a *descendant* of X (and X is an *ancestor* of Y) if and only if there is a directed path from X to Y. The full set of direct causes of X are also called the *parents* of X—abbreviated as *par* (X).

An important part of the appeal of directed graphs is that they provide a perspicuous representation of different systems of causal relationships of arbitrary complexity—one can represent (i) common-cause structures ($X \leftarrow Y \rightarrow Z$), (ii) chain structures ($X \rightarrow Y \rightarrow Z$), common effect structures (iii) ($X \rightarrow Y \leftarrow Z$), and structures built out of combinations of these. From an interventionist perspective each of these corresponds to a distinctive set of claims about what will happen under various combinations of interventions—each structure has a unique interventionist signature. For example, if (i) is the correct structure, then there will be an intervention on Y that will change X, an intervention on Y that will change Z, no interventions on X that will change Z, and so on. If (ii) is the correct structure, there is an intervention on X that will change Y and an intervention on Y that will change Z but no interventions on Y will change X. Thus (ii) is associated with a different set of claims about what will happen under interventions on X, Y, and Z and similarly for (iii): any distinct graph will have a distinct interventionist signature.

In the directed graph framework the notion of a hard intervention captured by (IV) has a simple representation: an intervention I on X with respect to Y "breaks" all other arrows directed into X, replacing these with an arrow from I to X while leaving all other arrows in the graph undisturbed. For example, in the

[39] A reminder: of course "variables" understood as representations should be distinguished from what they represent in the world, but in order to avoid pedantry I gloss over this in what follows.

$(X \leftarrow Y \rightarrow Z)$
(i)

Figure 2.2

$I \rightarrow X \quad Y \rightarrow Z$
(iv)

Figure 2.3

common-cause structure (i) in Figure 2.2, an intervention on X replaces (i) with (iv) in Figure 2.3.[40]

2.5.2. Equations

In the literature on causal modeling and econometrics causal relationships are also commonly represented by means of sets of so-called structural equations. These describe effect or dependent variables as functions of their direct causes—for example, in

$$Y = F(X_1 \ldots X_n), \tag{2.7}$$

$X_1 \ldots X_n$ represent the direct causes of Y. Systems of causal relationships are represented by systems of equations, one for each dependent (or endogenous variable). For example, the equations

$$Y = aX \tag{2.8}$$

$$Z = bY + cX \tag{2.9}$$

[40] When causal relationships are represented by equations as described immediately below, the corresponding representation of interventions is in terms of "wiping out" equations. An intervention on Y that sets its value to $Y = k$ in the equation $Y = F(X_1 \ldots X_n)$ wipes out this equation, replacing it with the equation $Y = k$ while leaving all other equations in the representation undisturbed. Values of other variables are then updated by substituting k for occurrences of Y in those equations. See Pearl 2001 for more details.

An alternative representation that some may prefer is this: the intervention variable I has an "off" value and one or more values corresponding to its being "on." When I is off, it is present in representation (i), but there is no arrow from I to X. When I is on, (i) is replaced by (iv). I see these representations as communicating the same information.

represent that X is a direct cause of Y and that X and Y are direct causes of Z. (In this particular case, the causal relationships are all linear, but in the more general case, the functions relating X to Y and X and Y to Z can take any form one likes.)

Within the interventionist framework, an equation like (2.9) is regarded as a correct representation of the causal relationships between X and Y and Z if and only if it correctly describes what would happen to the value of Z under interventions on the right-hand side (direct cause variables X and Y). For example, (2.9) is interpreted as claiming that an intervention that changes Y by ΔY will change Z by $b\Delta Y$—the equation is correct if and only if this is the case. Note that under this interpretation the equations (like directed graphs) do not just represent regularities or claims about patterns of association—they are understood as having modal or counterfactual import since they describe what would happen if various interventions were to be performed.

Although directed graphs can be used to represent causal relationships that are fundamentally or irreducibly chancy or indeterministic, most uses of such graphs as well as structural equation models assume that the underlying causal relationships governing the system of interest are deterministic. When the value of Y is not determined by its *known* direct causes $X_1 \ldots X_n$, it is usual to introduce an additive error term U, writing

$$Y = \mathrm{F}(X_1 \ldots X_n) + U.^{41}$$

Here U can be interpreted as representing the combined effect of all additional or omitted (unknown or unobserved) causes of Y besides $X_1 \ldots X_n$. U is typically assumed to be a random variable, taking different values for different individuals in the population of interest in accordance with a well-defined probability distribution. This allows us to treat Y as a random variable as well, with a well-defined probability distribution. Similarly we can add such error terms to directed graphs if we wish.

What is the relationship between the use of directed graphs and the use of equations to represent causal relationships? Assuming that the underlying structure is deterministic, arrows from $X_1 \ldots X_n$ to Y tell us that Y is some function of $X_1 \ldots X_n$. Usually it is also assumed that each of the $X_1 \ldots X_n$ is non-redundant in the sense that for some combination of values of the other variables, there are some changes in each of the $X_1 \ldots X_n$ that when produced by interventions will be associated with change in Y. However, beyond this the directed graph representation does not tell us specifically what this function is. By contrast,

[41] Structures of this sort are called pseudo-indeterministic in Spirtes, Glymour, and Scheines 2000.

the equational representation does tell us explicitly what functions relate the dependent variables to their direct causes. Although graphs thus represent less information than equations, this is not (for all purposes) a defect. As we will see, the topology or connectedness of a graph alone suffices for a number of inferences—one does not have to know the precise functional relationships among the variables. This can make such inference problems simpler and more transparent.[42]

One way of thinking about directed graphs is that they represent certain qualitative aspects of causal relationships—roughly the fact that *E* directly causally depends on *C*, without representing more quantitative aspects of those relationships or specific functional forms—for example, exactly how *E* depends on *C*. (Note that such qualitative relationships correspond to what is captured by (TC) and (DC). It is an interesting psychological question to what extent people's representations of causal relationships take this sort of qualitative form and to what extent such representations embody more specific assumptions relating to functional form and other matters.[43] As subsequent chapters show, the answer seems to be that sometimes people learn and reason in ways that suggest that they are operating with more qualitative representations of a sort that might be captured by directed graphs, and sometimes they seem to be operating with representations that embody more specific assumptions about functional forms. It is also an interesting psychological question to what extent people's causal representations embody a contrast between direct and more indirect causal relationships.[44] (See Chapter 4 for an affirmative answer to this question.)

Finally let me emphasize that although graphs and equations have a natural interpretation in interventionist terms, I do not at all mean to suggest that this is the only way of interpreting them causally or that their use commits one to any particular interpretation of causal notions. One might take $X \rightarrow Y$ or $Y = F(X)$ just to be a way of representing that *X* causes *Y* without any particular commitment to how the notion of "cause" is to be understood. Similarly one might take these devices to represent the presence of a causal relationship that is not to be understood in interventionist terms, although then some alternative interpretation needs to be supplied.

[42] This feature of graphs is systematically exploited in Pearl's work.

[43] Of course these specific assumptions need not be maximally detailed—for example, people might represent that *Y* is some monotonically increasing function of *X* without representing the exact form of the dependence of *Y* on *X*.

[44] Of course some people's causal representations obviously embody such a contrast—those who use directed graphs as public representational devices. My question has to do with the extent to which this is true of ordinary, lay causal thinking.

2.6. Varieties of Causal Relationships

So far I have suggested how the interventionist framework might be used to characterize notions of direct cause and total cause. One of the strengths of that framework is that it can be used to characterize and distinguish among a number of other causal notions, including notions that are not always clearly distinguished in either the philosophical or the non-philosophical literature. As an illustration, return to equations (2.8)–(2.9) and suppose that $a = -bc$. In this case, the influence of X on Z along the two different routes (the direct one from X to Z and the indirect one from X through Y to Z) cancels. Applying (TC) to this case, we see that X is not a total cause of Z since the overall or net effect of any intervention on X will not be to change Z. Nonetheless, there seems to be an obvious sense in which X is a cause of Z (indeed both a direct and an indirect cause of Z). This suggests that in addition to the notion of a total cause (or total effect) we also possess and make use of some notion of a cause along a route (or a contributing cause) or of a path-specific effect that allows us to think in terms of the effect of X on Z along the indirect route that goes through as well as the direct route from X to Z.[45] In fact, as Pearl (2000, 2001) shows, it is possible to characterize such a notion in interventionist terms and indeed to do so even when the relationships involved are nonlinear. As noted earlier, many other sorts of causal notions (e.g., actual causation) can also be given interventionist characterizations.

It is an open question to what extent alternative normative frameworks (e.g., regularity theories or theories that appeal to Lewisian counterfactuals) can capture distinctions among causal notions like those just described. To the best of my knowledge proponents of these frameworks have not tried to do this. The distinctions among these different causal notions is also psychologically interesting in at least two respects. First, there is the obvious question of the extent to which various subjects actually use these different notions in causal reasoning and distinguish among them. Second, verbal probes designed to elicit causal judgments in psychology experiments as well as the intuitions about various scenarios elicited by philosophers typically simply ask whether, for example, X causes Y (or, in the case of some psychological investigations, about the "strength" of the causal relationship between X and Y). But if there are a number of different causal notions—some of them different in subtle ways—it may be

[45] From a functional perspective we need this distinction because the total effect of an intervention on X corresponds to what happens under an intervention on X alone, while the notion of contributing cause or a path-specific effect of X corresponds to what will happen when there is an intervention on X and independent interventions on other variables. In particular, one might characterize the notion of X's being a contributing cause to Y in terms of their being some intervention on X that changes Y when other variables not on the route from X to Y are held fixed at some values via independent interventions (cf. Woodward 2003). These two kinds of information are relevant to different kinds of possible manipulations.

that according to some of them *X* causes *Y* and according to others *X* does not cause *Y*. (For example, *X* may be an indirect cause of *Y* along a path but not a total cause of *Y*.) So one needs to worry that an undifferentiated verbal probe that just asks whether *X* causes *Y* may be ambiguous or interpreted differently by different subjects—we will see examples subsequently. This is another example of a way in which normative theories that make distinctions among causal notions can have implications for descriptive investigations.

2.7. The Markov and Faithfulness Conditions

The discussion in Section 2.6 made no assumptions about the relationship between graphical structure and probabilistic relationships. It is sometimes reasonable to make such assumptions, and when this is the case, they can be exploited in powerful ways in causal inference, as well as being both philosophically and psychologically interesting in their own right. Suppose that there is a probability distribution *P* over the vertices **V** in the directed graph *G*. *G* and *P* are said to satisfy (one version of) the Causal Markov condition (CM) if and only if:

> (CM) For every subset **W** of the variables in **V**, **W** is independent of every other subset in *V* that does not contain the parents of **W** or descendants (effects) of **W**, conditional on the parents of **W**.

(CM) is a generalization of the familiar "screening off" or conditional independence relationships that a number of philosophers (and statisticians) have taken to characterize the relationship between causation and probability. (CM) implies, for example, that if two joint effects have a single common cause, then conditionalizing on this common cause renders those effects conditionally independent of each other. It also implies that if *X* does not cause *Y* and *Y* does not cause *X* and *X* and *Y* do not share a common cause, then *X* and *Y* are unconditionally independent—this is sometimes called the principle of the common cause.

A second useful assumption, called Faithfulness by Spirtes, Glymour, and Scheines 1993, 2001 is

> (F) A Graph *G* and associated probability distribution *P* satisfy the faithfulness condition if and only if every conditional independence relation true in *P* is entailed by the application of CM to *G*.

(F) says that independence relationships in *P* only arise because of the structure of the associated graph *G* (as these are entailed by (CM)) and not for other reasons. This rules out, for example, a causal structure like (2.8)–(2.9) with

$a = -bc$ in which the effects of *X* on Z along two different routes just happen to exactly cancel each other, so that *X* is independent of *Z*, although this independence does not follow from the graphical structure.

Although I will not discuss details here, these two assumptions can be combined to create algorithms that allow one to infer facts about causal structure (as represented by G) from the associated probability distribution *P*.[46] In some cases, the assumptions will allow for the identification of a unique graph consistent with *P*; in many other cases, they will allow the identification of an equivalence class of graphs that may share some important structural features. Thus, in the general case, even given assumptions like (CM) and (F), probability information in the form of independence and dependence information underdetermines causal structure to the extent that distinct graphs represent distinct causal structures. Indeed, an important theorem characterizes the extent of this underdetermination (Spirtes, Glymour, and Scheines 1993, 2000, Theorem 4.2, p. 90). Say that two directed acyclic graphs *G* and *G′* are faithfully indistinguishable if and only if every distribution faithful to *G* is faithful to *G′* and vice versa. Then:

> Two directed acyclic graphs *G* and *H* are faithfully indistinguishable if and only if (i) they have the same vertex set, (ii) any two vertices are adjacent in *G* if and only if they are adjacent in *H*, and (iii) any three vertices *X*, *Y*, and Z such that *X* is adjacent to *Y* and *Y* is adjacent to Z but *X* is not adjacent to *Z* are oriented as $X \rightarrow Y \leftarrow Z$ in *G* if and only if they are so oriented in *H*.

Thus, given a probability distribution *P*, graphs that are consistent with that distribution and that share the same vertices and adjacencies and the same colliders (where a collider is a structure of form $X \rightarrow Y \leftarrow Z$) are indistinguishable by inference procedures relying (just) on (CM) and Faithfulness. For example, the structures $X \rightarrow Y \rightarrow Z$, $Z \rightarrow Y \rightarrow X$, and $X \leftarrow Y \rightarrow Z$, are indistinguishable on the basis of probabilistic independence and dependence information, assuming (CM) and (F).

Probabilistic theories of causation like those briefly described in Section 2.2 commonly assume a connection between causal claims and probability along the lines of (CM) and little or nothing more. (Extant versions of probabilistic theories generally do *not* assume Faithfulness, and without this assumption the underdetermination problem is greatly compounded beyond what is described

[46] See Spirtes, Glymour and Scheines 1993/2001 and subsequent papers by these authors. These conditions—particularly (F)—can be weakened in various ways so that they fit a wider variety of circumstances, and one can then explore which inferences they justify. Of course one can also ask to what extent ordinary subjects use inference procedures like those developed by these authors in causal inference—see, e.g. Gopnik et al. 2004 for an affirmative answer to this question for children involved in simple inference tasks.

by the theorem just described). If we think of causal structure as something that can be represented graphically or by means of structural equations (whether or not these are understood in interventionist terms), then in many cases the apparent underdetermination of causal structure by the probability information (independence and dependence information) on which such probabilistic theories rely (assuming (CM) as a connecting principle) is enormous, which is just another way of saying that the project of trying to *define* causal relationships in terms of (or to *reduce* causal relationships to) facts about patterns of conditional and unconditional dependence along the lines of the probabilistic theories in Section 2.2 looks hopeless (or at least it looks hopeless in the absence of additional connecting principles).[47]

This is a fact that is of considerable potential interest for psychology. Consider the common proposal among psychologists that people's judgments of the causal strength of the relationship between *C* and *E* (understood as binary variables) tracks the quantity $\Delta p = Pr(E/C) - Pr(E/not\ C)$—the so-called contingency between *C* and *E*—or the suggestion that people represent the existence of a causal relationship from *C* to *E* by assigning a nonzero value to this quantity. This is an "associationist" model of causal judgment in the sense that it attempts to model such judgments purely in terms of information about conditional probabilities involving the cause and the effect. It is thus, as already noted, a kind of simplified psychological analogue of the probabilistic theories of causation described in Section 2.2, applied to a case in which there are just two variables *C* and *E*, presumably with the possibility of confounders and the possibility that *E* causes *C* somehow excluded. There are also more complex proposals about the empirical psychology of causal representation and judgment according to which people compute functions of the joint probability distribution of the variables involved, such as partial regression coefficients, with these functioning to represent causal relationships. These proposals can be applied to systems involving larger numbers of variables. However, the problem still remains that such probabilistic information typically underdetermines facts about causal structure. So one faces a choice—either (i) laypeople's[48] causal representations and judgments do not make the kinds of distinctions one can make with causal graphs or systems of

[47] Let me underscore again that this is just a claim about the relation between (in)dependence information, conditional and otherwise, in the probability distribution and causal structure, assuming (CM) and (F). Of course, it may be that in some or all cases there is additional information in the probability distribution (besides independence information) that in conjunction with additional connecting principles (besides (CM) and (F)) suffices to pin down a unique causal structure. At present we have no general account of how this might work although, as mentioned earlier, there is work showing that in the case of non-Gaussian distributions, higher moments of these distributions contain information relevant to the assessment of causal direction (Janzing et al. 2012).

[48] Again some people (e.g., researchers who use these devices) do make the kinds of distinctions among causal possibilities that are made by graphs and equations. So the claim must be that lay or naive subjects do not typically do so.

equations, so that people instead operate with less discriminating representations that can be fully characterized in terms of probability information alone (Δp or more sophisticated variants) or, alternatively, (ii) people do make the kinds of discriminations embodied in graphs and equations, in which case Δp and more sophisticated variants will not be empirically adequate models of human causal judgment. It is an empirical question which of these alternatives is correct, but as later chapters will show, the evidence seems to strongly favor (ii).

2.8. Model-Free versus Model-Based Accounts of Learning

It is common in accounts of learning (both causal and non-causal) to distinguish *model-free* from *model-based* approaches. The distinction is not always clear, but accounts according to which learning causal representations is a matter of learning associative relationships like Δp (or relationships characterized by the Rescorla-Wagner model described in Chapter 4) are generally thought of as model free, the idea being that the learner is not guided by a model that systematically tracks structural features of the external environment—or at least nothing beyond the acquisition of associative links.[49] By contrast, accounts according to which learners have causal representations that are structured like directed graphs or systems of equations are thought of as model based.[50] This is discussed in more detail in Chapter 4, but in general we can think of model-based representations as representing relations among environmental structures, often in compositional or maplike way. (By contrast, in model-free accounts the structures and computations that guide behavior need not be interpreted as representing anything in the external environment, or at least they need not do so in a systematic way.) For example, causal representations in terms of directed graphs are model based because they represent not just connections between pairs of variables but much more complex structures concerning how individual causal connections are related to each other—for example, that X causes Y which causes Z or alternatively that Y causes both X and Z.

Another way of putting this is that models of causal relationships in terms of directed graphs and systems of equations differ from regularity theories like

[49] In the literature on reward learning, this contrast is often understood in terms of the difference between a model-based system that represents the external reward environment, including relations between states, actions, and outcomes, and a model-free system in which all that is learned is the value of actions in different states, with no representation of outcomes—see, e.g., Dayan and Balleine 2002. Values are "internal" to organism rather than representations of the state of the environment.

[50] Similarly Bayesian and hierarchical Bayesian models with explicit distinctions between likelihoods and priors or among hierarchies of priors are usually thought of as model based, at least to the extent that likelihoods and priors are understood as representing facts about the external environment—e.g., facts about the frequencies with which various states occur.

Mackie's and probabilistic theories like those discussed in Section 2.2 in the following important way: the latter are what might be called one-level theories, that is, they work with the resources provided by just one level of information, which is taken *both* to represent the evidence (regularities, probability relations) on which causal claims are based *and* to provide the resources used to define or characterize causal relationships themselves. By contrast, when directed graphs or systems of equations are used to represent causal relationships, we have two distinct levels. Regularities or probability relationships are still thought of as evidence (i.e., one kind of evidence) for causal relationships, but the representation of causal relationships is seen as requiring a distinct level of structure corresponding to the graphs or equations, and there is no attempt to define features belonging to the second level solely in terms of the first level. Instead, the facts represented at the second level are seen as undetermined by the information at the first level, and the problem of causal inference/learning is seen as an issue of solving or partly solving this underdetermination problem. This in turn requires additional assumptions: possibilities include (CM), (F), invariance assumptions of various sorts (cf. Chapter 6), assumptions connecting causal claims to what happens under interventions, and more specific background information about the existence or nonexistence of various causal relationships. That the devices developed in various areas of science for causal representation and for the analysis of causal inference problems often have this two-level structure is one reason for thinking that such a structure (and particularly the second level) is doing useful work. It is plausible to guess that a similar point might hold for the representations of causal relationships by ordinary subjects (i.e., that the two-level structure accomplishes something that cannot be accomplished just by the one-level structure), although as I have said, whether in fact humans and other subjects operate with such a two-level structure in ordinary life is of course an empirical question—and one that will be addressed in subsequent chapters.

Before leaving this topic, I would be remiss if I did not emphasize another point. Directed graphs and structural equations are often useful devices for the representation of causal relationships. This does not mean, however, that such devices are useful for the representation of *all* causal relationships. Like all representational devices, directed graphs and structural equations have limitations that make them ill-suited (or at least imperfectly suited) for the representation of some causal relationships or some features of such relationships.[51] We have already noted, for example, that directed graphs do not represent information about functional form. In addition, directed graphs, at least in the forms

[51] I did not emphasize this fact as strongly as I perhaps should have in Woodward 2003, although there are plenty of examples in that book of causal representations that do not make use of structural equations or directed graphs.

currently employed, do not represent information about spatial relationships (between cause and effect and among different causes), and such information can be crucially important in causal cognition. Moreover, although temporal information can be represented in Bayes nets (e.g., through the use of time-indexed variables), the resulting representations are sometimes rather unperspicuous, in part because they involve discrete time steps rather than describing a continuous evolution of the system of interest. Many of the same points can be made about structural equations, at least as standardly used. Other forms of representation (most obviously, differential equations) are often better suited for the representation of causal relationships undergoing continuous temporal evolution or for representing the causal role of spatially distributed or extended factors. At least in many cases, when such equations have a causal interpretation, this interpretation also can be understood along interventionist lines—for example, when a violin string is modeled in terms of the wave equation, this equation in conjunction with suitable initial and boundary conditions describes how the string responds to an intervention that consists in plucking it.

That said, the pragmatic orientation I have defended throughout this book leads me to resist the suggestion that because graphs and structural equations are sometimes inadequate ways of representing causal relationships, they are always inadequate for this purpose. I deny that there is some single representational format that is best for all causal relationships. Instead, different representations will be most useful, depending on the characteristics of the causal relationships represented. This suggests (given the functional approach I have advocated) that, as a matter of empirical psychology, subjects may use different structures for the representation of different sorts of causal relationships.

2.9. Geometrical/Mechanical Theories

All of theories considered so far in this chapter have been difference-making theories. Standing in contrast to such theories is a group of theories that I will call geometrical/mechanical or GM. These theories are most naturally understood as applying to token-causal claims (or at least this is where they begin, with any treatment of type-level claims understood as derivative from token cases).[52] The basic idea of such approaches is that, contrary to what difference-making theories assume, whether *c* causes *e* does not have to do with a contrast or comparison of some kind between a situation in which *c* and *e* occur and an alternative situation in which *c* does (or would) not occur, but rather just has to do with what

[52] This is because spatiotemporal and mechanical relations apply most directly to individual token events.

happens in the situation in which *c* and *e* occur—in this respect such theories are "actualist"; causation is characterized just in terms of what actually happens. To distinguish the mere co-occurrence of *c* and *e* from cases in which *c* causes *e*, GM theories appeal to spatiotemporal and/or "mechanical" features that are present when the relationship between *c* and *e* is causal. One of the best-known philosophical theories of this sort is the causal process theory developed by Salmon (1984) and then elaborated by Dowe (2000). There are several different versions of this account (see Dowe 2009), but in one influential formulation, causation involves the "exchange" of some conserved quantity from cause to effect, with at least in many cases the quantity in question being energy or momentum. A paradigmatic causal interaction on such a theory is a sequence in which a moving billiard ball strikes a stationary one, and there is an exchange of energy and momentum with the result that the second begins to move.[53] The "intuition" that we are encouraged to have about this and other similar examples is that whether the impact of the first ball caused the second to move just depends on the nature of some "actual" relationship between these two events and is not a matter of any kind of comparison between the actual sequence of events and some alternative (e.g., an alternative in which the first ball fails to touch the second). Of course it is recognized that we may be willing to make such a comparison, at least when no other causes of the movement of the second ball are present, judging that the second ball would not have moved if the first had not struck it. However, this is regarded as an ancillary or derivative judgment, that follows from the original judgment of causation but is not "part of" or "constitutive" of the original judgment—see, for example, Bogen 2004.

In psychology an account of causal representation and judgment that is similar in spirit has been developed by Philip Wolff under the rubric of "force dynamics." (e.g., Wolff and Thorstad 2017). I will not attempt to describe Wolff's theory in any detail, but the basic idea is that the representation of causal relationships (and the judgments that result from these) involve the representation of "forces" of various sorts that are associated with entities that are represented either as force "generators" or as "patients" that are recipients of the forces. For example, when subjects judge that a moving billiard ball strikes a stationary ball and causes it to move, this involves a representation in which the first ball acts as a force generator that "transfers" a force to the second ball, which "receives" the force, in its role as a patient. Wolff elaborates this framework so that it can be

[53] Salmon and Dowe think of their accounts as tracking what physics tells us about the underlying nature of causation, while Wolff, a psychologist whose views are discussed later, motivates his account in part by claiming that ordinary subjects represent causal relations in the way that (he supposes) physics does. In this connection, it is worth noting that Wolff's idea that some objects acts as agents that exert forces on others that are patients or recipients with the latter not exerting forces on the former may or may not describe how people think, but it has no basis in the underlying physics.

used to distinguish among different causal concepts such as "cause," "help," and "prevent."

The causal process theory applies most naturally to a subset (see below) of certain physical phenomena that, for want of a better description, might be described as "mechanical"—indeed Salmon's name for his model was the "causal mechanical model," and Dowe describes his subject as "physical causation." The Salmon/Dowe theory is thus a theory about what causation *is* rather than (at least directly) a psychological theory about how people think about causation, as Wolff's is. Nonetheless, there are a number of questions of psychological interest that we might ask about the theory. First, consider the range of phenomena to which the Salmon/Dowe theory applies. Psychologists and others (e.g., Muenter and Bonawitz 2017 and the references therein; Tomasello and Call 1997) have explored aspects of the causal cognition of different kinds of subjects (adult humans, children, nonhuman primates) regarding phenomena that we might ordinarily think of as "mechanical"—some examples include phenomena involving support or lack of it (as when a book falls from a table despite being in contact with the table because its center of gravity is off the table), phenomena associated with impenetrability (it would be anomalous for a solid object to move through a solid barrier), and the causal role of rigidity and shape in tool use (a rake with a rigid hook on the end is more effective at retrieving out-of-reach food than a flexible string).

Interestingly, despite its "mechanical" character, the Salmon/Dowe theory does not seem to apply (at least straightforwardly and literally) to the causal interactions that are present in such cases. In the example of the book resting on the table, and taking the usual physics analysis as the correct account of what is really going on, the book exerts a downward force on the table due to gravity and the table exerts an equal and opposing upward force on the book so that the net force on the book is zero. (Thus, contrary to what Wolff's account apparently suggests, neither the book nor the table is treated as an agent and the other as a patient.) As a matter of physics, there is no "transfer" of energy/momentum (or other conserved quantity), at least a macroscopic level, even though the example is naturally described as one in which a causal interaction is present. Similarly, it is also not clear how the invocation of conserved quantities helps to illuminate the relevant aspects of mechanical phenomena associated with rigidity and impenetrability: touching the end of a flexible string to an object may "transfer" some small amount of energy/momentum to the object, but understanding this is not the kind of causal understanding that is relevant if one wants to retrieve the object and must choose an appropriate tool for doing so.[54] (This is not just

[54] Suppose that all mechanical interactions conform (at some level of analysis) to conservation laws—"at some level of analysis" means that even if, e.g., a collision between two balls appears to be inelastic at some coarse-grained level, energy/momentum conservation is respected at some

a thought experiment; the results of some actual experiments having to do with problems of this sort will be described in subsequent chapters.)

As noted previously, even if there are many "mechanical" phenomena that are not, as a matter of physics, well captured by the Salmon-Dowe theory, it is a separate question whether subjects represent or reason about phenomena involving mechanical interactions in terms of the resources present in that theory. But, in view of the considerations just described, it seems doubtful that subjects who succeed in such tasks involving support, shape, and rigidity considerations reason about them in terms of the Salmon/Dowe framework, given the limited range of applicability of that framework. For example, if they were to use only the resources of the Salmon/Dowe framework, it is unclear how they would be able to reason successfully about what would happen if, for example, support were to be removed or a nonrigid (but momentum transferring) tool were to employed. Adult humans typically do reason successfully about such cases, although in many cases apes do not. Successful reasoning in such cases seems to require appreciation of difference-making or dependency considerations—that whether putting an implement in contact with some object will allow one to retrieve it depends on whether the implement is rigid, has a hook at the end, etc. This information goes beyond information about whether spatial contact and energy/momentum transfer is present.[55] A psychological theory about how subjects reason about such cases will need to employ richer resources as well.

With regard to Wolff's theory, thinking about mechanical interactions in terms of forces often more accurately reflects what is relevant in the underlying physics than does talk of energy/momentum transfer, at least taken in itself, as we see in the case of the book resting on the table, although Wolff's notion

finer-grained level, although there may be dissipation into the environment or the interior of the balls. Nonetheless there seems no reason to believe that all of mechanics can be derived just from conservation laws—laws and other constraints specifying the detailed nature of the mechanical interactions are also required. To the extent that which causal relationships hold depends on these other laws and constraints, conservation-based considerations alone will fail to specify them.

55 This is not to say the notion of a causal process that Salmon and Dowe attempt to capture is scientifically unimportant or resists a clear characterization. As Wilson (2017) notes, this notion is best captured by the contrast between systems that are governed by different kinds of differential equations. When a system is governed by a hyperbolic differential equation (as opposed to an elliptical or parabolic partial differential equation) and admits of a well-posed initial value formulation, the solution domains for these equations will have "characteristic surfaces" or cone-like structures that characterize an upper limit on how fast disturbances or signals can propagate. In the case of the equations of classical electromagnetism, these correspond to the familiar light cone structure. Causal processes are what propagate along such surfaces. However, this notion of causal process is clearly connected to difference-making or dependency ideas (rather than standing in contrast to them) since differential equations are representations of difference-making relations. See Woodward 2016 for additional discussion. Thinking of causal processes in terms of hyperbolic differential equation allows one to avoid having to make use of unclear notions such as "intersection," "possession" of a conserved quantity, and so on that are employed by Salmon and Dowe. A pseudo-process can then be characterized just by the fact that its behavior is not described by a relevant hyperbolic partial differential equation.

of "force transfer" is foreign to physics. However, it is notorious that in many cases involving mechanical interactions such as billiard ball collisions the underlying physics (i.e., the internal changes in the balls in response to the collision and the forces involved in them) is extremely complicated (cf. Wilson 2006, 2017)—this is why elementary physics textbook treatments do not try to model this. Whatever it is that lay subjects do when they reason about such cases, they do not make use of this sort of physical information about underlying internal forces. To this it might be replied that force dynamics can at least yield a qualitative (if not quantitative) representation of causal relationships (in terms of force vectors), and this can serve as a basis for judgment about them. Perhaps so, but it would be desirable to have a more precise characterization of how this works than Wolff provides. An alternative account that predicts fairly well how adults judge regarding collisions and trajectories of small ball-like objects is the Noisy-Newton theory (Sanborn et al. 2009; see also Gerstenberg et al. 2012), which is discussed in Chapter 4. This models such judgments in terms of a noisy version of Newtonian mechanics assuming inelastic collisions and achieves a fairly good fit to people's actual judgments. As far as I am aware, Wolff's theory does not achieve this level of predictive accuracy.

Switching to another example, suppose that we want to understand how people are able to correctly judge that a book has been partially placed on a table in such way that it will fall off it. To do this people need something more than just a representation that the table exerts an upward force on the book, gravity exerts a downward force on it, and so on. Maybe when people see the book fall, they think about this in terms of the gravitational force being stronger than the force exerted by the table, or something similar, but this does not help us to understand how people can successfully judge ahead of time that the book will fall just from looking at how it is placed on the table—presumably this instead has something to do with spatial relations between the book and the table conjoined with perhaps implicit assumptions about the book possessing a relatively uniform mass distribution so that if its geometrical center is off the table, it is like to fall. Wolff's theory does not seem to incorporate considerations of this sort.

So far I have been arguing that the scope of the Salmon/Dowe theory is rather narrower than the full range of phenomena that we think of as "mechanical." Another common philosophical complaint about this theory is that it is also too narrow at, so to speak, the other end of the scale—there are many causal relationships that are not mechanical in any relevant sense and which don't seem (at least in any obvious sense) to involve transfer of conserved quantities. (Certainly we don't know how to represent/describe them in terms of conserved quantities.) These include causal relationships involving many "higher level" entities and properties that figure both in common-sense reasoning and in the so-called special sciences: causal claims relating mental events and behavior, causal claims relating economic variables, and so on. Presumably Salmon's view was that underlying, true, upper-level causal claims are lower-level facts about the transfer of energy and momentum, with the truth of the former being "grounded" in the latter. However, since the details of these lower-level facts and how they connect to upper-level causal claims are in many cases unknown (either to scientists or to lay subjects), it is hard to see (at least in many cases) how his theory can usefully be brought to bear on such upper-level claims. Certainly it is not obvious how to use the theory to capture ordinary people's causal reasoning about such claims.

In contrast to Salmon, Wolff attempts to extend his account well beyond what one ordinarily thinks of as mechanical interactions. He suggests that people also represent causal relationships involving psychological and social phenomena in terms of forces—economic, psychological, and social forces, for example. However, even if this is correct as a psychological claim, the notions of force involved seem largely metaphorical. It is unclear how attributing such force representations to people can explain *successful* causal cognition in connection with these variables. How exactly do such representations allow me to reason successfully about how you would respond if I were to choose one course of action over another or whether an increase in the interest rate by the central bank will cause an increase in inflation? Game theory and economic theories that do not traffic in forces are much better at providing predictions about these matters. If I am trying to reason about whether my action will cause you to become angry, I may or may not think of anger as a "force," but it is hard to see how whether or not I do so contributes to the reliability of my reasoning. Something more than force representation must be going on when I engage in such reasoning.

There is another line of objection that applies to the Salmon/Dowe theory and indeed to any theory that tries to completely dispense with the difference-making aspects of causation. Consider someone who throws a tennis ball against a solid brick wall, which of course remains intact, with the ball rebounding from the wall. This is a paradigmatic causal interaction according to the Salmon-Dowe theory, with the impact of the ball "transferring" energy and momentum to the wall. Nonetheless the continued existence of the wall (and its remaining intact and upright) and so on is not an effect of the impact of the ball. Similarly (cf. Hitchcock 1995), consider someone who hits the cue ball with his cue stick, transferring blue chalk from the end of the cue stick to the cue ball and from that to the eight ball, which drops into the corner pocket. The blue chalk carries a very small amount of energy and momentum from its initial position on the cue stick to the eight ball. Still the presence of the blue chalk and the energy/momentum transfer associated with it do not (in realistic cases) cause the eight ball to go into the pocket.

What examples like this show is that the Salmon/Dowe theory fails to make certain fine-grained distinctions that are important in causal thinking. What one would like to say is that the energy and momentum associated with the chalk transfer are not "large enough" to "make a difference" to whether the eight ball goes into corner pocket. Similarly the energy/momentum transferred to the wall is sufficiently small that it makes no difference to whether the wall stands up or falls down. But saying this reintroduces the "difference-making" aspect of causation that causal processes theories were trying to avoid. Of course a number of cases in which *c* causes *e* will be cases in which there is a causal process in the Salmon/Dowe sense connecting *c* to *e*, but it looks as though the presence of the process by itself does not guarantee that *e* depends on *c* or that *c* causes *e*. This is another case in which the normative inadequacy of a philosophical theory has implications for a corresponding thesis in empirical psychology. Even if ordinary people have internal representations of energy/momentum transfer, reasoning successfully about cases like those described here requires additional resources having to do with the representation of difference-making information.

Despite these limitations I do not think that we should simply dismiss causal process or other approaches to causation that emphasize the role of geometrical/mechanical considerations, either as components of a normative theory or as components of an empirical account of causal cognition. Trying to produce a general account of what is involved in *c*'s causing *e* in terms of the existence of a connecting process (or some other appropriate geometrical/mechanical relationship) between *c* and *e* is unlikely to be successful, but facts about such a connection, when present, can nonetheless play an important role in causal learning and reasoning.

This can happen in a number of ways. First, there are many cases in which there are spatiotemporal requirements on the relation between a candidate cause and effect that are necessary or near-necessary conditions of the existence of a causal relationship between them. (We can think of such spatiotemporal conditions as a subset of the broader class of geometrical/mechanical conditions.)[56] Sometimes but not always these have to do with immediate spatiotemporal contact or proximity. When Billy and Suzy throw rocks at a bottle, the fact that both rocks follow continuous spatiotemporal trajectories makes it easier to tell which rock is associated with which thrower. If Suzy's rock comes into spatial contact with the bottle first and it shatters immediately after, then, unless something very unusual is going on, we can conclude that it was the impact of Suzy's rock that caused the

[56] Some conditions that figure in geometrical mechanical accounts (such as whether there is contact) seem characterizable in spatiotemporal terms. Other properties that seem "mechanical" (such as rigidity or impenetrability) are not purely spatial or geometrical, although, e.g., changes in shape seem relevant to their presence.

shattering.[57] Note, though, that this does not mean that we can *define* or fully characterize what it is for Suzy's rock to cause the shattering in terms of facts about the spatiotemporal contact, continuous causal processes, and energy/momentum transfer. (Correspondingly, the story we tell about the empirical psychology of causal cognition regarding such cases will need to include more than subjects' representations of spatiotemporal relations and energy momentum transfer.) Indeed, the inference I have described concerning Suzy's rock seems to require a number of additional (and more specific) assumptions—that the shattering had some cause (the bottle does not shatter spontaneously), that a rock that hits a bottle can cause it to shatter if it has enough energy/momentum (a difference-making consideration, as noted previously), and that a rock that does not touch an intact bottle cannot cause it to shatter (which allows us to eliminate Billy's rock as a possible cause). Rather than using spatiotemporal information to define what it is for Suzy's rock to be a cause, it seems more correct to think of the situation as one in which we use such information in conjunction with other assumptions to infer the existence of a causal conclusion understood as a difference-making relationship, with this having implications both for normative theorizing and for empirical psychology.

A related point is that the presence of energy momentum transfer (or evidence for it) or the presence of some spatio-geometrical relationship can function as a clue that some dependency or difference-making relationship in which we are interested may be present or alternatively could not be present, even if we cannot define causation in terms of energy/momentum transfer. If there is no energy/momentum transfer between Billy's rock and the bottle, we can conclude that it is almost certain that his throw did not cause the shattering even though it is also true that the presence of energy/momentum transfer between Billy's rock and the bottle is not *sufficient* for that rock to cause the shattering.[58] Similarly the presence of the chalk on the eight ball after its being present on the cue stick strongly suggests that the ball or some cause of its motion interacted causally with the cue stick. Although the presence of the chalk (or the fact that the chalk was communicated by the cue stick to the eight ball) is (we are assuming) not causally relevant to whether the eight ball goes into the pocket, it suggests where we might look for a factor that is causally relevant to the eight ball's motion—something

[57] It is relevant to note in this connection that capturing the causal role of Suzy's rock by means of structural equations or directed graphs seems to require a variable that corresponds to whether Suzy's rock hits the bottle or not—see Halpern 2016. Of course this variable describes a spatiotemporal relationship, even though structural equation / directed graph accounts are paradigmatic difference-making accounts.

[58] As Paul and Hall 2013 note, even if Billy's rock does not touch the bottle, it may cause the motion of molecules in the air to impact on the bottle and light rays striking the rock along its trajectory may be reflected onto the bottle. So there will likely be some energy/momentum transfer from the rock to the bottle even if the rock does not cause the shattering.

about the cue stick, such as its prior movement. More generally, that some c^* is involved in energy/momentum exchange with some e^* can suggest that that there may be a dependency relationship between some c associated with c^* and some e associated with e^*, even if the $c \rightarrow e$ relationship cannot be characterized just as a matter of energy momentum exchange.[59] Put differently, information about the presence of an appropriate geometrical and/or energy/momentum transference relation can be a clue that some dependency relation or other is present, even though this information typically does not pick out which dependency relation is present or even tell us that any dependency relation of a sort we might be interested in is present.[60]

Thinking about matters this way suggests a somewhat different role for geometrical/mechanical considerations than the role they are assigned in causal process theories. Rather than trying to use these considerations to define or characterize what it is for a relationship to be causal (or what it is to have a causal representation), I suggest that for the purposes of understanding causal cognition, we should focus instead on better understanding the relationship between geometrical/mechanical information and difference-making relations, on the assumption that the former can convey information about the latter. A distinctive feature of adult human causal cognition is that we can often use geometrical mechanical information, in conjunction with other information or assumptions, to rapidly reach conclusions about difference-making relationships: we integrate geometrical/mechanical information into our causal representations and reasoning, along with other sorts of information having to do with difference-making. We immediately "read off," for example, from the shape and apparent

[59] I will not try to characterize what "associated" means. The kind of relationship I have in mind is the sort that obtains when the impact of a rock *R* on someone's head causes death *D*. *R* is associated with energy/momentum transfer even though the mere fact that such transfer has occurred does not by itself capture the fact that *R* caused *D*.

[60] It is worth noting the existence of another (related) role for "causal process" considerations that is relatively unexplored in the philosophical literature and that may be of considerable interest for the empirical psychology of causal cognition. Often when there is a causal relationship between two relata *C* and *E* there is either (i) a connecting physical structure associated with *C* and *E* or (ii) some "mark" or recognizable structure that is "transferred" or moves from *C* to *E* or both. For example, *C* may have to do with the movement of one end of a rigid rod and *E* with the movement of the other end of the rod, with the rod itself being the connecting structure. Or a neuron may connect one part of the brain to another so that information and/or a material trace such as some radioactive material can be transferred along this connection. In cases of this sort the existence of such a physical connection (even in conjunction with some mark transference) may not be *sufficient* for the existence of a causal relationship of any sort that we are interested in. For example an anatomical connection from brain region 1 to brain region 2 will not by itself show that region 1 causally influences region 2 in any relevant way (that there is a "functional" connection). Nonetheless it may be very plausible that *if* there is no physical connection from region 1 to region 2, then there is no causal connection—the physical connection is *necessary* (or nearly so) for causation, so that we can exclude candidates for causal relationships on this ground. Information of this sort about connections and evidence for transfer can play an important role in learning about causal relationships in science and, one suspects, in nonscientific contexts.

rigidity of an implement whether it can be used to make a difference for the retrieval of some object, and we can see at a glance that removing a box that supports another will make a difference for whether it falls. I say that this is a distinctive feature of adult human causal cognition because, as we shall see (Chapter 4), it seems to be lacking to a surprising extent in other primates—these animals do not appear to use geometrical/mechanical information to infer difference-making relationships in the way that we do. I conclude from this that a satisfactory account of human causal cognition needs to assign an important role to geometrical/mechanical (as well as temporal) information but that we cannot use this information to fully characterize what it is for a relationship to be causal (or for a representation to be a representation of a causal relationship). I believe that a similar conclusion holds for the role of the force dynamics described by Wolff.

My discussion in this section has mainly focused on logical/philosophical differences between difference-making and geometrical mechanical accounts of causation and, in connection with the latter, differences between causal process accounts and those that focus on "mechanical" properties like support and rigidity. Various examples show that one can have connecting processes without difference-making and conversely. Moreover, there are "mechanical" relationships that are not well understood in terms of connecting processes. (These various elements can and do "dissociate" from one another, both in terms of logical relationships and, sometimes, empirically.)

As will become apparent in subsequent chapters, the fact that these elements are independent in this way has important consequences for the psychology of causal cognition. Suppose, as I have argued, that the full suite of adult human capacities for causal reasoning and judgment involves the appreciation of difference-making relations and how these connect to spatiotemporal and other kinds of "mechanical" information. It follows from the preceding discussion that, both as a matter of logic and as a matter of what is true empirically, sensitivity to difference-making information and the ability to reason with it is not just a matter of sensitivity to geometrical/mechanical relationships. Evidence for the latter does not by itself establish the presence of abilities associated with appreciation of difference-making information. By contrast, some psychologists have claimed that the fact that subjects like babies are sensitive (e.g., as revealed in looking-time experiments) to various geometrical/mechanical cues and recognize anomalous geometrical/mechanical behavior (e.g., they are sensitive to the difference between collisions that "look causal" and those that do not or to changes in trajectories like those that occur in collisions but without spatial contact or to the difference between scenarios in which solid objects do or do not appear to pass through each other) shows that these subjects possess a "concept" of causation. However to the extent that the full suite of abilities associated with causal reasoning involves an appreciation of the role of difference-making relationships (and even putting aside, for the sake of argument, misgivings about "concept of causation" talk), this conclusion does not follow. It does not follow because the sensitivities to geometrical/mechanical information just described do not automatically bring in their wake, either as a matter of logic or empirically, the other abilities associated with causal reasoning, including those associated with difference-making relations.[61] This is one of many cases in which getting clear about the logical relationships among various elements that are associated with causal thinking can help to clarify what empirical evidence does or does not show.

I conclude this chapter by commenting briefly on a related question raised by an anonymous referee: I've been skeptical about geometrical/mechanical views as complete accounts of causation, instead favoring difference-making (and more specifically interventionist) accounts. But why not adopt a "pluralist" or non-unified framework instead? The idea presumably would be that there are several distinct but equally legitimate ways of thinking about causation, one organized around difference-making ideas and the other around geometrical/mechanical considerations. That the latter does not build in difference-making features does not indicate any inadequacy; it just reflects the fact that the geometrical mechanical account captures a *different* causal notion. I will not try to discuss this issue in the detail that it deserves. I will just remark, as I argue in this and subsequent chapters, that adult humans do seem to have relatively unified ways of thinking about causation. Furthermore it seems normatively appropriate that they should. Even putting aside the inadequacies of talk of distinct "concepts" of causation (see Chapter 3), it seems clear, as argued in this chapter and in Chapter 4, that rather than operating with distinct and unconnected notions of difference-making and geometrical mechanical causation, we connect these two, expecting, for example, that geometrical/mechanical information will bear on difference-making relationships. Moreover, given that we want a unified

[61] There is thus an important empirical question: when babies exhibit sensitivity to geometrical mechanical considerations of the sort found in looking-time experiments, to what extent do they integrate or connect this with difference-making considerations, including those related to manipulation? In principle one might imagine that at some early stages in development such integration has not yet occurred, so that as an empirical matter a dissociation is present. Designing a real experiment to explore this will not be easy because some appropriate test for sensitivity to difference-making information is required and it not obvious how to test for this—looking-time information by itself seems insufficient. But if the babies are capable of simple manipulations such as using a tool to retrieve a toy, one possibility might be to test whether they choose an appropriate tool based on geometrical mechanical considerations (i.e., a tool in spatial contact with the toy in contrast to one that is not). Success in doing this might be taken as some evidence that they understand the relevance of geometrical mechanical considerations to what makes a difference to retrieving the toy. To the extent that a baby understands the relevance of such considerations—integrating them with a grasp of difference-making relations—we have something that begins to look (at least in some respects) like an "adult" understanding of causation.

account, difference-making seems a better candidate for a central organizing principle, since we obviously recognize difference-making relations even when geometrical/mechanical considerations do not apply, as in cases of social and psychological causation.

Appendix: Actual Causation

In Section 2.1 I invoked a familiar distinction between type and token or actual cause claims.[62] I suggested that these are less closely related to one another than is often assumed (in the sense that an account of token causation is not required if we want to provide accounts of varieties of type causation), and used this claim to motivate my decision to focus largely on the former. This appendix attempts to provide some additional background in support of this claim. It is not intended as a full account of actual causation—that would be a very large project—but rather tries to locate this notion within the larger geography of causal claims[63].

First, although the distinction between claims like

Smoking causes lung cancer (2.1)

and

Jones's smoking caused his lung cancer (2.2)

is real and important, one needs to take care that the terminology of "type" versus "token" causes does not mislead. In most if not all cases, "type" causal claims are not plausibly interpretable as claims about causal relations between abstract types of events (whatever that might mean), at least in any sense of "type" that contrasts with "token." Instead the units, so to speak, over which type-causal claims range are really "tokens." For example, (2.1) should *not* be understood as

[62] There is an additional complication regarding the type/token distinction that is worth noting, although I will not discuss it in detail. This is that the distinction (as treated in this book) is not exhaustive: there is a further category having to do, roughly, with the causation of variation in particular populations, either cross-sectionally or over time. (These are called "causal role" claims in Woodward 1995.) As an illustration consider the claim that changes in the incidence of cigarette smoking are causally responsible for most the changes in the incidence of lung cancer in the US population in the twentieth century. On the one hand, this is a causal attribution claim involving effect-to-cause reasoning and in this respect like actual cause judgments. On the other hand, it does not have to do with the causation of any particular person's lung cancer in the way that actual cause judgments do.

[63] For additional discussion of some of the issues that arise in modeling actual cause judgments, see Woodward, forthcoming d.

the claim that the abstract type "smoking" somehow causes another abstract type ("lung cancer"). Instead, what (2.1) means or implies is something like this: individual people, who are left unspecified, either would or may be caused to develop lung cancer by their smoking if they were to smoke. In other words, the causal relation claimed by (2.1) is causation involving actual or possible particular (but presumably prolonged) episodes of smoking by individual people and occurrences of lung cancer, again in individual people.

This observation might seem to support claims about token causation being the "primary" or "central" causal notion, but I think that this would be a mistake. The basic point is that we need to distinguish the issue of what "units" are involved in some causal relationship from whether the judgments we are making about that relationship are "token" or "actual cause" judgments and if so, what conditions the latter judgments must satisfy in order to be true or correct. The units (individual people, individual episodes of smoking, occurrences of lung cancer in individuals) are the same for (2.1) and (2.2)—what is different is that different conditions need to be satisfied in order for (2.1) and (2.2) to be true. In other words, we have two sets of practices or standards or causal questions, one involving claims like (2.1) and the other involving claims like (2.2)—the issue is what these are and how they are related. This is the issue that is raised by the claim that token/actual causal judgments should be understood as foundational with respect to type-causal claims. It isn't settled merely by observing that both (2.1) and (2.2) are in some sense "about" individual occurrences of smoking and lung cancer. If (2.1) and (2.2) correspond to very different causal questions that are embedded in different practices with different goals and hence different standards of correctness, it isn't obvious that token cause judgments can serve as a basis for type cause judgments.[64]

What then are the standards of correctness for (as I'm going to call them from now on) actual cause claims? I won't try to answer this large and difficult question here, but I do want to emphasize some of the respects in which these seem to be different from the standards we employ in assessing more type-level claims.

One difference is this: when one makes an actual cause judgment, one is typically[65] presented with a particular outcome whose occurrence is taken as known—for example, it is known that Jones has lung cancer and (at least usually) it is assumed that something or other caused this outcome. In making an actual

[64] This is one reason for preferring the terminology of "actual cause" to that of "token." The latter does not really get at the difference between (2.1) and (2.2).

[65] Although this is typically the case, it isn't always so. As noted in Hitchcock 2017, I can "look forward" in wondering whether some event will be an actual/token cause of an effect. But even when the event and the effect occur, it is still an additional question whether the event was an actual cause of the effect. A doctor may wonder whether a particular injection will save a patients life, give the injection with the patient surviving and yet there is still the additional question of whether the injection was the actual cause of survival.

cause judgment, the typical task is to reason "backward" from this occurrence to what caused it (from a known effect to a not-yet-known cause)—was it Jones's smoking or something else, such as his exposure to asbestos? The goal is to discriminate among these alternatives by identifying the actual cause.

In the case of so-called type-causal claims, the direction of inference is often (but not always) the opposite—from some known event or factor to its possible effects.[66] That is, one begins with a candidate for a cause, characterized generically as a type of event or factor, and then reasons "forward" from this candidate to ask whether it causes some type of effect. For example, one begins by focusing on smoking and asks whether it has some carcinogenic effect such as lung cancer. Experimentation or statistical investigation using population wide covariational data is often directly relevant to establishing whether type-causal claims are true, as (2.1) illustrates. By contrast, establishing an actual cause claim like (2.2) requires information of a different sort, including detailed information about a particular event or individual, although other kinds of information may also be relevant. At least in most cases one can't establish whether a claim like 2.2 is true by conducting an experiment or running a regression on population data. This is closely related to the general view within the causal inference literature outside of philosophy (in statistics, epidemiology, econometrics) that effect-to-cause reasoning and establishing actual cause claims is often (not always but often) harder and requires more information (or information that is often less readily available) than cause-to-effect reasoning and establishing type-level claims. Although it is well established that smoking causes lung cancer, it is typically very difficult to establish that Jones's lung cancer was caused by his smoking if some other potential cause of cancer was also present.[67] Inference to actual causes is thus a different computational problem than inference from causes to effects.

Another complicating factor in discussion of actual cause judgments, in both the philosophical and the psychological literature, concerns the role of *causal selection*. As I will use this notion, causal selection problems arise when there are a number of factors that are known to be causally relevant to an effect and the

[66] This difference in orientation is illustrated by the different questions associated with the fielder and wall example discussed in Chapter 3. In this example, a window is protected by a solid brick wall. A ball is thrown in the direction of the window but is caught by a fielder before it reaches the wall. If we ask whether the catch had the effect of preventing the window from breaking (a question that calls for cause-to-effect reasoning), we are inclined to say no since there was never any possibility of the ball breaking the window. If we ask what caused the ball not to reach the window (effect-to-cause reasoning), we are likely to say "the fielder's catch." Note that both answers appeal to the same underlying causal facts.

[67] On the other hand, as emphasized in Hitchcock 2017, one can sometimes be in a position to know that *c* was an actual cause of *e* while lacking a great deal of information about associated counterfactuals—e.g., what would have happened if *c* had not occurred. Woodward 2020a discusses some implications of this observation in the context of disease prevention.

immediate goal is to select one (or some small number of these) as "the cause" or "a cause" of the effect. For example, when a short circuit contributes causally to a fire, the presence of oxygen will also be causally relevant (and commonly known to be so). Nonetheless, under normal circumstances our usual practice is to settle on the short circuit as the cause of the fire and to treat the presence of oxygen differently, perhaps as a mere background condition.

Empirical studies of causal selection show (unsurprisingly) that such judgments are influenced by such factors as whether the occurrence of the candidate actual cause is "normal" or "abnormal" (in both the statistical sense and in the sense of whether there is a violation of non-statistical norms).[68] As an example of the latter, in a scenario in which staff are permitted to take pens from a secretary's desk and faculty members are prohibited from doing so, if both take pens and the secretary is unable to take a message as result, the faculty member's behavior is more like to be described as the cause (or a stronger cause) of this outcome. Although it is true, as noted in Chapter 5, that considerations having to do with departures from what is actual, normal, or a serious possibility also play a role in some type-causal judgments, they don't seem to play the *same* role.[69] For example, violations of norms don't seem to play much of a role in many type-causal judgments, nor does causal selection when understood in the way described previously. In addition, the actual distribution of values for a cause variable seems to play less of a role in type-causal judgments than in token-causal judgments—see Chapter 5. This casts further doubt on the idea that actual cause judgments can serve as a foundation for type cause judgments or that all of the features of the former are also found in the latter. More generally, it suggests that it is a mistake to motivate claims about causation in general by focusing only on actual cause judgments, with their many idiosyncratic features. Causation ≠ actual causation.

An important additional question has to do with whether all actual cause judgments share the features present in judgments influenced by causal selection. Although the issue deserves more attention than I can give it here, it is not obvious that the problem we face in assessing (2.2) should be thought of as a

[68] For a discussion of some of the complex considerations that, as an empirical matter, influence actual cause judgments see Kominsky et al. 2015 and Icard et al. 2017. Actual cause judgments in tort law provide one important illustration of the role played by the normative considerations since such judgments are often influenced by the incentive effects of assigning fault to one party rather than another: when the joint actions of several parties contribute to a harm, the actions of the party that can most easily and cheaply avoid contributing to the harm are often taken to be the actual cause. This fits well with the efficiency of intervention function for such judgments described subsequently.

[69] As noted in Chapter 5, to the extent that normality-like considerations play a role in type-causal judgments, these are heavily influenced by generic subject matter considerations—e.g., fundamental (and hence abnormal) changes in human psychology are not treated as serious possibilities in economic modeling. This is rather different from the role played by "normality" in causal selection in actual cause judgments.

causal selection problem or at least as the same kind of problem as the one we face in the short-circuit/oxygen example. In connection with (2.2), what we want to know is whether it was smoking that caused Jones's cancer rather than something else like exposure to asbestos or, alternatively, perhaps smoking and asbestos acting together. The background assumption is that if one of these is the token/actual cause of the effect, the others will not be. In other words, if it is true that smoking was the actual cause, then asbestos exposure is naturally viewed as causally irrelevant to Jones's cancer, rather than a relevant factor that is not selected, as is the case with oxygen in the previous example.

Nonetheless in the recent literature on models of actual cause judgment, both empirical and normative, it is common to find both kinds of cases—those like (2.2) and those like the oxygen/short-circuit example—described as actual cause judgments. Several of the most prominent recent treatments (e.g., Kominsky et al. 2015; Halpern 2016) attempt to capture both sorts of judgments in a common account that appeals to considerations having to do with the "normality" or not of candidate causes and variable settings—a project that, as already noted, seems very plausible in the case of the oxygen/short-circuit example. The upshot of such formal treatments is (in effect) to assimilate examples like (2.2) to straightforward cases of causal selection. I do not claim (at least here) that such common treatment accounts are mistaken. I do think, however, that to the extent that such accounts are correct, they support my earlier judgment that the causal relations reported in actual cause claims are not metaphysically or scientifically fundamental. This is because, as noted previously, selection is clearly made on the basis of factors that in many cases are different from the sorts of considerations that figure in causal reasoning, particularly of a generic or type sort, in many areas of science.

In addition, even if one does not accept such common treatment, normality-based accounts, they point to a serious empirical issue in modeling actual cause judgments. I suggested earlier that it may seem intuitively plausible that (2.2) is different from straightforward causal selection cases. However it is far from clear (at least at present) how to provide a principled basis for distinguishing actual cause judgments that involve selection from those that do not or whether there is any such distinction to be had—again, this is reflected in the fact that prominent formal models do not embody such a distinction. What I mean by such a principled basis is a probe of some kind (presumably verbal) that would allow us to tell whether we are dealing with an actual cause judgment that involves causal selection or one that does not. The verbal probes that are currently used to investigate actual cause judgments, which usually are causal strength probes, do not accomplish this. To the extent we are unable to find a way of characterizing and recognizing this distinction, we seem forced to an account of actual cause judgment in which normality considerations play a central role and hence an account in which actual cause judgments behave in some respects rather differently from other sorts of causal judgments. On the other hand, if there is such a distinction to be made, it follows that many current approaches mistakenly try to model rather different kinds of judgments in a way that does not distinguish them.

I alluded earlier to the absence of a generally agreed-upon account of the goal or function of actual cause judgments. As remarked earlier, one promising "functional" story is that proposed by Hitchcock and by Knobe and colleagues.[70] Their idea is that actual cause judgments often have the function of guiding us toward those interventions on a causal factor that in the actual circumstances are the most appropriate target (in terms of either their ready availability as a means of manipulation or their normative desirability) for altering the effect. This makes sense of many judgments involving causal selection, such as judgment that the occurrence of the short circuit was the actual cause of the fire—in most circumstances this is a better target for intervention than the presence of oxygen if one wishes to produce or prevent the fire. Arguably, however, this rationale works less smoothly for some other actual cause judgment, such as late preemption. Intervening in Suzy's throw won't prevent the vase from being shattered if, under this intervention, Billy's throw would have shattered it instead.

Another natural and not sharply distinct approach to actual cause judgments associates many of them with the attribution of blame, responsibility, or fault[71] —concerns that again are not operative (at least to the same extent or in the same way) with more type-level causal judgments, particularly in science. By "blame," "responsibility," and "fault," I have in mind extended senses of these notions that encompass both attributions of moral and legal responsibility, as when we ask whether the blow struck by Jones caused (= was responsible for) Smith's death and attributions of fault in which biological structures and designed objects don't behave as they "should," as in "the failure of the O-rings at low temperatures caused the *Challenger* disaster" (presumably a case of causal selection). Very roughly, the idea is that when we make an actual cause judgment regarding some occurrence, we attempt to trace back the causal history or etiology of the occurrence with an eye to finding some action or natural occurrence that we can hold "responsible" for the occurrence. The "natural home"

[70] For Knobe and colleagues, see, e.g., Icard et al. 2017. A more detailed version of such an account has been developed by Hitchcock (2017), who links actual cause judgments to situations involving planning in which we care about consistent exploitation of path-specific effects.

[71] I don't mean that people are in general unable to distinguish between actual cause claims and responsibility claims or that they always interpret queries about actual causation as queries about responsibility. I suggest that they *sometimes* do the latter, particularly but not exclusively when human agents are involved.

of many actual cause judgments thus includes, for example, contexts involving legal reasoning (did A's action cause the damage to B's property?), moral and prudential reasoning, medical diagnostic or forensic reasoning (what caused the patient's/victim's death?), and fault detection in engineering contexts (e.g., in circuit design).

If this is one of the things that we do or aim at when we make actual cause judgments, several consequences follow that make for differences from type-causal differences. One, discussed briefly in Chapter 8, is that we should expect that the characterizations of actual causes often legitimately incorporate information that seems irrelevant to the causation of the effect when this serves to distinguish the actual cause from other candidates for the actual cause. We judge that the impact of Suzy's rock caused the window to break even though the fact the rock was thrown by Suzy (rather than some else) is irrelevant to the breaking. We do this because (or to the extent that) it is important to attribute responsibility for the breaking to Suzy rather than some other cause. Type-causal judgments are often less influenced by this sort of consideration.

One obvious possibility suggested by these observations is that the category of actual cause judgments may turn out to be rather heterogeneous in the sense that such judgments can have a number of different functions. Some of these may be closely connected to fault and responsibility ascription, but this may be less true for others. Some of these may be closely connected to causal selection problems and others less so. If something like this is correct, it would explain why capturing the full range of such judgments within a single account has turned out to be such a difficult problem.[72]

Finally, let me note that if actual cause and type cause judgments are relatively independent in the way described, this raises some interesting empirical questions. For example, does a capacity for one kind of judgment precede the other developmentally, or do both develop together at the same time? Presumably even if actual causation is logically or metaphysically primary, it doesn't automatically follow that a capacity to make such judgments will emerge before the capacity to make type-causal judgments (or to engage in behavior that reflects such judgments), but perhaps it would be natural to expect something

[72] Glymour et al. (2010) describe a large number of examples involving actual cause judgments and argue convincingly that none of the available normative theories returns judgments about all these examples that accord with "intuition." (I'll add that I at least have no clear "intuitions" regarding some of their cases.) They also argue, again convincingly, that given the enormous number of possible examples, especially when one considers cases involving large numbers of variables, there is no reason to suppose that any method that proceeds inductively, attempting to constructing a theory that covers just the examples that happen to occur to the theorist, is likely to be successful—there will be no grounds for confidence that there do not exist additional counterexamples that have not yet been thought of. They note as well, as I have, the difficulty of making progress with this problem in the absence of a generally accepted normative theory.

like this. A related developmental question has to do with the role of norms and abnormality in actual cause judgments. Is it true, as some current accounts seem to imply, that subjects will lack the capacity to make actual cause judgments (or will have a greatly restricted capacity) if they lack understanding of the relevant norms? Is language required for this? Can nonverbal children and animals make (what we can interpret as) actual cause judgments? What evidence would show this?

3

Methods for Investigating Causal Cognition

Armchair Philosophy, X-Phi, and Empirical Psychology

3.1. Introduction: Some General Features of Empirical Research on Causal Cognition and How These Contrast with Intuition-Driven Methodologies and Experimental Philosophy

Subsequent chapters will describe in more detail various kinds of empirical research on causal cognition (which I will sometimes abbreviate as ERC) and how this might be brought to bear on normative theorizing (and conversely). But before digging into these details, I want to step back and consider some general features of ERC and what it can show. I also want to compare this research to the intuition or judgments-about-cases-based methodologies used in traditional armchair philosophy as well as to research in experimental philosophy (X-phi). My motivation for proceeding in this way is several-fold. First, armchair-based methods are widely used in philosophical discussion of causation, and so it is important to understand just what they can (and cannot) establish. A second motivation is that I suspect that many philosophers will fail to appreciate just how different some ERC is from armchair philosophy, both in the methods employed and in the kinds of conclusions it can establish. For this reason, the opening sections of this chapter underscore some of these differences—part of my goal is to help philosophers to see how many other sources of information besides armchair-based methods are available for the problems that interest them. Another set of themes has to do with various aims one might have in constructing a theory of causation and how these relate to intuition-based methods. I see much of the philosophical literature as focused on one of two possibilities—either intuition-based methods are used with the aim of capturing "our concept" of causation, or they are used with the goal of discovering facts about the "nature"[1] of causation itself. I claim that intuition-based methods are not very apt for either purpose. In addition, the aims themselves are either unclear or (in the case of organizing inquiry around "our concept of causation")

[1] Non-philosophical readers may wonder what sorts of facts have to do with the "nature" of causation. Different philosophers have different candidates for such facts: it has been claimed to be part

Causation with a Human Face. James Woodward, Oxford University Press. © Oxford University Press 2021.
DOI: 10.1093/oso/9780197585412.003.0004

overly limiting. Subsequent sections in this chapter then describe aims for both normative and empirical theorizing that are more fruitful.

Although methods involving intuition and judgment about cases are of limited usefulness in the roles just described (i.e., as sources of information about the nature or concept of causation), I stress that they nonetheless can provide genuine information (or evidence) that is useful in theorizing about causal reasoning and causation. Very briefly, these methods can provide information about or evidence concerning *shared practices of causal judgment*.[2] This in turn can play a useful role in both descriptive and normative theorizing about causal reasoning. Note that this conception of the significance of intuition-based methods requires that we think of the relevant information they provide as *not* having to do with their ostensible subject matter. Suppose that philosopher X reports the judgment that the gardener's failure to water the flowers caused their death in some hypothetical scenario. Although this judgment is "about" a causal relationship that would exist if the scenario were realized, I contend that the evidential role of this judgment is most naturally understood[3] in terms of a claim about how X and others judge about a case of this kind. Thus, despite some of the differences I describe subsequently between judgments about cases in the philosophical literature and ERC, there is an important respect in which the former potentially can have an epistemic or evidential status that is similar to the results of an ERC experiment—in both cases we are being provided with (what is claimed to be) information about how people judge. In this chapter I explain why this understanding of the evidential role of judgments about cases is, contrary to what philosophical readers might initially think, not counterintuitive at all.

Since the structure of this chapter is somewhat complex, with a number of interrelated themes, I begin with a brief road map to what follows. Section 3.1 draws attention to various ways in the methods and results in ERC differ from intuition-based methods in philosophy. Section 3.2 takes up questions about what intuition-based methods can (and cannot) tell us, and Section 3.3 relates this material to issues having to do with the goals that a theory of causation or causal reasoning might legitimately have. Rejecting the suggestion that theorizing should aim at describing our "concept" of causation or, alternatively, at characterizing what causation "is," under various interpretations of this, subsequent sections explore and defend other possible goals.

of the nature of causation that it is an "intrinsic" relation, that it is transitive and asymmetric, that involves transfer of energy, etc.

[2] I don't claim this is the only legitimate role for judgments about cases—see subsequent discussion.

[3] As suggested earlier, "most naturally understood" means that this is the construal that assigns an evidential role that is defensible.

3.1.1. Judgments about Cases, X-phi, Concepts, and What Causation Is

We noted in Chapter 1 that a very standard procedure in philosophy is to assess philosophical theories by comparing these with "judgments" or "intuitions" about particular cases covered by the theories in question. (I will interchangeably speak of "intuitions," "judgments," or "intuitive judgments" in what follows.) Appeals to judgment/intuition can take many different forms, but in one very common version, a case or scenario is described verbally, a judgment is generated about whether the case is or is not an instance of some concept or category X, and then this judgment is compared with what some philosophical theory implies about how the case should be classified, with the theory gaining or losing support depending on whether the results that it yields agree with such judgments. (The judgment thus typically involves a "binary" act of classification rather than something more graded.) The philosophical literature on causation is full of examples that can be understood in terms of such a strategy: a scenario is described in which Billy and Suzy throw rocks, each of which will shatter a bottle if contact occurs; Billy's rock hits first, and the bottle shatters. Then an intuition/judgment is elicited—for example, that Billy's throw caused the bottle to shatter and that Suzy's did not. Philosophical theories of causation are assessed at least in part in terms of whether they capture or reproduce this intuition.

As I will understand this procedure, the intuition/judgment itself is taken to have some evidentiary role or to serve as a source of information of some kind.[4] This raises several obvious questions. If the intuition (or its content) is evidence, what is it evidence for? Several possible answers will be discussed in what follows. One is that the intuition provides evidence/information about "our" concept (or concepts) of causation. A second possibility is that the intuition provides information about what causation "is" or its "nature," perhaps by providing some sort of rationalistic grasp or insight into this. Another question, also explored later, has to do with why we should believe that intuitions have any such evidentiary role—what the source or basis of their epistemic authority is.

Experimental philosophy (X-phi) often presents itself in opposition to the "armchair" methodology just described, and in some important respects it *is* different. In X-phi, typically the responses of a number of subjects (rather than the responses of a single philosopher and perhaps a few colleagues) are collected, and the subjects are often non-philosophers. Moreover, at least in many cases, experimental philosophers do not treat these responses as having the properties some philosophers assign to intuitions (e.g., as a source of insight into the

[4] Some philosophers deny that judgment about cases should be understood in this way. I discuss their views briefly in what follows.

nature of things). Nonetheless, it is striking how much continuity there is between some kinds of X-phi research and the more traditional intuition-driven methods.[5] These similarities include the fact that in both sorts of investigations, the subjects are adult humans and commonly what is recorded are their verbal responses or judgments regarding verbally described (as opposed to materially realized) scenarios. Moreover, in both kinds of investigation these responses are often (although by no means always) taken to be evidence concerning features of the *concepts* subjects possess, even if in some cases of X-phi research the claimed result is that the folk don't possess the concepts that philosophers attribute to them.[6]

By contrast, a great deal of ERC does not follow the traditional intuitions-about-cases philosophical methodology *and* departs from the methodology that is frequently adopted in X-phi research. Although a substantial amount of research in both X-phi and traditional philosophy has as its declared target discoveries about concepts, I will argue in what follows that much ERC is not helpfully viewed as aimed at making discoveries about concepts of causation at all. Instead the research goals are often much broader, having to do with features of causal reasoning and inference that are not plausibly viewed as "part of the (or a) concept of causation." For example, these research goals may have to do with understanding the underlying computations and representations involved in learning and reasoning. They may also be relatively focused, having to do with, for example, uncovering particular assumptions, including "default assumptions" (see Section 3.3.2) that underlie various of causal learning strategies. As I will argue later, even on a very expansive conception of what is involved in mastery of the concept of "cause," not every interesting feature of causal cognition that is a target of empirical research reflects such mastery.

Still less is ERC designed to arrive at conclusions about what causation "is" (at least in any sense that is likely to be of interest to metaphysicians) in the way that some intuition- or judgment-driven methodologies aspire to do. Moreover, the evidence or empirical results to which ERC appeals need not be (and often are not) "intuition-like" or judgment-like in the way that philosophers think of these notions. For one thing, the evidence may not take the form of a verbal report or judgment at all—instead it may take the form of nonverbal behavior,

[5] As I note in what follows, these similarities are *not* characteristic of all X-phi research, including a number of recent developments, which instead have much more in common with ERC. See, for example, Knobe 2016 for a general defense of X-phi research that is continuous with ERC and Kominsky et al. 2015 for an example of such research.

[6] Presumably this is in part a reflection of the origins of the experimental philosophy program. This arose in part as an empirical challenge to the claims of traditional philosophers about the extent to which their intuitions/judgments about cases are widely shared and about the extent to which these provide information about folk concepts. This focus on concepts and what verbal judgments can tell us about them is retained from more traditional philosophy models, even if the conclusions are different.

as described in more detail in what follows. Even when the evidence does take the form of a verbal judgment, it may not involve a simple classificatory judgment of a case as falling under or failing to fall under some concept of interest. For example, as noted earlier, it may involve a graded, nonbinary judgment, as is the case with causal strength. More fundamentally, the evidential significance of such judgments in ERC is usually taken to be quite different from the evidential significance philosophers often assign to their judgments/intuitions. Among other considerations, the verbal judgments in ERC are often the dependent variables in experiments, with other variables being manipulated by the experimenter, with the goal of determining how these (causally) affect the verbal judgments. That is, the goal is to first establish conclusions about the judgments that subjects make and then to identify the factors affecting why the subjects judge as they do. There need not be (and there often is not) any assumption that the judgments themselves reflect some special access or insight into either the nature or the concept of causation—the judgments are instead simply treated as evidence for (or explananda of) some model of underlying representations and computations, where this may not be interpretable as a claim about the subject's causal concepts.[7]

In addition, as I will argue in what follows, although reports of intuition/judgment of the sort that philosophers rely on will sometimes provide evidence for what the intuiter and perhaps others judge (or, more broadly, for the existence of certain shared practices of causal judgment), in many cases they do not by themselves provide reliable evidence for conclusions about the factors *causally* influencing such judgments, which is a typical goal of ERC. A similar assessment holds for the survey-like research frequently conducted by experimental philosophers. These may tell us what judgments people make, but they usually do not—indeed are not designed to—provide reliable evidence for *why* they judge as they do. Finally, although some ERC does involve eliciting subjects' verbal judgments about verbally described hypothetical scenarios, in other cases even though verbal responses are involved, the scenarios themselves are materially realized rather than merely described verbally, and this also can make an important difference to the content and significance of the responses.

In what follows, I elaborate on some of the points made in the preceding paragraphs.

[7] An additional difference between traditional reports of intuitions by philosophers and verbal responses in ERC is that in the former case, the reports are usually (although not always) shaped by the expectation that the reported judgment will be shared by others—that is, the philosopher does not just report what she judges but often takes her report to reflect how others will also judge. By contrast, in a typical ERC experiment subjects are asked to merely report their own judgment (although they may of course also be influenced by a desire to report judgments with which others or the experimenter agrees).

3.1.2. Experiments in Which the Result Is Not a Verbal Report

A number of the experiments (and other observations) described in subsequent chapters do not involve eliciting verbal responses at all, so it is hard to assimilate them to "an intuitions-as-evidence-for-a-concept or an underlying nature" paradigm for this reason alone. (They are also in this respect unlike the verbal responses elicited in much X-phi research.) Instead, ERC may involve nonverbal subjects (infants, nonhuman animals) with the dependent variable being some form of nonverbal behavior—for example, whether a primate can successfully use a tool to retrieve food. Moreover, even when the subjects are verbal humans, the experimental result again may involve nonverbal behavior—the dependent variable may be whether the subjects (e.g., young but verbal children) are able to carry out an action that will cause a machine to light up,[8] with this being the measure of whether there is successful causal learning. (Again we see a focus on success, which can sometimes be measured by nonverbal behavior.) In addition, when subjects are verbal, a combination of measures involving verbal responses and nonverbal behavior may be employed. One of the advantages of using nonverbal measures is that it allows for experiments with subjects like animals and infants. This permits comparisons of causal cognition across different species and for the tracking of developmental trajectories (e.g., from infants to toddlers to adults) that otherwise would not be possible. In addition, as some of the experiments described subsequently will illustrate, verbal responses can be ambiguous and difficult to interpret—sometimes nonverbal behavior can provide clearer evidence.

In experiments in which the dependent variable involves nonverbal behavior, it may be tempting to try to interpret this as judgments about cases. However, in many cases this seems forced. Consider an ape who chooses a stick without a hook at the end in preference to a hooked tool in an attempt to retrieve an out-of-reach food item. This may tell us something interesting about the ape's "causal cognition," but it is not clear that it is fruitful to interpret this as a case in which the ape has certain intuitions or makes certain judgments and then to use these to provide evidence about the ape's "concept" of causation (or of proto-causation or whatever). It is presumably even less plausible to treat the ape's behavior as evidence bearing on the nature of causation.

[8] As in Gopnik et al. 2004.

3.1.3. Verbal Reports That Do Not Follow the "Intuitions/ Judgments as Evidence for Concepts" Paradigm

In the intuitions/judgments as evidence for concepts paradigm, the philosopher considers examples and then makes a "binary" judgment about these, classifying them as either falling under some concept or not—the case is either a case of justice or not, either a case of causation or not. Although some investigations in ERC take this form, many do not, even when a verbal response is asked for. Instead, verbal probes calling for more graded, nonbinary responses are employed. For example, subjects may be asked to rate on a sliding scale (running, say, from 0 to 7 or from 0 to 100) how strongly they would agree that it would be appropriate to say that a causal relationship is present when they are presented with certain stimuli. These ratings are taken to be judgments of "causal strength," a notion introduced in Chapter 1 and discussed at some length in subsequent chapters.

It seems problematic (or at least not obviously appropriate) to assimilate such graded strength judgments to the responses that philosophers typically try to elicit when they ask about intuitions. This is partly because of their graded, nonbinary character but more fundamentally because they play a role somewhat different from the role philosophers assign to intuitions. To repeat an earlier observation: while philosophers often treat intuitions as a source of access to facts about causation or our concept of it, psychologists treat causal strength judgments as explananda, variations in which they are trying to explain in terms of some set of independent variables (perhaps with an associated computation).[9] That is, the experimental paradigm is one in which some stimulus or body of information is varied or manipulated (e.g., contingency information about the association between a putative cause and effect)—this is the independent variable—and the goal is to record subject causal strength judgments and see how these change in response to changes in the stimuli. The role of causal strength judgments is thus similar to the role of subject responses in a psychophysics experiment.[10]

To this it may be responded that philosophers who appeal to intuition often *try* at least to do something similar to what has just been described—they vary different features of their scenarios and consider how their intuitive judgments

[9] To be fair, philosophers sometimes talk as though the goal of their theorizing is capturing or "explaining" intuitions. But to the extent that their theories are not computational and capturing intuitions means something like describing them in a compact way rather than providing causal explanations of judgments, their theories don't look much like standard psychological theories. I say more about this later.

[10] When a subject in a psychophysical experiment judges that one experienced sound is twice as loud as another, this judgment is not treated as a source of insight into, say, the nature of loudness or into our concept of sound. Instead, the goal is to identify other variables, variation in which explains variations in judgments of loudness.

change in response, in this way hoping to learn how the features causally influence those responses. (In effect, they attempt to experiment on themselves.) I discuss this response in more detail in Section 3.2.3. Here I will just observe that such a procedure requires some very strong assumptions about, for example, the philosopher's ability to mentally control for potential confounding influences that may undermine the reliability of such causal inferences.

3.1.4. Material Realization

Another important way in which some ERC differs from typical X-phi survey experiments is that in the former, the "scenarios" may, so to speak, be realized materially rather than just described verbally. For example, subjects may be presented with various physical objects (toy airplanes, blickets (a made-up name for objects used in psychological experiments), and blicket detectors—see Chapter 4), rather than just with verbal descriptions of these. In some cases subjects may be asked to make verbal judgments about causal relations involving these objects but (as noted previously) they may also be asked to engage in nonverbal behavior—for example, to "activate" the objects (e.g., Gopnik and Sobel 2000), with this behavior being the dependent variable in the experiment.

This difference between verbal description and material realization matters for several reasons. First, there is substantial evidence that even adult humans sometimes respond very differently to stimuli that take the form of verbal descriptions of scenarios than to stimuli that are materially realized. As an illustration mentioned earlier, subjects often perform differently and sometimes more optimally on statistical inference problems when presented with ecologically realistic data in the form of materially realized frequencies than when given word problems involving probabilities, particularly when these involve explicit reasoning. Saffran et al. 1996 found that eight-month-old children can learn to segment words from fluent speech just based on statistical relationships between neighboring speech sounds, even though a number of well-known studies report that adult humans are often remarkably bad at verbally presented reasoning problems involving explicit calculation with probabilities. As a causal reasoning illustration, Danks et al. (2014) found that although certain causal judgments about verbally described scenarios are influenced by information about norms, this effect largely disappears when subjects are asked to base causal judgments on materially realized frequencies.[11] More generally, as remarked in Chapter 1,

[11] Another illustration is provided by the fact that adult humans without special training are often bad at constructing explicit experimental designs that eliminate confounding. By contrast, even young children can take behavioral steps to avoid confounding when faced with concrete causal inference problems—see, for example, Kushnir and Gopnik, 2005.

results from explicit reasoning tasks, particularly with verbally described hypothetical scenarios, can dissociate from results from tasks that rely on implicit processing with information presented in other, more material formats, with the latter sometimes being more reliable and accurate than the former. We get a misleading picture of human abilities if we focus only on the former, but this is what methods employing intuition/judgments about verbally described cases do.

A second point is that materially realized experimental designs often naturally allow for additional behavioral (and physiological/neural) measures besides the verbal responses on which both traditional intuition-based approaches and X-phi often focus.[12] Most obviously, researchers can record whether subjects succeed or not at the experimental task, and this can tell us something about the structure and extent of their causal understanding, beyond what is suggested by their verbal behavior. Other measures of nonverbal behavior and processing can also be employed—for example, reaction times, looking0time measures, eye-tracking,[13] neuroimaging techniques,[14] and so on. This can help to deal with some of the problems with exclusive reliance on verbal responses—problems that are nontrivial, as we will see subsequently. (Also see Chapter 7.) A related important advantage of such "material" designs is that they allow for the use of subjects who cannot speak, such as preverbal children and nonhuman animals. Including such subjects enables comparisons of causal cognition across species and also allows one to trace developmental trajectories of aspects of causal cognition among humans, as several of the experiments described in what follows illustrate. This greatly broadens the range of issues that can be explored.[15]

I argued in Chapter 1 that an important part of what theories of causal cognition should seek to understand is how subjects learn about causal relationships and how successful (in contrast to unsuccessful) performance is achieved. Here

[12] Just to be clear, I am not claiming that the use of verbally described scenarios precludes the use of such other measures. For example, Joshua Greene's well-known experiments on moral reasoning measure, in addition to subjects' verbal responses to hypothetical scenarios, both neural hemodynamic response via fMRI and reaction times. But in many or most cases experimental philosophers do not employ such alternative measures.

[13] See the results of the eye-tracking experiments by Gerstenberg et al. (2017) discussed in Chapter 4.

[14] As an illustration, imaging of subjects involved in so-called causal perception (as when one "sees" one billiard ball cause another to move) shows that the neural areas involved in causal perception (e.g., superior temporal sulcus) are different from the neural areas involved in causal judgment tasks not involving causal perception, with these being more frontal (e.g., dorsolateral prefrontal cortex). This, along with other behavioral experiments, strongly suggests that different sorts of processing are involved in causal perception and causal judgment not involving causal perception and that, contrary to what some have claimed, the former is unlikely to be the sole source of whatever is involved in the latter. See Woodward 2011a for additional discussion.

[15] Of course, verbal responses to vignettes also can sometimes be evaluated according to whether they are correct of not according to some normative theory. My point is not that only "material" experiments allow for such evaluation (or that all material experiments automatically do), but rather that such experiments often can be designed in such a way that they provide uncontroversial benchmarks for successful performance.

normative theories (which make claims about what success involves) play a central role. Although some kinds of verbal response can be assessed for correctness or success, materially realized experiments calling for nonverbal responses often have the virtue that they can make use of rather straightforward and uncontroversial criteria for success or correctness such as whether or not a "blicket detector" activates when a subject places a blicket on it. Here the assessment that the subject has successfully caused the machine to activate does not depend on the adoption of any particular controversial theory of causation. Verbal responses to verbally described scenarios in which there are no clear criteria for correctness are often harder to interpret, as will become apparent from some examples discussed in subsequent chapters.

3.2. What Can Reports of Intuitions/Judgment about Cases Tell Us?

With this as background, I turn next to some more detailed discussion, beginning with some remarks about the role of appeals to intuitive judgment and what it can show. Again part of my justification for this focus is the central role such appeals play in much of the philosophical literature on causation. Another goal is to distinguish what judgments about cases can tell us from other sorts of information about causal cognition that is unlikely to be reliably accessible via judgments about cases. My focus in this part of my discussion will be on the idea that intuitions/judgments about cases are evidence or sources of information about something else—about either shared practices, concepts, or extra-mental items like causal relationships as they are in the world. Appeals to intuition/judgments arguably have other uses (besides this evidential one) in philosophical argument, but I will not focus on these in what follows.[16]

As philosophical readers will be aware, there is substantial current debate about the nature and role of intuition and appeals to judgments about cases.

[16] One such "other" use that is important both in philosophical discussion and elsewhere (e.g., in science or mathematics) is simply to illustrate a concept (that is, concept$_n$—see Chapter 1) or distinction that a writer introduces. To return to an example mentioned earlier, in his discussion of various notions of direct and indirect effect, Pearl (2001) provides several examples to illustrate the distinctions he has made. These examples might be understood as "cases," and no doubt Pearl hopes that his discussion of them will seem "intuitive" in the ordinary sense of that term—that is, that readers will find the distinctions clear, well motivated in the sense that they can see why the distinctions are useful, and so on. But this is different from claiming that our intuitive responses to the examples are evidence that our judgments about them are correct or evidence that the distinctions are normatively appropriate. Similarly some well-known thought experiments in philosophy might be construed as illustrations of distinctions or as ways of drawing attention to the fact that "we distinguish" between such and such and so and so (e.g., knowledge and justified true belief). Obviously one can make this claim without also claiming that we are correct to so distinguish.

Some writers not only defend this practice but are happy to use the word "intuition" in describing it. Many different accounts are offered about what intuition involves and why it has epistemic authority. Some favor pictures according to which intuitions involve "intellectual appearances" that provide a sort of rationalistic grasp of mind-independent "natures" or "essences" that are the objects of the intuitions. Sometimes this is accompanied by the suggestion that the knowledge so provided is modal knowledge of some kind—knowledge of how matters must be (with respect to causation, freedom, justice, etc.) rather than merely how they are. Others suppose that what is delivered is instead knowledge about concepts. Some but not all philosophers take intuitions to have a special sort of phenomenology or character that distinguishes them from ordinary judgments—they are fast, automatic, pre-theoretical in some sense, and often accompanied by strong feelings of certainty or correctness.

Recently several philosophers (e.g., Williamson 2007; Cappelen 2012) who are sympathetic to armchair philosophy and to appeals to judgments about cases, have argued that we should not understand these in terms of appeals to anything like intuition. Instead they claim that such judgments amount to reports or assertions of generally accepted facts about extra-mental items such as knowledge, justice, or causation. For example, Gettier's famous discussion is understood not as appealing to intuitions that certain cases do not involve knowledge but rather as drawing our attention to certain facts about knowledge—in particular the fact that Gettier's cases are not instances of knowledge.[17]

In what follows I will use "intuition" and "judgments about cases" in an inclusive way that covers the views of both philosophers who think these have special properties (a special phenomenology, a capacity to provide rationalistic insight etc.) and those who think of them instead as a form of ordinary judgment. At least as far as the philosophical literature on causation goes, I think that Williamson, Cappelen, and others are wrong, as an exegetical matter, in claiming that philosophers don't often argue for their conclusions by appealing to intuitions, where these are understood as having some special status or authority.[18] However, little will turn on my being right about this—the reader can replace talk of intuition with talk of judgment about cases in what follows and understand the latter as "ordinary" judgments that do not involve the exercise of some special faculty.[19]

[17] If you find this maneuver unhelpful, I'm sympathetic. I'm just reporting what these writers say.

[18] For a recent example of the explicit use of intuition-talk in connection with causation, see Paul and Hall 2013.

[19] I will, however, assume that the category of intuitions/judgments about cases as this figures in philosophical discussion should not be understood so broadly as to cover all cases of judgment or knowledge claims. For example, an ordinary perceptual judgment (that there are three water glasses on the table) should not be construed as an appeal to intuition or a case judgment in the relevant sense. Similarly, when I judge that modus ponens is a valid inference form, we don't have to think of this as somehow depending on an appeal to intuition—the usual analysis in terms of truth tables

One point that I will insist on, however, is the following: whatever it is that intuition/judgments about cases are claimed to provide information about, we need some basis for assessing the accuracy or reliability of such judgments. We need an answer to the question of why (if ever) we should trust such judgments or what the source of their epistemic authority is. We should resist accounts that fail to address this question. This is so whether or not one follows Williamson and others in thinking of judgments about cases as claims about generally accepted facts. Even if this is the way most philosophers think about such judgments, we still need an answer to the question of why we should believe that such judgments do report facts.[20] The most obvious way of establishing reliability consists in showing that such judgments are well calibrated with respect to their

suffices to show that this inference is truth-preserving and thus has a means/ends justification. I mention this because there is some tendency in the philosophical literature to use the notion of intuition/judgment about cases so broadly that it covers virtually all sources of knowledge. This makes it look as though skepticism about appeals to intuition commits one to much more general forms of skepticism.

[20] For what it is worth, I see Williamson, Cappelen, and others as failing to address these issue about reliability in a convincing way. As noted previously, these writers hold that we should understand judgments about cases as appeals to mind-independent facts that in turn typically reflect widely shared background knowledge. Sometimes this is put in terms of a contrast between *intuitings* and the *intuited* (the latter being the content of the judgment). It is said that it is this content (at least to the extent it reports a fact, which seems to be presumed) that is doing the argumentative work and not the fact that the content is the upshot of an intuiting. From my perspective, this fails to come to terms with legitimate concerns about reliability and calibration. If one is concerned about the epistemic authority of judgments about cases, proceeding on the presumption that these can be treated (at least often) as reports of extra-mental facts or reflections of background knowledge seems question-begging. We can grant the distinction between intuitings and the content of what is intuited, but it is still in order to ask why, in various cases, we should believe that the latter describes a fact. Similarly whether or not it is true that the content of typical judgments about cases has to do with mind-independent items (rather than, e.g., claims about how we think), we still face the question of why we should take such contents to reflect truths about such mind—independent items.

Cappelen suggests that to the extent that we are unsure about whether a judgment reflects facts of background knowledge, we can appeal to additional evidence and standardly available procedures of assessment. But at least in many appeals to judgments about cases in philosophy, this is *not* what is done, and it is hard to see what it might involve. When Gettier says that the subjects in his scenarios do not know, he does not back this up with further *evidence* of any ordinary sort. If he were to provide additional grounds in support of his claims, it is hard to see what this could consist in besides either (i) a normative theory of what knowledge consists in (a claim that we shouldn't treat Gettier's examples as cases of knowledge) or (ii) a claim that it is not our practice to apply the word "know" to his cases. The latter is an appeal to a shared practice of judgment and thus accords with the construal of the significance of intuition that I advocate. If instead the judgments about cases he reports are being supported by some normative epistemological theory, then again this is not just a fact about what is in the world.

My view is that similar conclusions hold when the judgments in question have to do with causation. Consider someone who reports his judgment that the failure of the gardener to water the flowers caused them to die. He may think of himself as reporting a fact (or what would be a fact if the scenario were realized). Nonetheless we are entitled to ask what the justification is for treating this causal claim about flower death (with its implication that omissions can be causes) as a fact. Here too appeal to background knowledge or ordinary empirical procedures seems question-begging. No ordinary empirical investigation into how matters stand in the world is going to tell us whether or not it is a fact that omissions can be causes. As with the Gettier cases, it is hard to see how one might

targets X—that is, they have good error characteristics in connection with classifying these targets.[21] Obviously this requires a specification both of the target (is it a concept, some worldly item, something else?) and of some independent standard (independent of the intuitions being assessed) for whether the target is correctly classified. In this respect, appeals to intuition/judgment should be assessed just as we would any other putative source of information. If, say, I claim to be able correctly distinguish cats from dogs on the basis of my visual perception, then establishing that I am reliable at doing this requires some independent standard for identifying cats and dogs and evidence that I can classify in accord with this standard. Similarly for the intuitions expressed in causal judgments and elsewhere in philosophy. One reason for thinking that intuitions/judgments can sometimes be sources of information about the judgments of others (as I have claimed) is that we can independently assess whether such intuitions are well calibrated with respect to such targets. That is, we have independent ways of checking what others' judgments are—for example, we can ask them or observe their behavior—and in this way we can assess to what extent we are accurate trackers of the judgments of others. In the case of other possible targets for intuition—for example, intuitions as possible sources of information about the nature of causation, it is much less clear how to assess calibration. Presumably this is why philosophers fall back in such cases on standards of assessment other than calibration, such as the coherence of various intuitive judgments with one another—a standard that I claim is inadequate.

In what follows I explore various possibilities about what appeals to intuitive judgment might show, beginning with the least ambitious. Those who are interested in what can be learned from intuitions/judgments about cases can think of

go about trying to argue for or against this claim except by reconstruing it as a claim about shared practices or as a normative claim about how we ought to judge.

[21] Some philosophers, particularly those who are critical of appeals to intuitions, try to bypass calibration considerations and to argue that intuitions are often unreliable simply on the grounds that people's intuitions can disagree or that they are sometimes influenced by "irrelevant" factors. I regard these arguments as problematic. If people's intuitions disagree, this may just reflect differences in the interpretation of the question associated with the intuition—see the ensuing discussion. Moreover, even if disagreement shows that some are not well calibrated, this is consistent with others being well calibrated. This last fact may be of considerable philosophical and psychological interest. As an illustration, a third of the subjects in an experiment described by Walsh and Sloman 2011 (see Chapter 7) report the judgment that a marble that strikes a domino that then falls over does not cause this outcome when a backup cause is present. Whatever is going on with such judgments, they do not show that the remaining two-thirds of subjects, who disagree with this judgment, are unreliable, either about the generally accepted pattern of judgment in such cases or about what is normatively appropriate. Similarly, the fact that intuitive judgment can be influenced by allegedly irrelevant factors is consistent with such judgments often being reliable, as we see in the case of visual illusions. The upshot (in my opinion) is that if the issue is the reliability or not of intuitive judgments, there is no good way of getting at this that avoids considerations of calibration—extent of agreement and irrelevant influences are not good substitutes for this.

what follows as a partial vindication of the idea that these can serve as a source of information, where the information in question has to do with shared practices of causal judgment. At the same time they should think of my discussion as a challenge to explain why we shouldn't move beyond appeals to intuition and judgments about cases in the ways indicated, since there are other, richer sources of information and many interesting questions both about causation and about causal cognition that cannot be settled by appeals to what people intuit or judge.

3.2.1. Intuition as a Source of Information about What the Intuiter Judges

One might think that reports of intuition/judgment (at least) yield information about what the intuiter judges. But even this may not always be true if we impose some constraints on what counts as an intuition or judgment. One might think, for example, that to count as a (report of) an intuition/judgment, what is reported should not be too unstable over time and not subject to certain kinds of order or framing effects, the rationale for these conditions being that in their absence it is not clear what the subject's judgment is or even if anything is present that deserves to be called a "judgment" (or at least a judgment that has any kind of authority or that is worth capturing or explaining). Unsurprisingly, there is evidence (e.g., Machery 2017 and the references therein)[22] that some reports of intuition show effects of this nature, including instability even for a single subject.

A related but more subtle issue is this: there is considerable evidence that the verbal probes employed in psychology experiments, including those discussed in what follows, can be ambiguous in ways that are not obvious, either to the experimenter or her subjects.[23] Different experimental subjects may interpret the same verbal probe in different ways, and the experimenter may interpret the probe differently than her subjects. Both experimenter and subjects may be confused about the question being asked and fail to recognize different possible ways of interpreting it. Although it is easy to see that this is possible in an experiment in which the experimenter and subject are different people, it also seems entirely possible that the same thing can happen when the philosopher puts questions to herself and reports her intuitive responses. (I'm assuming here that it is plausible to think of a report of an intuition/judgment as a report of an answer to a question that the philosopher puts to herself.) That is, the intuiter may be confused about the question she is putting to herself or may think that slightly different

[22] But see Cole 2016 for evidence that subjects are often able to recognize when their intuitions are stable.

[23] See especially the disputes surrounding Cheng's causal power theory discussed in Chapter 7.

variants are really the same question when they are not and in fact call for different answers.[24] Interpreting the question differently, other philosophers may conclude that they do not share the intuition when in fact they would have the same intuitive judgment if they had understood the question similarly. Or perhaps they mistakenly conclude that they do share the intuition but this is only because each interprets the question differently. In all of these cases, although there may be a verbal report that is identified by the speaker as reporting an intuition, it again may be unclear which judgment is being reported.

Although I believe these (both order and framing effects and ambiguous questions) to be real possibilities when appeals are made to intuitive judgments, the issues they raise can sometimes be adequately addressed through a combination of normative analysis and empirical investigation. Empirical investigation can reveal when order and framing effects are present and sometimes suggest strategies for lessening their impact—in principle these are available not just in experiments and multi-person surveys but also in armchair judgments about cases due to a single philosopher. As illustrated in Chapter 7, normative analysis followed by empirical exploration can help to reveal when questions are ambiguous and suggest less ambiguous formulations—another reason why normative theory is essential to empirical investigation in this area. In principle, effects, if

[24] Here is a possible example drawn from the philosophical literature. McDermott (1995) considers a case in which a ball is headed toward a window. Between the ball and the window are (first) a fielder and then a very solid brick wall. The fielder catches the ball. McDermott invites us to consider our "intuitions" (his word) about whether the fielder's action "prevented" the ball from striking the window. He observes that most of the people to whom he put this question—call it (i)—respond at first with a "no" on the grounds that even if the fielder had failed to stop the ball, it would not have reached the window. In other words, the fielder's action did not "make a difference" to whether the window was struck. McDermott suggests, however, that this initial reaction is mistaken on the following grounds: he asks his responders, "Well, something prevented the ball from striking the window. So between the fielder and the brick wall, which was it?" He reports that when presented with this question—call it (ii)—most of his respondents change their mind and agree that the fielder was a preventer. Obviously this argument depends on the assumption that questions (i) and (ii) are alternative ways of eliciting the same intuition—that (ii) is just a clearer version of (i), so that the answer to (ii) can be used to cast doubt on the initial answer to (i). I think this is far from obvious. Question (i) is about whether a putative cause has a certain effect (preventing the window from breaking) where it is assumed to be an open question whether the cause has that effect. It thus involves cause-to-effect reasoning. Question (ii) instead involves effect-to-cause reasoning—the question assumes that a certain effect has been caused (with the causation of the effect being further characterized as a case of prevention of window hitting) and then asks for which of two competing alternatives is the cause of this, in line with the description of actual cause judgments in Chapter 2. To the extent that it is treated as an open question whether the catch has the effect of preventing, a negative answer to (i) seems reasonable. By contrast (ii) assumes that this negative answer is mistaken—that is, that this *is* a case of prevention, with the only issue being whether the fielder or the wall was the preventer. No wonder the two questions elicit different answers!

Let me add that I'm not trying to argue that one of these answers is right and the other wrong or even that my particular analysis of why we are pulled in different directions is the correct one. I do suggest, however, that this is a case in which it is not straightforward that the two different verbal probes used to elicit an intuition correspond to the same question and that confusion can result from assuming otherwise. It is also worth noting how this example illustrates some of the differences between judgments of actual causation and judgments of the effects of causes.

any, of variations in wording can be explored experimentally.[25] One can test experimentally whether disambiguation has been successful by seeing whether it leads, as expected, to reduction in variation in subject response. So while skeptics and critics are right to raise the worry that framing, order effects, and ambiguities may be present in appeals to intuitions/judgments about cases (and for that matter in X-phi and in ERC), these need not be insuperable problems in principle.

3.2.2. Intuition as a Source of Information about the Judgments of Others

Another natural thought, endorsed in Chapter 1, is that reports of intuitions or judgments can sometimes tell us something about the judgments of others (or more generally about shared practices of reasoning and inference, which may extend well beyond judgments—see the later discussion). Philosophers who report intuitions often suppose that others will share these intuitions and/or that they are good judges of whether others will share them. Such assumptions are reflected in the common use of such tropes as "people think that such and such," "we think," "we would (or would not) say . . ." in connection with intuition-like reports. Put differently, we might think of philosophers who use intuitions in this way as attempting to use themselves and their judgments as sources of information about how others will judge in broadly the same way that I might, in non-philosophical contexts, use my own response to an actual or possible situation to predict how others will respond to a similar situation.

Obviously whether (and to what extent) the intuiter's judgments are representative or shared by others is an empirical issue.[26] A number of experimental philosophers claim to have found examples (most not involving causal cognition) in which the intuitive judgments of philosophers seem to depart significantly from those of the folk or at least subpopulations of the folk. (See Machery 2017 for a summary.)[27] Other experimental philosophers claim instead that

[25] Thus in my view the proper response to order and contextual effects or ambiguity involving verbal probes is not to eschew the use of verbal responses (whether in experiments, surveys, or reports of intuitions) but rather to recognize that these are most likely to be valuable (i) when accompanied by testing for order and contextual effects and the use of designs than minimize these, careful analysis of the verbal probe itself, how it is likely to be interpreted by subjects, and how the results it elicits compare with the results of other verbal probes and (ii) when the verbal response is combined with other sorts of evidence, including evidence about nonverbal behavior.

[26] Put differently, the reliability of intuitive judgments in the role just described is closely bound up with the issue of how good philosophers are at descriptive psychology (regarding their own and others reasoning and cognition).

[27] For additional, more optimistic discussion of the conditions under which philosophers are good judges of the judgments of others, see Nagel 2012.

in some cases philosophers are fairly good judges of how others will judge and moreover that they have a fairly good sense of when their judgments about the judgments of others are likely to be accurate—see Nagel 2012 for the case of knowledge claims. My sense is that philosophers (and psychologists) are often reasonably good judges (prior to detailed empirical investigation) of how others will judge in cases involving causal judgment, even if they are not such good judges about other sorts of cases. For example, as discussed later, Lewis's (1986) claims about the role of sensitivity, Hall's (2004) claims about the distinction between dependence and production, and Yablo's (1992) claims about the role of proportionality all seem to track patterns that, as an empirical matter, are present in people's judgments.

When philosophers do make mistaken claims about shared patterns of causal judgment and reasoning, it seems to me that this is often because they underestimate the amount of variability in these, rather than being completely off the mark. That is, the philosopher's claims may correctly characterize the judgments of *some* nontrivial group of people, but such judgments may be less widely shared than the philosopher supposes. This may be the case for judgments involving the causal role of absences, for example, with some subjects being more willing to attribute a causal status to absences than others, so that neither the claim that absences are almost always judged to be causes nor the claim that they almost never are is correct.

Several additional points are worth emphasizing. First, there is nothing epistemically mysterious about the two roles assigned to intuition so far. Claims about how others will judge are ordinary empirical claims and can be assessed accordingly; X-phi surveys may be particularly useful in this regard. The abilities involved in successfully judging how others will judge are likewise unmysterious.

Second, note that on the view just described, we don't have to see judgments/reports of intuition as providing information only about some special subset of facts about causation or causal cognition—for example, only information about what belongs to the *concept* of causation. In principle, such reports, either of one's own judgment or of judgments one expects others to share, might provide information about many different features of practices associated with causal cognition—for example, patterns of reasoning and learning strategies—that are not naturally viewed as having to do with concepts of causation. In other words, since these are empirical psychological claims, we need not think of them as claims about conceptual truths about causation or anything of that sort.[28] I stress this point because many philosophers hold that intuition or intuition-based

[28] Note also that this construal of the role of intuitive judgments allows us to avoid puzzles about how such judgments can provide us with anything more than information that is already contained in our concepts.

methodologies should be understood as sources of information about the structure of our concepts (see, e.g., the subsequent discussion of Goldman) and perhaps mainly or only that.[29] At least at the level of principle, I see no reason for this restriction. I might, for example, have access to certain features of how I reason about some causal inference problem and use this, correctly, as a basis for thinking that others also reason in the same way, even if I don't think of the reasoning in question as built into the concept of causation.

Let me also emphasize again that insofar as we think of intuitive judgment as a possible source of information about some aspects of the causal cognition, on the parts both of oneself and of others, there is no reason to think that such judgments are the *only* source of such information (or even, in many cases, the best source of such information). In addition to other examples mentioned previously in which the evidence is not intuition-like because it does not involve a verbal report, information about shared practices can come from anthropologists and historians (in the case of practices in the past), among others. If our particular interest is causal reasoning in science (rather than lay causal reasoning), then cases studies of particular episodes of such reasoning, both successful and unsuccessful, may also be very informative about shared practices. Thus, as far as providing such information about self and others goes, there is nothing *special* about the intuitions or reports of judgments about cases produced in philosophical contexts.[30]

[29] Two further points: First, presumably the idea that intuition is primarily or entirely a source of information about concepts is appealing to some philosophers because claims about concepts are regarded as true a priori, if true at all. This thus allows (it is thought) the truths delivered by intuition to have an a priori status and not to be in need of empirical support. But of course even if truths about concepts are a priori, the claim that the intuiter can correctly discern these truths via intuition is an empirical claim. Second, putting aside possible misgivings about the notion of "concept" itself (e.g., problems having to do with deciding what does or does not belong to some concept), we should distinguish (i) the claim that intuitions/judgments about cases can sometimes furnish information about what many will find it natural to describe as "concepts," from (ii) the claim that intuition and other sorts of empirical information about causal cognition provide only or mainly information about concepts rather than information about other features of cognition. My target in the preceding is (ii) rather than (i)—the latter seems so weak as to be relatively uninteresting. Thanks to Thomas Blanchard for raising this issue.

[30] I noted earlier that a number of philosophers (Williamson, Capellen, and others) have claimed that we should not understand philosophers who make judgments about cases as appealing to "intuitions." Instead we should understand them as appealing to something like generally accepted background knowledge. They argue that this shows the irrelevance of X-phi investigations since (supposedly) these are premised on the assumption that the traditional philosopher does appeal to intuitions. This is a very puzzling line of argument. Suppose it is true that when the armchair philosopher reports a judgment about a case, this should be understood as a claim about generally accepted background knowledge. Why then can't the same thing be said about the subjects in an X-phi experiment—that is, that these subjects are relying on generally accepted background knowledge in their reports? Certainly if armchair philosophers don't appeal to intuitions, it seems even less plausible that naive subjects in X-phi experiments do this. Moreover, why doesn't the same conclusion also follow for the reports of subjects in ordinary psychological experiments? If armchair philosophers are really reporting generally accepted background knowledge, shouldn't lay subjects as well as philosophers have access to it? And if this is so, why aren't the judgments of subjects in X-phi or psychological experiments just as relevant to whatever inquiry the philosopher is engaged in as the judgments of the armchair philosopher?

So far I have suggested that judgments about cases can be a source of information about the judgments of oneself and others, but I have said nothing in this chapter about why such information might be valuable. Insofar as we are interested in the empirical psychology of adult human causal cognition, the answer to this question should be obvious. To the extent that we are interested in normative theorizing, the answer is the one that I gestured at in Chapter 1—generally accepted patterns of judgment and reasoning are sometimes and perhaps often ones that work and that may have an independent normative justification. So we have good reason to pay attention to such patterns. I elaborate on this idea later in this chapter.

It is also worth noting that the rationale for intuition/judgment just described provides little support for the common philosophical practice of appealing to intuitions about outlandish or science-fiction-type cases that are far removed from situations with which the intuiter and others have any experience—for example, cases involving magical spells, worlds governed by laws that far different from any contemplated as governing our world, and the like. Or at least this is so if the goal is to get information about actual practices of reasoning and judgment that have been shaped by feedback (hopefully in reliability-enhancing ways) from real-world experience. If we are to use information about actual practices in a normative argument along the lines sketched in Chapter 1, there is no reason to suppose that whatever factors contribute to the reliability of practices of reasoning and judgment applied to familiar and common cases or to cases encountered empirically will also carry over to whatever practices are suggested by science-fictional examples.

Finally, two more general observations: some philosophers, concerned to defend traditional armchair methods, have claimed that what they are doing when the employ such methods (and the information such methods elicit) is completely different from what is elicited from X-phi investigations. They conclude that the latter are irrelevant to their concerns. Presumably they would make a similar claim about empirical investigations conducted by psychologists. However, as noted earlier, in the case of causal judgment there is often substantial congruence between some of the conclusions reached by armchair methods and conclusions reached by psychologists. This suggests that at least to some

For these reasons arguments about whether armchair philosophers are correctly described as appealing to intuitions strike me as a red herring. It seems to me that if the views like those of Capellen and Williamson are to be successful, they must establish two claims. First (i), the armchair judgments of philosophers are importantly different from the judgments of subjects in X-phi or psychology experiments. Second (ii), because of this difference, the judgments of armchair philosophers have some status or authority that those of other subjects do not have. Arguments that armchair philosophers are not reporting intuitions do not establish either of these claims. I don't see that there is any *general* case for either (i) or (ii), although philosophers may have relevant special expertise in particular cases.

extent and in some respects armchair philosophers and psychologists concerned with causal judgment are engaged in a common enterprise. Viewing the armchair philosopher as making claims about shared patterns of judgment makes sense of this commonality and suggests that the investigations pursued by these two groups are not as completely disjoint as sometimes claimed.[31]

A closely related consideration is this: the relationship between armchair philosophy and X-phi is fairly frequently antagonistic: the X-phi philosopher attempts to show that people don't judge in accord with what the armchair philosopher claims, and the armchair philosopher dismisses the X-phi results as irrelevant. Of course when armchair claims and X-phi results are discordant, this is worth knowing. But there is no reason why we should expect the interactions between these two groups to always take this negative, debunking form. As noted previously, in some cases, including cases involving causal cognition, armchair claims and the results of empirical investigations support one another. One would think that this outcome should be welcomed by both groups. A substantial part of the research discussed in what follows illustrates this more cooperative possibility.

In the two uses of appeals to intuition/judgments about cases described so far, we are thinking of it as a source of information about people's judgments—what they judge about cases—and perhaps as a source of information about other overt and readily accessible features of practices surrounding causal cognition. Information about these matters should be distinguished from information about the *causal factors* that affect such judgments. It is a further question whether (and what) intuition or armchair judgment can tell us about this. In saying that this is a further question, I do not mean to claim that intuition/judgment about cases is always unreliable about such matters but only that there is an obvious difference between (i) the claim that people judge that p and (ii) the claim that such and such are the factors that causally affect whether they judge that p (and/or the claim that the computations and representations that underlie judgment about p are such and such). Additional evidence or argument that goes beyond what is needed to establish (i) is required to establish (ii). In particular, since (ii) is a causal claim, we need to conform to the usual standards of causal inference in establishing (ii)—this includes, for example, control of potential confounders. To explore this theme, I turn to a contrast between surveys and experiments.

[31] This is not to deny that there are also important differences between what these two kinds of investigation can tell us. As argued later, controlled experiments are likely to be a more reliable source of information about the causal factors influencing judgments than armchair reflection. On the other hand, armchair analysis may be a better source of ideas about distinctions among different sorts of causal judgments than surveys of lay responses.

3.2.3. Intuitions and Surveys versus Experiments

As I will use these terms, the goal of an "experiment" is to determine which factors causally affect some dependent variable *Y* of interest—in the present context this will be some feature of causal cognition, which might be measured by a verbal response or in some other way. In an experiment this is accomplished by *manipulating* some feature of the experimental situation—for example, some presented stimulus *X*—and then attempting to ascertain the effect, if any, of this variation in *X* on *Y*. For example, the dependent variable might be subjects' judgments of causal strength and the experimental question might be how these are affected by variations in a stimulus that consists of contingency information. The experiment involves manipulating this stimulus and observing any change in verbal response. In doing this, appropriate experimental control is essential—one needs to eliminate the possibility that whatever variation in *Y* is observed is due to some other factor besides *X*, so that the variation can instead be reliably assigned to *X*.

Contrast this with a case that merely involves a description of a scenario (or the presentation of a stimulus) and then a request for a response or judgment, but with no investigation into the factors that affect this judgment. A substantial number of the surveys conducted in X-phi experiments have this sort of structure—that is, they involve a recording of responses but with no systematic effort to investigate the causal factors that affect these responses.[32] Of course such surveys are distinguished from traditional philosophical armchair methods in that they involve a number of subjects rather than just one, and the subjects are typically nonexpert. Perhaps also the philosophers think more carefully about their responses. But to the extent that surveys do not investigate (or their design does not permit the investigation of) factors that causally influence the reported responses, they are more like (mere) reports of judgments than real experiments.[33] In this respect, at least, the appellation "experimental philosophy" seems a misleading description of this sort of research.

In saying this I am *not* claiming that *only* experiments can provide reliable evidence for factors causally affecting subjects' judgments. Experiments are one way of providing such evidence but by no means the only way. If analysis of the factors causally affecting judgment and reasoning is the goal, the alternative to experimentation is causal modeling based on observational data of subject responses and other covariates (or some combination of this and experimentation). Such

[32] For examples of experimental philosophy in which survey-like results play an important role (and which are described by their authors as surveys) see Schaffer and Knobe 2012 and Nahamias et al. 2005.

[33] Of course this does not mean that surveys are without value—merely that they don't provide certain sorts of information.

methodologies have been employed by some investigators, including some experimental philosophers, and can be used to establish reliable conclusions about the causal factors affecting subject responses.[34] My point is that some additional steps and argument, whether experimental or non-experimental, beyond mere reporting of judgments are required to support conclusions about the causal factors affecting such judgments. Usually, although not always, the methodology employed for this purpose in psychology is at least in part experimental, and in this respect psychology experiments are importantly different from those X-phi surveys that are not designed to warrant causal conclusions.

A second issue, already mentioned in passing and addressed in more detail subsequently, bears on the role of intuition/judgment (as opposed to X-phi surveys) in establishing causal conclusions. It might be argued that the intuiter can, so to speak, do controlled experiments on herself, varying features of cases, noting variations in her intuitive responses, and in this way establish which features cause the responses. I address this argument later in this section.

All of this has been rather abstract, and some more detailed illustrations may be helpful. Suppose, first, that a subject or group of subjects is presented with a scenario in which a gardener omits to water some plants and the plants die and where it is also stipulated that if the gardener had watered the plants they would not have died. The subjects are then asked whether the gardener's omission caused the plants to die (or whether it would be "appropriate" to say this or some other similar question) and the response is recorded. Such responses can tell us in principle about how the subjects and (if the responses are representative) how others judge. They thus can be used to refute or confirm claims about the causal judgments about absences that people sometimes make. Note, though, that if this is all that is done, we have no information about the factors causally influencing the responses.

Suppose now that the subjects are presented with different scenarios (e.g., some involving absences and some involving "presences," as when cases in which non-waterings and sprayings with poison are both followed with plant death) and there is variation in response with, say, the poisonings being judged as causal and the non-waterings not. Can we conclude that the factor that causes this variation in judgment is variation along the absence/presence dimension—that is, that the subjects judge that non-waterings are not causes *because* they involve absences or omissions and that the poisonings are causes *because* they involve the presence of a "positive" factor? In my view this conclusion has not been established. Unless other factors that may be influencing the judgments have been

[34] For an example of research on causal cognition involving a mixture of experimental inferences and causal modeling, see Icard, Kominsky, and Knobe 2017.

adequately controlled for, we don't know that the absence/presence contrast is what causes the variation in judgment, rather than some other factor.

This is not just idle skepticism. In my 2006 (see also Chapter 7) I proposed that people are less likely tend to judge relationships of intervention-supporting or non-backtracking counterfactual dependence as causal (or are less likely to judge them as "paradigmatically causal") to the extent that those relationships are relatively non-invariant and are more likely to judge them as causal to the extent that they are relatively invariant, with invariance understood along the lines described in Chapter 2. I suggested that this applies both to causal claims involving presences or positive causal factors and to causal claims involving absences or omissions: claims that the absence of *C* causes some effect *E* tend to be accepted (regarded as uncontroversially causal) when the absence of the *C* → *E* relationship is relatively invariant (as in "absence of oxygen causes death") and not when those claims involve relationships that are relatively non-invariant. In fact there is substantial empirical evidence that such invariance-based considerations do influence judgments of causal status, including judgments about the causal role of absences in the way described (cf. Vasilyeva et al. 2018). It is thus a real possibility that in our imaginary study the subjects are judging that the non-waterings are non-causal not because they involve absences per se but because of invariance-based considerations: they regard the non-watering → death relationship as relatively non-invariant and judge its causal status accordingly. To establish that non-waterings are judged as non-causal *because* they involve absences (and reflect a general tendency of people to judge absences as non-causal), one would need to rule out alternative explanations like the one just described. This is what a well-designed experiment attempts to do—by, for example, varying the extent to which the relationship in question is invariant and whether the candidate cause is a presence or absence independently and seeing whether each of these has an effect on judgment.[35]

Let me now turn to the possibility mentioned earlier—that reliable inferences about the causes of judgments and perhaps other aspects of causal cognition can be accomplished from the armchair, just by thinking. The idea goes something like this: the philosopher finds two cases that (it is supposed) match exactly in all possibly relevant respects except for the presence or absence of a single feature X. The philosopher takes this to be a case in which she is controlling "in her

[35] For a study that makes related points, see Henne et al. 2019. These researchers are interested in the common tendency to judge actions as more causal than omissions. One standard philosophical account of why people exhibit this tendency goes as follows: people judge actions to be more causal because in such cases there is a connecting process between the action and its effect, while such a process is absent in the case of omissions, even when there is counterfactual dependence. (This in effect is the diagnosis provided by causal process theorists such as Dowe 2000.) Henne et al. provide evidence that this is not what is causally influencing judgments about the causal role of actions, at least in some cases.

mind" for all the other relevant differences between the two cases besides X. The philosopher finds that her intuitive response to the two cases is that one is an instance of concept C and the other is not, and concludes that it is recognition of the presence or absence of X that causes this differential response.[36] Generalizing, the philosopher infers that the same is true for the similar judgments of others. The use of this procedure is, I believe, fairly common among philosophers who employ judgment-about-cases methods. That is, it is common for philosophers to make claims not just about their own and others' judgments but also about the factors that influence such judgments on the basis of their own responses to cases (perhaps supplemented by the deliverances of introspection), even if this extra layer of claims is not always explicitly recognized as an additional commitment.[37]

I don't want to claim that this armchair method never leads to correct conclusions about the causal factors affecting judgment. Nonetheless, several observations seem in order. First, as already remarked, claims about what causes judgments are stronger and more exposed to inductive risk than mere claims about how people judge. Similarly for claims about the causation of other features of causal cognition. Second, the mental matching procedure described here seems to require that the philosopher have access to all (has thought of and represents mentally all) of the other factors besides X that might influence her differential response, and, moreover, that she can recognize when these are present and influencing her response, and that she can also tell when she has successfully removed or controlled for such influences, so that any variation in her judgments can be attributed to the causes she considers. Again, I don't claim that this is impossible. However, there are many cases in which real experiments seem to show that what is actually influencing people's differential responses is not what they judge to be influencing them—our access to the causes of our behavior, including our verbal behavior in reporting intuitive judgments, can be quite unreliable.[38]

[36] This methodology is explicitly endorsed in Kamm 1993: she thinks she can tell from the armchair which factors influence her intuitive responses. As noted in Chapter 1, Lewis's various discussions of causation contain many claims about the factors that causally influence judgments about cases, and the same is true of many other philosophers who report judgments about causal relationships. As argued previously and subsequently, some of Lewis's claims about causal influences on his judgements are supported by experimental investigation and are arguably correct; this is not the case for other claims. Again the moral that I want to draw from this is not that it is illicit to propose hypotheses about influencing causal factors or that such hypotheses are virtually never correct; it is rather that such hypotheses require independent confirmation.

[37] It is also worth noting that to the extent that philosophers make claims about the factors that causally influence their own and others' judgments about cases, these seem to be straightforward empirical claims. They are not candidates for a priori or conceptual truths, even if some other intuition-based claims are.

[38] Indeed this feature is emphasized by a number of the philosophers who are sympathetic to a nontrivial role for appeals to intuition—they observe that even when our intuitions seem clear, the processes generating those intuitions seem opaque to us. See, for example, Nagel 2012. Wilson 2002 is a classic discussion of our lack of insight into the causation of our behavior.

Note also that if we always had reliable access to what is influencing our responses via the mental matching procedure described earlier, real experiments would appear to be unnecessary—we could reach reliable conclusions about the factors influencing our judgment and reasoning just by using the matching procedure. Relatedly, if the matching procedure was highly reliable, one would not expect results from real experiments about the causal factors influencing judgment to be surprising—the experimental results could be reliably anticipated through the matching procedure. That real experiments do not seem to be superfluous (and sometimes suggest surprising conclusions, particularly about underlying causal influence and processing) suggests that there are some limitations on the reliability of the matching procedure. It also suggests a role for experimentation (or detailed causal modeling) that goes beyond standard philosophical methods involving judgments about cases.[39] Finally, note that even if the matching procedure can sometimes reliably establish which factors are influencing judgment, we often aspire to something more than this—for example, a computational account that maps input stimuli onto judgments and other responses. I assume that even the most enthusiastic proponent of the matching procedure is unlikely to think that this is a reliable way of establishing or testing such a computational account.

Let me also add that even if my skepticism about the matching procedure is misguided, one would think that philosophers who employ the matching procedure would welcome its supplementation by real experiments and more conventional causal analysis. If the method is highly reliable, it would be nice to have that fact confirmed by other means. If it is not so reliable, that would be worth knowing too. More generally, armchair philosophers who infer to causal influences on their intuitions and empirical psychologists are, to an important extent, involved in very similar enterprises and would benefit from mutual interaction.

3.2.4. Other Limitations of Appeals to Judgments about Cases

There are additional limitations of appeals to verbally expressed judgments, whether these are intuition-like or collected in surveys. In humans abilities

[39] I have received a fair amount of pushback from philosophers outside of philosophy of science regarding my skepticism about the reliability of the matching method as a source of information about the factors causally influencing armchair judgment. This does not persuade me that my skepticism is misplaced, but it does suggest to me that many philosophers see no in principle difference between psychology experiments and what might be delivered by the mental matching procedure and judgments about cases. If so, this might explain why those philosophers tend to neglect experimental results that seem relevant to their concerns—it is as though they think that everything relevant to their concerns is already present in responses to cases and reflection on these.

associated with causal cognition develop over time, and this is something that we wish to understand. Methodologies based on adult judgments about cases (including X-phi-style surveys) often are not helpful in investigating such developmental trajectories, both because this requires evidence from nonverbal subjects and because adult judgments about cases are likely to be of limited usefulness in understanding the judgment and cognition of small children, verbal or not. A similar point holds for explorations of the continuities and differences between causal cognition in humans and in other animals.[40]

3.3. What Causation Is and Our Concept of It as Goals of Inquiry

In Chapter 1, I distinguished two different possibilities that are popular within philosophy as goals for a "theory" of causation (as opposed to an account of the empirical psychology of causal cognition)—an account of what causation "is" and an account of "our concept" of causation. Intuition and judgments about cases have been treated as a source of information relevant to both of these goals. I begin with the first.

3.3.1. What Causation "Is"

Philosophers who invoke this as a goal of inquiry often have in mind a contrast with a (mere) investigation into our concept of (or how we think about) causation. The idea is that the philosopher should be interested in causation "as it is in the world" or in its "nature" and not merely in how we conceptualize or think about causation. (Similarly, it should be justice and not merely our concept of justice that is of interest to moral philosophers and knowledge rather than our concept of knowledge that is of interest to epistemologists.) Indeed some philosophers (e.g., Paul 2010, 2012) suggest that not just appeals to judgments about cases/intuition but also other empirical results from X-phi or cognitive science can provide information about extra-mental phenomena—about the "natures" of causation or knowledge, as they are in the world rather than merely how they are conceptualized by us.

How should we understand such "what X is" claims when X is something like causation? How should we understand the contention that philosophers should

[40] Some readers may be inclined to deny that nonhuman animals are capable of anything that deserves to be called causal cognition—issues related to this are explored in Chapter 4. But even if you are inclined to this view, some account is needed of what abilities such subjects have and how they differ from those present in adult humans.

focus on such claims, as opposed to claims about how we think about causation? I take it to be uncontroversial that claims about causal relationships themselves, as they exist in the world, are not to be confused with claims about the psychology of causal cognition, so that is not what is at issue. However, if we take seriously the approach advocated in Chapter 1, according to which part of our project is to understand why (in virtue of what strategies and reasoning patterns) human causal cognition is successful, then, as argued earlier, this inevitably leads to a focus on *both* psychological facts about people and facts about causal relationships as they exist in the world. We are interested, after all, in both (i) the discovery of true causal relationships (and among these, relationships having certain other objective, worldly features like invariance) and (ii) an understanding of how we do or could accomplish such discoveries. If (i) is understood as involving us in questions about "what causation is," the idea that we might fruitfully pursue an investigation into human causal cognition without considering (i)-type questions is in my view a nonstarter. As argued earlier, we need a view about what is "out there" before we can even begin to talk about *success* (or failure) in causal reasoning. In this sense, an investigation into causal cognition inevitably leads us to questions about how we are to understand causal relationships as they exist in the world. Conversely, although, perhaps apart from cases of mental causation, thinking does not create causal relationships, our thinking about causation (the "concepts" we employ and so on) is obviously something that we humans have created. Inevitably, then, when we seek to describe causal relationships as they exist in the world, we make use of categories and patterns of thought that are human constructions. In these respects there are bound to be close connections between what causation is and how we think about it.

I suspect, however, that this does not really get at what motivates those who advocate a focus on what causation is, as opposed to how we think about it. So let's explore what this might involve in a bit more detail. It is possible to understand "what causation is" in a somewhat deflationary way, according to which how we think about causation maps onto what causation is in a fairly straightforward way—indeed, this seems to characterize many of the accounts in the philosophical literature. For example, one might interpret David Lewis as claiming that causation "is" the ancestral (transitive closure) of non-backtracking counterfactual dependence (or, alternatively, the worldly relation corresponding to such dependence), where the latter is understood along the lines described in Chapter 2. If one also thinks, as Lewis does, that one of the adequacy conditions on an account of causation is that it fit a range of common-sense causal judgments or intuitions (reflecting common-sense ways of thinking about causation), then there is an obvious sense in which such judgments might be interpreted as giving us information about what causation is. Other familiar theories of causation might be interpreted along similar lines.[41] For example, the interventionist account that I favor might be construed as an account of what causation is in the sense that the account claims that (roughly) causal relationships "are" those relationships in the world that support interventions. Indeed, as observed earlier, any approach that focuses in part on methodological questions in the way that I recommend will need at a minimum to provide some specification of the target (as interventionism does and as I take Lewis to be doing) we are trying to learn about or discover. So if *that* is all that is meant by a concern with what causation is, it seems unobjectionable and indeed unavoidable. Again, however, on this sort of interpretation, it is assumed that how we think (learn and reason) about causation does not sharply come apart from what causation is.

It seems to me that if the contrast between "concept of" (or better, how we think about) and "is" is to have more substantial significance, the idea must be that it is possible for how we think about causation, both in ordinary life and perhaps in parts of science, to come apart from what it is or its underlying "nature." There are philosophical accounts of causation (and of how concepts work) according to which something like this happens or can happen. For example, according to some accounts, physics tells us about the underlying "nature" of causation—perhaps this has to do with the transfer of energy and momentum or, alternatively, with the instantiation of underlying physical laws of some distinctive sort such as "fundamental laws of temporal evolution," applied to a Cauchy surface, as suggested in Mauldin 2007 and Paul and Hall 2013. Now add to this the further idea that this underlying nature is sharply distinct from how ordinary people mostly think about causation, which may be in terms of transmission of "umph," certain stereotypes and paradigm cases, manipulation, and so on. If so, these features of how we think tell us little or nothing about the underlying nature of causation. (The folk certainly don't explicitly think about causation in terms of conservation laws and hyperbolic differential equations.) This idea might seem to be motivated by theories of meaning and concept possession regarding natural kind terms of the sort defended by Kripke and Putnam: just as the way ordinary folk think about gold (as a yellowish, shiny, precious metal) is distinct from a correct conceptualization of its underlying nature (given in terms of its atomic number or some similarly "scientific" characterization), so also for causation.[42] Just as with gold, ordinary lay thinking about causation may

[41] As another illustration, the treatment of causation (or causal reasoning) in terms of causal power and the connection between causation and invariance due to Cheng (1997) that is discussed in Chapter 7 might also be understood (I suppose) as a proposal about what causation "is." But when understood in this way, Cheng also does not assume that what causation is can sharply come apart from how we think about it. Her view is that we think about causation in terms of causal power *and* that this captures important aspects of features of causal relationships as they exist in the world.

[42] Of course this assumes that "causation" is itself or behaves like a natural kind term. This is explicit in some writers such as Strevens 2019. See also Kistler (2014, 81): "Causation is a natural kind of relation (or process), in the sense in which gold, water, cats, or humans can be conceived of as natural

be largely or completely ignorant of its underlying nature, which it is the task of the theorist to discern.

"Dualistic" theories of this sort face a number of difficulties. I won't discuss these in detail but rather want to draw attention to an obvious dialectical puzzle: the more sharply one distinguishes between what causation is and how we think about it, the harder it is to see how information provided by intuition or ordinary judgment (or for that matter information provided by ERC) can provide us with information about what causation is. In other words it is hard to see how this sort of dualism can be combined with the idea that intuition/judgment about ordinary cases can be a source of information about the underlying nature of causation. Instead, results from intuitions/judgments about cases (as well as ERC) seem at best to bear on how people think about causation, with claims about the underlying nature of causation requiring support in some other way—presumably through fundamental physics or perhaps metaphysics.[43] After all, in the case of gold, we don't think that we get at its underlying nature through an examination of how ordinary people think (or their intuitive judgments) about gold but rather through scientific investigation. If "what gold is" is sharply distinct from how ordinary people think about it, why doesn't a similar conclusion hold for causation? Of course this does not by itself undercut the "sharply distinct" thesis, but it does seem to create difficulties for someone who wants to combine that thesis with a reliance on intuitions/judgments about cases—a not uncommon position among philosophers who discuss causation.

One way of trying to get around this difficulty is to suppose that we have some intellectual faculty (a faculty of rational insight) that puts us directly in touch with facts about what causation is, with our intuitions/judgments reflecting the operation of this faculty. Again I assume it is very hard to tell a naturalistic story about how this faculty works and why it is reliable. Moreover, to the extent that the correct story about the nature of causation is to be found in fundamental physics (rather than, say, metaphysics) it is even harder to see how the armchair intuition of most philosophers could be a reliable source of information about *that*.[44]

kinds of substances or individuals." For reasons described in the appendix to Chapter 1, I do not think that "cause" behaves like a kind term. It is also worth adding that in the case of gold, questions like "What is it made of or composed of?" have straightforward answers that may not be known to ordinary users. It is not clear that the corresponding question for causation makes sense.

[43] A similar objection against the use of intuition to answer questions about what X is, where this is sharply distinct from our concept of X, is advanced by Goldman 2007.

[44] Could the judgments/intuitions of a subset of philosophers—for example, philosophers of physics—be a reliable source of information about what causation is? Well, for one thing, they disagree among themselves both in their understanding of causation and in the role they assign to intuitions—compare Maudlin 2007; Kutach 2013; and Frisch 2014. More importantly, to the extent that the intuitions in question are based on genuine physical knowledge, it is the latter that is doing the work, not the having of the intuitions.

Moving beyond appeals to intuition and judgments about cases to empirical psychology, we should note that ERC researchers typically do not present their results as direct or explicit claims about what causation "is" and they certainly don't present claims about an underlying nature of causation that is sharply distinct from how people think about it. (After all, they are psychologists.) So to the extent that we are interested in engaging with ERC, it appears that its philosophical significance should not be understood in terms of claims about what causation is that adopt the sharp distinctness thesis.[45]

Of course even if judgments about cases is not a good source of information about the nature of causation when this is understood as sharply distinct from how we think about causation, one might still wish to pursue an inquiry into the nature question, relying on some other source of information. I won't comment further on this project, but I do want to briefly consider its implication for methodology and normative assessment. I have agreed that if what is meant by "what causation is" is some characterization of the target—some useful specification of what it is that we are trying to discover when we evaluate patterns of reasoning and judgment in terms of whether they conduce to successful discovery of causal relationships—then we need such a characterization for normative assessment. However, at least when sharply divorced from "how we think" about causation, substantive accounts of what causation is don't seem to be very helpful for such methodological purposes. Consider methodological issues like the following: in a complicated causal structure with many variables and many potential causal connections among them, which variables should we control for or condition on in order to determine the causal effect of one of these variables on another? Are variables like race and gender legitimate causal variables? If either (i) X causes Y or (ii) Y causes X and time order information is unavailable, what other information, if any, might be used to determine whether (i) or (ii) is correct? I claim that even if an oracle were to tell us (truly) that causation is transfer of energy or involves instantiation of a hyperbolic differential equation or a relationship of necessitation between universals, this would help us little if at all in connection with these methodological questions. It just isn't clear how to connect the methodological questions with claims of the sort described earlier about the underlying nature of causation. Moreover, there seem to be few if any cases in which anyone has seriously tried to do so.[46] Of course connections with the psychology of causal cognition are equally difficult to discern.

One common defense of the idea that judgments about cases should be understood as providing information about or evidence for claims about the world or

[45] If results from ERC are not a source of information about what causation is, this of course raises the question of why we should suppose intuition/judgment about cases is such a source. Why is the latter epistemically privileged?

[46] For additional argument along these lines, see Woodward 2015a.

at least mind-independent facts appeals to the observation that the overt subject matter of such judgments has to do with such worldly mind-independent items rather than with concepts or shared practices. For example, a judgment like "The impact of the rock caused the window to shatter" is overtly about the impact, the shattering, and the causal relation between them. However, although this is true, it does not follow that such judgments are appropriately treated as evidence for or sources of information about either the truth of particular causal claims or more general claims about the nature of causation.[47]

Consider first the possibility that such judgments are (at least) evidence for the truth of the particular causal claims reported in the judgments (or when the case is hypothetical) evidence that the reported claim would be true if the hypothetical scenario were to be realized. Thus when Billy witnesses a certain scenario unfold and judges that Suzy's rock caused the window to break, it is claimed that his judgment is evidence for (or a source of information bearing on the truth of) what he judges, Although philosophers commonly say things like this, I think that it is hard to make sense of this suggestion. The evidence for the truth of the claim that Suzy's rock caused the window to break has to do with such facts as that the window was intact until the rock came in spatial contact, that it broke at that moment, that no alternative cause of the breaking was present, and so on, perhaps combined with general facts about what rocks can do to windows.[48] Billy's judgment may be based on this evidence/information, but neither the content of his judgments nor the fact that he makes this judgment is in itself evidence for the causal claim about Suzy's rock any more than my judgment that smoking causes cancer is evidence for or a source of information about the truth of this judgment. A similar point holds for the judgments of the armchair philosopher who merely imagines the scenario. Causal judgments are not self-certifying about the truth of their contents in this way.[49]

[47] Consider a religious person's intuition/judgment that God exists. This is not evidence for God's existence or about the nature of God. It may well be evidence for something else—for example, this person's religious beliefs.

[48] Here I agree in part with Williamson 2007. Of course one might treat Billy's judgment as testimony (Clark Glymour's suggestion) but this just raises the issue of the basis on which we are entitled to treat that testimony as reliable. This problem is particularly acute when the intuitions to which philosophers appeal are controversial as is the case with intuitions about the causal status of absences and double prevention relations.

[49] At the risk of belaboring the obvious, let me emphasize that the issue is not whether Billy's (or the armchair philosophers) judgment is true or correct. We can stipulate that it is (or would be, if the scenario were realized). The issue is rather whether their judgment (either its content or their making of it) is evidence for or a source of information either (i) about what is true of the episode or (ii) perhaps, more generally about the nature of causation. It is these claims that I deny. I mention this because some philosophers seem to think that if one is skeptical about intuition/judgment about cases as a source of information or evidence about whether, say, various causal relationships hold, one must also be skeptical about whether those relationships in fact hold or whether the judgments about them are correct. This is a non sequitur.

Could Billy's judgment somehow be evidence for more general claims about what causation is or its nature without being evidence for the particular causal claim reported? That is, could it be evidence for more general claims about causation that are somehow not based on more particular claims about causation in particular cases? It is hard to understand how this is supposed to work (at least without strong further assumptions),[50] and in any case the overt content of Billy's judgment does not have to do with the nature of causation, so we can no longer run the argument that because Billy's judgment is about p, then, if it is a reliable source of information about anything, we should construe it as a reliable source of information about p.

3.3.2. Judgments about Cases, ERC, and Our Concept of Causation

Rejecting the idea that judgments/intuitions about cases (or for that matter ERC) give us some sort of direct access to an underlying nature of causation that is sharply distinct from how we think about it leads to an obvious alternative suggestion, which is that intuitions and perhaps ERC should be interpreted as (or at least have as their goal the production of) evidence bearing on "our concept" of causation. In the case of standard philosophical appeals to intuition, there is a prima facie appealing story to be told about this, which has been nicely articulated by Goldman (2007). In outline the story goes like this: someone who masters a concept learns how to apply the concept correctly to a range of cases and to recognize incorrect applications—this is part of what is meant by mastering (or possessing) the concept in question. As Goldman puts it, "Possessing a concept makes one disposed to have pro-intuitions toward correct applications and a con-intuition toward incorrect applications" (2007, 14–15).[51] Thus when we consult intuitions about particular cases (of putative causal relationships or anything else), basically what we are doing is using our mastery of the concept to determine whether it correctly applies to such cases. In this way intuitions/judgments understood as answers to questions about whether the concept correctly applies to various cases provide us with information about the concept itself, perhaps making explicit what was previously only implicit in subjects' use of the concept. Extending this idea, we might attempt to tell a similar story

[50] If one is willing to assume that the nature of causation is given by whatever best systematizes ordinary judgments about causation, then it is obvious how Billy's judgments is relevant to the nature of causation. But I see no reason to accept this assumption—see 3.3.3.

[51] The passage continues with a dash, followed by an important qualification: "—correct, that is relative to the content of the concept as it exists in the subject's head." Thus, according to Goldman, whether this content is shared by others is a further issue. This passage is cited in Stich and Tobia 2016, 9.

concerning (at least) some verbal judgments about scenarios produced by experiments in ERC—perhaps these too should be understood as evidence regarding our concept(s) of causation.

In assessing this suggestion, let me again note a deflationary way of interpreting it, which parallels the deflationary interpretation of what causation is. If "evidence about our concept" just means evidence bearing on how people think about, represent, reason about, and learn about causal relationships (i.e., their causal cognition) then (as suggested earlier) I agree that people's intuitive judgments can sometimes provide information about this. However, for reasons described previously, I also claim that such judgments tell us less about many aspects of causal cognition than is often supposed. Thinking of the goal of a theory of causation or causal cognition as (just) having to do with capturing our *concept* (or concepts) of causation is overly restrictive. There are many important features of causal cognition that are not naturally viewed as having to do with "our concept of causation" at all—instead they have to do with such matters as computations underlying causal learning, possible forms of causal representation, and so on. Trying to interpret all of these as just having to do with features of our concept of causation seems to require an implausibly expansive notion of "concept" and also seems unnecessary.[52]

To expand on this, consider the questions, much discussed in the ERC literature, of (i) whether people represent (or sometimes or often represent) causal relationships by means of causal Bayes nets (or at least in terms of directed graphs), (ii) whether they reason in accord with the Causal Markov condition and (iii) whether, in learning about causal relationships from statistical data, they use constraint-based methods (of the sort described in Spirtes, Glymour, and Scheines 1993/2000) or some alternative. Is a commitment to (i), (ii), and (iii) built into our concept or concepts of causation? Some subjects seem to conform in their behavior to (i)–(iii), at least in connection with some inference problems, but others do not. Does this show (assuming that both sets of subjects share our concept of causation) that (i)–(iii) is not built into this concept? That (i)–(iii) *is* built into "our" concept but the second set of subjects does not have our concept? Rather than getting embroiled in such questions, it seems preferable to focus directly on the extent to which (i)–(iii) characterizes people's reasoning and not try to express the results as claims about concepts. Questions (i)–(iii) can be highly relevant to understanding people's (or some people's) causal reasoning without being part of their concept of causation.

As another illustration of the same general point, consider the role of *default* assumptions of various sorts in causal reasoning. These reflect assumptions that

[52] Here I agree with writers like Williamson and Cappelen who deny that what emerges from the method of cases is exclusively or even primarily information about what is in concepts.

reasoners tend to assume, defeasibly, as a starting point in inquiry, but which then can be modified or rejected as additional evidence accumulates. For example, Kushnir and Gopnik (2007) show that small children tend to assume as a default that many causal relations require spatial contact between cause and effect (or at least they learn such relations more readily) but they also give up this expectation when the evidence so indicates, and learn causal relationships in which there is no spatial contact. One might ask whether it is part of the children's "concept" of causation that causes must be in spatial contact with their effects. On the one hand, one may be tempted to answer yes since children do have the default. On the other hand, one might be inclined to answer no since the children do learn causal relations where there is no spatial contact (although if one adopts this line one needs some other way of recognizing the role of the default in causal reasoning).[53] I take this to suggest that this "Is it part of the concept?" question is unhelpful and fails to capture the role the default assumption plays in the children's reasoning.

A closely related issue concerns how we should think about variability or differences between subjects. If we are to characterize such differences in terms of concepts those subjects possess, we need principled answers to questions about the individuation of concepts and whether subjects are operating with the same or different concepts. An illustration: in experiments discussed in Chapters 4 and 7 subjects are given information about the conditional probabilities with which an effect *E* occurs in the presence and absence of a cause *C*, where other background causes of *E* may also be present. Some subjects give judgments of causal strength (measured by a particular verbal probe) that track $\Delta p = Pr(E/C) - Pr(E/not\ C)$ (see Chapter 2), while others produce judgments that track "causal power" in the sense of Cheng 1997, where causal power $= \Delta p\ /\ [1 - Pr(E/not\ C)]$. Moreover, given certain natural assumptions, the response that tracks Cheng's measure can be shown to be the normatively correct one and the use of Δp normatively incorrect, roughly because it fails to appropriately adjust for the presence of the other background causes besides *C*.

How should we interpret these results in terms of the subject's concepts of causation? Do subjects who provide different strength judgments in response to the same stimuli have different concepts of causation? Do they have the same concept, with one group (the Δp group) misapplying that concept, by failing to adjust appropriately for the presence of background causes? Does each group have the same two concepts, one corresponding to what is measured by Δp and the other corresponding to causal power, but with the verbal cue employed in the

[53] Of course one might also hold that when the children learn causal relations that do not involve spatial contact, they change their concept of causation, but this again raises the question of how such concepts are to be individuated.

experiment triggering one of these concepts in one group and the other concept in the second group[54]? Again, it seems to me that there is much to be said in favor of not getting entangled in such questions if we can avoid them. In fact, as noted earlier, researchers in ERC tend not to express their results in terms of claims about people's concepts of causation. Instead, the focus is on modeling and understanding much more specific effects and the processes that underlie them, including issues about computations and representations associated with such processes. For example, in the case of effects expressed as causal strength judgments, the focus is on understanding the factors that influence certain patterns in such judgments and the representations and computations that underlie these.[55] I suggest that philosophers too would be well advised to not try to force all of their claims about causation and causal cognition into claims about causal concepts. At the very least it is not necessary to do this. We should particularly avoid assuming that all relevant or interesting features of causal cognition somehow follow from or are explained by our possession of some concept of cause.[56]

Finally, let me underscore an additional reason for avoiding framing conclusions about causation or causal cognition as (just) claims about our concept of cause or about what causation is. Recall one of the themes of Chapter 1: the value of thinking about causation and causal reasoning in functional terms—that is, thinking in terms of the point(s) or purpose(s) of causal thinking and what we want to accomplish. One limitation of focusing attention entirely on questions like "What is causation?" or "What belongs to our concept of causation?" is that this functional/normative dimension tends to be pushed into the background. I think it important to keep it in the foreground.

3.3.3. Capturing/Explaining Judgments of Causation

Another suggestion that one sometimes finds in philosophical discussion is that the goal of a theory of causation should be to "capture" or perhaps "explain"

[54] If we want to describe the experimental results in terms of concepts, this last possibility is by no means a crazy suggestion, as we shall see in Chapter 7.

[55] Knobe (2016) argues that this increasingly true of research in X-phi as well—recent results in this area often are not expressed in terms of claims about concepts.

[56] I assume that one reason why Goldman and others are attracted to the idea that intuitive judgment is a source of information about concepts is that it fits with a standard picture of philosophy as centrally concerned with conceptual analysis and intuition as providing the raw material for this. But this view of philosophy as having to do primarily with conceptual analysis is now rejected by many philosophers and, in my opinion, for good reason. We can retain Goldman's view that intuitive judgment can be one source of information about some aspects of causal cognition without accepting the idea that this primarily provides information about concepts.

(where "explain" may mean something like systematize or reproduce)[57] particular judgments about causal relations. Depending on how it is understood, this goal may be closer to the discovering the nature of causation goal discussed earlier or the articulating our concept goal. Alternatively it may be a kind of mixture of these two possibilities.

What might be meant by talk of capturing or explain in this context? One possibility is that the aim is reproducing typical judgments with no particular commitment one way or the other to whether those judgments are correct in the sense of corresponding to how things stand in the world. A possible analogy is with a theory of the grammar of some natural language where the goal is to reproduce or generate the judgments of native speakers about grammaticality; in a similar way a theory of causation might aim at reproducing causal judgments people endorse. This is one possible interpretation of the aim of the theory of actual causality presented in Halpern 2016, although Halpern does not seem to make use of empirical studies of actual patterns of causal judgment, relying instead, as many linguists do, on his "intuitions" about how people judge. Unsurprisingly, although the resulting theory captures some range of typical judgments, it does not capture all, in part because Halpern is also guided (as he acknowledges) by certain normative constraints that generate nontypical judgments. In any case, since Halpern uses the framework of structural equations, the resulting theory does not look much like most of the theories philosophers propose when they talk about capturing or explaining causal judgments.

A second possibility that may be closer to what many philosophers have in mind takes what is to be explained or captured to be something more like facts (or features of the world) reported by causal judgments, with the underlying nature of causation "explaining" those facts—perhaps we infer to this underlying nature as part of an inference to the best explanation of those facts. Here too some sort of analogy with explanations that appeal to underlying natures of ordinary material things may be operative: there are various particular facts about gold—that it is shiny, ductile, and so on—and the underlying nature of gold explains these. So also for the relation between the underlying nature of causation and various particular causal judgments about cases. One obvious problem with this proposal is that while the underlying structure of gold may figure in an ordinary causal or scientific explanation of its superficial properties, the nature of causation doesn't cause ordinary causal judgments or provide an ordinary scientific explanation of them. So we face the question of what "explain" (or "capture" etc.) means or involves in the latter context. In addition, we face many of

[57] Note that finding a way of systematizing or reproducing judgments is prima facie rather different from providing an explanation that cites the causal factors affecting those judgments. See Woodward 2003.

the same difficulties that arise for the "what causation is" proposal discussed in previous sections. The judgments we are trying to explain are likely to be at least somewhat inconsistent, and the resulting theory is likely not going to reflect the function of causal talk or be normatively useful. (For more on this topic see the remarks on reflective equilibrium in Section 3.8.)

3.4. Different Projects Connected to Understanding Causal Cognition and Their Connection to Intuition/Judgments about Cases

So far the discussion in this chapter has focused mainly on criticisms of the idea that the main goal of a theory of causation should be to uncover what causation is or features of our concept of causation (and that appeals to judgments about cases can accomplish this). In this section I discuss in more detail some other goals that accounts of causation and causal cognition can legitimately have. Here I expand on some of my remarks in Chapter 1.

3.4.1. Explaining Success

One such goal or project is explaining success. We may think of this as having the following schematic structure:

(3.4.1a) The causal cognition and associated practices in some population of subjects exhibit features F. (F may have to do with patterns of judgment, learning or reasoning strategies, use of certain representations of causal relationships, distinctions among causal relationships, and so on.)

(3.4.1b) The presence of F contributes to the achievement of goal G, where G is a goal associated with causal cognition.

(3.4.1c) Achievement of G counts as success in connection with causal cognition.

Conclusion: The presence of F explains why subjects' causal cognition is successful in some particular respect, related to the achievement of G.

Note that in this schema it is the presence of F that explains the subject's success. That is, the explanandum is this success and not the presence of F—the latter is part of the explanans.

Premise (3.4.1a) is an empirical claim that can be established by ordinary empirical investigation. As argued earlier, individual judgments about cases ("intuitions") can (in principle) be *one* possible source of information about the presence of some sorts of Fs (e.g., those having to do with judgments) and thus support some premises of form (3.4.1a). However, as also argued previously, there are many other sources of information about possible success-making features F. Moreover, we need not regard all these features themselves as built into our causal concepts.

Premise (3.4.1b) is of course independent of Premise (3.4.1a). Often premises of form (3.4.1b) are established by analyses that have a logical, mathematical, or conceptual component, although empirical (or quasi-empirical) evidence including the results of simulations and calibration studies may be relevant too.[58] To return to a possibility discussed previously, if feature F is some strategy for learning causal relationships from statistical data, mathematical analysis may be able to demonstrate that, given additional assumptions, the strategy is reliable in the sense of having good error rates in learning the relationships in question. It may even be possible to show that with enough data the strategy will identify the single relationship that is correct or, alternatively, an equivalence class that contains the correct relationship. Results of this kind for learning causal relationships are described in Spirtes, Glymour, and Scheines 1993/2000. Similarly, as described in Chapter 7, mathematical analysis can establish that, given a conception of causal relationships as relationships satisfying certain invariance requirements, the power PC measure of causal strength proposed by Cheng is superior to alternative measures and, moreover, better tracks certain features of causal relationships that generalize to new populations. To this extent we have an explanation of why subjects judging in accord with power PC are more likely to be successful in certain specified respects (such as generalization), in contrast to subjects who judge in accord with some alternative measure of causal strength. Other results about the optimality or reliability of various learning or identification strategies can be found throughout statistics, econometrics, and learning theory. To the extent subjects follow such strategies, this can explain why they succeed. An advantage of such mathematical analyses (in addition to the certainty they can in principle provide) is that they can support conclusions not just about the overall reliability of a strategy but about which particular features F are responsible for this reliability.

Mathematical proofs are not the only way of providing support for premises of form (3.4.1b). Another possibility is simulations of various sorts. For example,

[58] I see each of these—mathematical analysis, simulation, or empirical calibration—as an instance of the means/ends justification described in Chapter 1.

one can construct (perhaps virtually) examples of systems with various known causal structures, use these to generate data, and then see how reliable some learning strategy is in recovering this structure. Reliability can also be demonstrated by applying a learning strategy or analysis to real-life data where the truth about which causal relationships are correct is known on some independent basis.[59] However, although these last two approaches can provide information about the reliability of some procedure, they may leave it less clear which particular features F are responsible for reliability. To this extent, they may fail to fully explain the success of subjects who follow the procedures.

Because establishing the second premise, (3.4.1b), often involves logical or mathematical analysis, it will to this extent be something that philosophers and others can do from the armchair or by means of a priori reasoning—thus illustrating a legitimate role for this sort of reasoning. However, such reasoning does not seem to be what many philosophers have in mind when they appeal to intuition or judgments about cases. Again these typically involve classification, having to do with whether or not some case is an instance of X. Even if judgment/intuition gives people access to whether their causal cognition or shared practices of causal cognition possess some feature F, without further analysis, this access is not going to tell them whether F contributes to success in connection with goal G. This requires something different from intuitive judgments about cases as standardly conceived.

3.4.2. Normative Assessment

The project of explaining success described in Section 3.4.1 is obviously closely connected to normative assessment. In showing that certain features F present in causal cognition conduce to successful achievement of various goals, one also provides a basis for a positive normative assessment of those features. However, insofar as we are interested just in normative assessment, the extent to which, as an empirical matter, various features present in causal cognition are widely shared is not directly relevant but at best suggestive.[60] This is because what we are interested in is whether or not these features would conduce to success if they were present, and this is logically independent of the extent to which they are in fact present in people's cognition. In other words, to the extent that we are interested in normative assessment, we are interested in

[59] As with the previously mentioned results from Janzing et al. 2012. These authors tested various algorithms for learning causal direction from correlational data on real-life examples for which the correct causal direction was independently known.

[60] "Suggestive" because as explained earlier, when a feature is widespread this may suggest that the possibility that it may have some normative rationale is worth taking seriously.

claims of the sort associated with Premise (3.4.1b) but not with claims of form (3.4.1a). And of course we are also interested in identifying features of causal cognition that do not effectively contribute to our goals and replacing them with better alternatives. A great deal of work on causal learning and inference conducted in disciplines like machine learning, statistics, and econometrics has this character—it is about the discovery of techniques for reliable inference, invention of concepts and distinctions that behave in normatively appropriate ways, and so on, with no commitment to whether various populations use those techniques.

Another possible project that might be pursed in connection with causal cognition is proximal explanation of features of causal cognition.

3.4.3. Proximal Explanation of Features of Causal Cognition

Suppose it can be established that the causal cognition of some group of subjects exhibits feature F, where F might be, for example, the presence of some pattern of judgment or some learning or reasoning strategy. Then we may ask for a proximal explanation of why this is the case—that is, about the proximal factors or causes that influence whether or not F is present. ("Proximal" here contrasts with an explanation in terms of function or "ultimate" causes—see Section 3.4.4.) One possibility is that this proximal explanation takes the form of a specification of inputs that are available to subjects, the computations they perform on those inputs, and the representations that are involved and how these lead to feature F. For example, in Cheng 1997 (discussed in Chapter 7), feature F involves subjects' causal strength judgments about the relation between a cause *C* and an effect *E*, and the input available to the subjects is information about the contingency between *C* and *E*. Cheng's model specifies computations subjects perform on this input to arrive at their causal strength judgments. In this sense the model purports to explain why subjects make the causal strength judgments they do (again this is different from explaining why those judgments are normatively appropriate, as in 3.4.1–2). In other cases, no formal computational model may be provided but there still may be claims about the proximal factors that influence some feature present in subjects' causal cognition. For example, Lombrozo 2010 (see also Vasilyeva et al. 2017) claims that variations in subjects' causal strength judgments regarding various scenarios are explained in part by variations in the invariance (in the sense of Woodward 2003, 2007, 2010) of the causal relationships in those scenarios. These authors provide experimental evidence in support of these causal claims.

3.4.4. Functional Explanation of Features of Causal Cognition

Ernst Mayr's distinction between "proximate" and "ultimate" causes captures a familiar contrast between two types of explanation. In addition to asking about the proximate causes for explanandum *E*, we can also ask a more "functional" or teleological question—is some sort of selective process responsible for *E*? For Mayr, genetic causes for a phenotypical trait are paradigmatic proximate causes, while an explanation for the presence of the trait in terms of natural selection involves an appeal to an ultimate cause. In introducing Mayr's distinction I don't mean to claim that it (or the use to which he puts it) is entirely unproblematic. I mention it here only because we can draw a parallel distinction in connection with causal cognition: We can ask about the proximate causes of some feature of causal cognition (as in 3.4.3), but we can also ask whether there is an explanation in terms of ultimate causes—that is, whether there is a selective explanation for the feature in question. To posit such a selective explanation is to claim that the feature is present in causal cognition *because* of consequences that it has, where this implies that some selective process is operative that selects for the feature in question because it has those consequences. For some features of causal cognition the selective process might involve natural selection, but for other features it may involve learning processes of various sorts (with feedback), both individual and cultural. For example, use of strategies that lead to success (understood as the achievement of some goal, such as the discovery of relationships that can be used for manipulation) over the course of an individual's lifetime may be reinforced because of this success, while less successful strategies are not reinforced, so that the successful strategies come to predominate. Many theories of learning postulate that it proceeds via some version of such a reinforcement process.

As in the projects discussed previously, while intuitive judgments about cases may sometimes furnish information about features possessed by causal cognition, these judgments will not by themselves provide information that might be used to vindicate claims that various features of causal cognition are present *because* of the operation of selective processes.

Note that in all of the argument patterns or projects in Sections 3.4.1–3.4.4, one does not just take judgments about cases or reports of intuitions at face value and then try to codify or systematize them. First, if the project is explaining success (or constructing a normative theory of what conduces to success, as in 3.4.1), reports of intuitions or judgments are taken to be of significance only insofar as they reflect features that we can independently identify as conducive to goals associated with causal thinking. If our judgments report features of causal thinking that we cannot make sense of or rationalize in this way, they carry no weight as far as explaining success is concerned or for any kind of normative justification of those features. In particular we don't try to argue that various features of our causal cognition or patterns of judgment are correct or normatively justified merely because they reflect our intuitions or judgments about cases (or some systemization of these, perhaps balanced against other sorts of considerations).

3.5. More on Success

I have emphasized the importance of explaining success in the project of understanding causal cognition. Of course this requires criteria for success and failure. As argued previously, these can be supplied by claims about particular goals associated with causal cognition and normative theories related to these, but it is also worth emphasizing that in some cases it is possible to design experiments in which the criteria for success are relatively independent of any particular normative theory and relatively uncontroversial. For example (cf. Gopnik et al. 2004), a subject may be presented with a task in which the goal is to get a machine to activate (e.g., by placing the appropriate block from a set of choices on it) and where the subject has previously been presented with evidence about whether various blocks do or do not activate the machine. Success is measured by whether the subject does in fact choose a block that activates the machine. This is taken to be a measure of whether the subject has learned a certain causal relationship (that placing some particular block on the machine causes it to activate). A setup of this sort is designed so that given appropriate background assumptions, any reasonable theory of causation will judge that when the block is placed on the machine and it activates, the former causes the latter.[61] So in taking this to be the criterion for correctness, we are not making question-begging assumptions that amount to presupposing the correctness of any particular theory of causation or even any very specific theory about goals of causal cognition.[62]

By contrast, at least in many cases, theories constructed around appeals to judgments about cases (or for that matter X-phi type surveys) are not structured in such a way that they allow us to assess via some independent standard whether those responses are correct or not. Instead, to the extent that there is a standard of correctness/success at all, it is supplied by whatever theory best systematizes

[61] There is a subtlety/complication here that should be acknowledged, although it does not affect the main point. This is that in the actual experiments there is (unknown to the participants) confounding since whether the detector activates is controlled by the experimenter, not by the block placed on the detector. It merely looks as though the blocks activate the machine. My point is that given these appearances, any reasonable theory of causation will judge that the block caused the machine to activate.

[62] Admittedly this experimental test for whether there is causal learning is one that is congenial to an interventionist since it involves (apparently) intervening to make the machine go. However, even if interventionism is mistaken as a general theory of causation, it is implausible that, in the circumstances of the experiment, interventionism is mistaken in its judgment that it is rational to conclude that placing the block on the machine causes it to activate.

the responses. This prevents our asking questions about whether the subjects are getting things right according to some standard that is independent of the responses themselves.[63] This also undercuts the possibility of providing nontrivial explanations of correct or successful judgment when it occurs.

3.6. The Significance of Variability

Although there is considerable uniformity in response in a number of causal judgment tasks among adult humans, there is also (as we shall see in subsequent chapters) significant variability in performance in other causal cognition experiments, both for adult humans and (unsurprisingly) even more so for other subjects (e.g., children of different ages and nonhuman animals). As noted previously, if we try interpret the results of such experiments as claims about subjects' concepts of causation, we face obvious issues about how concepts are to be individuated: when does a difference in performance indicate a difference in causal concepts, and when should it to be understood in terms of possession of the same concept but with variation along some other dimension (e.g., beliefs external to the concept or memory limitations)? Moreover, to the extent that the tasks involve verbal judgments, the presence of variability creates problems for interpreting the judgments as (reliable) "intuitions," especially if we have no standard for reliability other than agreement among intuitions.

The alternative framework I favor adopts a quite different stance toward the significance of variability. Since one of our main concerns is explaining success, the presence of variability, particularly when it involves some subjects succeeding and others not succeeding at some causal cognition task, can be quite informative: it leads us to ask why (in virtue of what learning strategies or forms of causal reasoning) some of the subjects are successful and others are not.[64] (Thus the flip

[63] If the difference between these approaches is unclear, consider the following. Suppose that in Gopnik's experiment, information about which blocks the children choose is recorded but not information about whether or not the children chose correctly (i.e., whether they activate the machine). This information is then systematized (e.g., it is observed that most children chose block 1 over block 2, perhaps certain kinds of inconsistency in judgment are cleaned up, and so on). The standard of correctness is then taken to be conformity or not to this systematization, so that an individual choice of 1 over 2 is taken to be correct because it coheres with what others choose, regardless of whether it is 1 or 2 that activates the detector. This is roughly analogous to the procedure that is followed by philosophical approaches that proceed by systematizing judgment about cases with no independent standard of correctness. Whatever might be said about this, it is different from the procedure followed in Gopnik's experiments.

[64] As many have observed, it is very common in psychology experiments to neglect individual variation, focusing instead on average behavior (e.g., mean differences between treatment and control groups). However, it is increasingly recognized that variation in individual performance can be quite informative and hence warrants more attention—this is so in connection with causal cognition and also more generally.

side of explaining success is also explaining failure.) Moreover, when the variability concerns differences among humans at various ages, this can serve as a source of information about the development of causal cognition. Often this puts the focus not so much on the question of whether all (or almost all) subjects judge or reason in a certain way, but rather on what subjects of various kinds "can do" or "do with some frequency" (that is, what capacities or abilities they have or fail to have, rather than on the details of the frequencies with which different judgments occur), particularly when it can be shown that other sorts of subjects rarely or never do these things.

As an illustration, discussed in more detail in Chapter 4, (many) adult humans are able to combine evidence from their own and others' interventions and from passive observation (not involving interventions) in making causal judgments. Young (two-year-old) children and perhaps nonhuman primates are apparently not able to achieve this integration, but four-year-olds are. This suggests the need for a learning/developmental story about how the older children acquire the ability to do this. Again, rather than trying to describe all of this in terms of claims about concepts possessed or not possessed by various subjects, it seems more fruitful to talk in terms of abilities and reasoning patterns subjects possess or fail to possess, and how these conduce to success or failure in various tasks. Such claims about what subjects of various sorts "can do" don't imply claims about universal or nearly universal behavior—the former claims are consistent with some considerable number of subjects failing at the task in question.[65] This focus on capacities is characteristic of a great deal of explanation in cognitive psychology and neuroscience.

As another illustration, many adult humans are able to recognize the difference between conditioning and intervening and use this distinction in guiding their inferences (cf. Chapter 4), but by no means all adults do. Adults who recognize this distinction are more successful or reliable in the inferences they make. This is interesting from the point of view of understanding causal cognition and when it is successful even though not all subjects behave in the same way. To the extent that talk of shared concepts implies something like near universality of response (or at least a very high level of shared response or judgment), it may not be an adequate vehicle for capturing claims about capacities possessed by many (or some) but not all subjects.[66]

[65] Although in the example under discussion, none of the two-year-olds and all of the four-year-olds completed the task

[66] Some readers may worry that there is a tension between my claims that human causal cognition is often "rational" and the existence of variability. In some cases this tension can be resolved by the observation that the variability is the result of the subjects being engaged in different reasoning tasks that are only superficially similar, with different criteria for success. In other cases, the tension is real, but here I remind readers that I regard the extent to which human causal cognition is rational as an empirical matter. I am not claiming that it is always rational but rather that it is often rational enough that it is illuminating to view in a rational analysis framework. Rational analysis can

Note also that even if it is true that not all or almost all subjects do X (where X may be, for example, judging or reasoning in a certain way), it is often of considerable interest to determine how those subjects who do X accomplish this—that is, what learning or reasoning strategies or computations they employ in doing X. For example, by no means all subjects behave as Bayesians when faced with causal inference problems, but there is evidence that some significant number, including children, do in connection with certain problems and that when they do so, they are able to make accurate causal judgments by taking base rate information into account in the way prescribed by Bayes' theorem—see Sobel et al. 2004; Griffiths et al. 2011 and Chapter 4. Here the ability to make certain causal judgments correctly is explained in terms of subjects calculating in accord with Bayes' theorem and the normative appropriateness of such calculations. Again this analysis is not just a matter of figuring out what concepts such subjects possess.

As yet another illustration, suppose that although a substantial number of subjects do judge in accord with Cheng's power PC model, a substantial number do not—a claim that is currently a matter of dispute (Chapter 7). Of course this does not in itself undermine the normative status of that model or the normative appeal of the invariance-based ideas on which it rests. It is also consistent with the model providing an adequate explanation of the judgments of those subjects who do conform to the model and also of why their causal cognition succeeds in certain respects. Here again just focusing on the extent to which judgments are shared leads us away from many interesting discoveries, both descriptive and normative.

The tendency in recent philosophical discussion of the role of intuition or judgments about cases to focus on the question of whether such responses are very widely or nearly universally shared is in some respects very natural and appropriate. After all, when a philosopher claims, in an unqualified way, that "people judge that so and so . . . ," it is entirely in order to assess whether such claims are true or false. On the other hand, it is also important to bear in mind that to the extent that our interest is in normative theory and explaining success, discoveries that judgments and other practices are non-universal may be less consequential than is sometimes supposed—a demonstration of non-universality need not undermine normative claims or explanations of success to the extent that it occurs.[67]

be illuminating, for those subjects who are rational, whether or not it is illuminating for nonrational subjects. Moreover, it can sometimes be illuminating for the latter by, for example, suggesting specific diagnoses for failures of rationality.

[67] It is true, as Edouard Machery has pointed out to me, that a significant amount of work in X-phi does focus on individual variation. However, to a large extent, this work focuses on *demographic* differences as a source of variation—for example, it might be claimed that Asians make different

Given these remarks about variability, one might wonder why it matters whether some set of causal judgments are shared to any very significant degree at all. One possible answer was suggested in Chapter 1: when judgments or other features of causal cognition are shared to some nontrivial extent rather than being completely idiosyncratic, this *may* provide some reason to take seriously the possibility that the judgments or features are well adapted to the circumstances in which they are found and conducive to normative success. Of course, as emphasized earlier, showing that this possibility is actual requires an independent argument, not just evidence that the judgment is widely shared.

3.7. Reflective Equilibrium to the Rescue?

I noted earlier that one problem with any approach that begins with subjects' judgments about cases and then attempts to systematize these, treating the resulting account as an account of either our concept of causation or the nature of causation, is the variability (indeed inconsistency) of these judgments: the same subject may report apparently inconsistent judgments, and different subjects may also endorse inconsistent judgments.[68] In the absence of a single theory that fully captures everyone's judgments, a standard philosophical response is to appeal to some version of *reflective equilibrium*—one looks for a theory that captures as many judgments as possible, but where this may involve rejecting some intuitions in favor of others in order to maximize overall systematic coherence.[69] Additional constraints/desiderata that are motivated on more general philosophical grounds, such as the demand that the resulting theory be

judgments about reference or free will than Europeans do. This work is valuable in debunking claims about universally shared ways of thinking about reference and free will. However, my own view, for what it is worth, is that understanding the differences in representation, computation, and learning strategies that lead to individual variation (rather than just focusing on demographic variables) is likely to be more illuminating from the point of view of psychology. This is partly because, as an empirical matter, there is a considerable amount of individual variation within demographic groups with respect to many judgments and cognitive strategies, and it is worthwhile to understand the sources of this. In addition, for reasons that I have discussed elsewhere and are rooted in an interventionist approach to causation (Woodward 2003), I don't think that factors like ethnic group membership and gender are good candidates for causes at all—hence not good candidates for causes of individual variation.

[68] Of course it is possible to merely report these judgments, in all of their variability and inconsistency, but I assume that virtually no philosopher or psychologist will be satisfied with that. Everyone wants a more unified theory of some kind about these judgments—one that systematizes or explains or rationalizes them in some way. In particular if the judgments are interpreted as telling us about a single coherent underlying concept or what causation is, one needs some way of dealing with inconsistent judgments.

[69] As noted earlier, this process may also involve deciding that certain judgments are correct on the basis of systematic considerations even when people have no clear intuitions, as in Lewis's "spoils to the victor" arguments.

"reductive" or that it reflect certain requirements coming metaphysics or from "fundamental physics," may be added to this mix—a prominent recent example is Paul and Hall 2013, and a similar program seems to underlie Lewis's work on causation (e.g., 1986).

A general problem with this approach is that there is no reason to suppose that there is a unique outcome that represents the best possible trade-off among the different intuitions and other desiderata (and no obvious way of telling whether we have found such an outcome, supposing that it exists). Instead there may be multiple equilibria, each corresponding to different ways of trading off or balancing the desiderata just described. Or there may be no equilibrium.[70] Indeed, the variety of different theories that have been produced by investigators claiming to follow something like (some version of) the method of reflective equilibrium seems to support the conclusion that either there are multiple equilibria or that all but one of the theorists have misapplied the procedure (and we can't tell which).

But even putting this aside, there are several additional difficulties. One is that it is unclear what the goal or point of enterprise just described is. It can't be intended just as empirical psychology or empirical description/explanation of aspects of causal judgment, since even if we accept that judgments about cases is a source of information about these, the project under discussion involves rejecting some of these judgments in favor of others, settling cases in which judgment is uncertain in definite ways (rather than just reporting the uncertainty) and in many cases subjecting the whole investigation to additional philosophical constraints that are not motivated by empirical psychology. A straightforwardly descriptive enterprise in empirical psychology would not take this form. But at the same time, the account that emerges from such a procedure will not have an obvious functional or normative rationale (in the sense described in Chapter 1 or earlier in this chapter) either. One way of seeing this is simply to note that the reflective equilibrium procedure as described assigns no role to what the end product is to be used for or whether it is well or poorly designed to achieve goals associated with causal thinking. Instead, the product looks like a hybrid, partly constrained by the goal of capturing intuitions and partly constrained by other sorts of considerations that seem to have little to do with functional/normative considerations.[71] The result does not fit into any of the categories—explaining

[70] In the many areas of science (thermodynamics, economics etc.) in which some notion of equilibrium is invoked, it is recognized that one can't just assume or postulate that an "equilibrium" exists. One has to first define what is meant by equilibrium, and then investigate, either analytically or empirically, whether any equilibrium exists, whether there is a unique equilibrium or many, whether the equilibrium is stable, and so on. Unfortunately, when the notion of reflective equilibrium is invoked in philosophy, nothing of this sort is provided.

[71] In assessing the role of reflective equilibrium in providing normative justification, it is worth considering one of the sources of this idea—Goodman's (1955) claim that the only standard for correctness in inductive judgment is agreement with the inductive judgments we actually make (or

success, providing functional or proximate cause explanations, etc.—that are described in Section 3.4. The question of what the point of the resulting theory is (and how it relates to other possible projects involving causation and causal cognition) thus looms large.

3.8. More on the Relationship between the Normative and the Empirical

It should be clear from the preceding discussion that the relationships between empirical results about causal cognition and normative theory are (or can be) complicated, subtle, and multifaceted. Let me underscore again that on my view empirical results (whether they derive from people's intuitive judgments understood as a basis for claims about how people judge and think or derive from some other source) are *not*, taken in themselves, "evidence" for any particular normative theory. Empirical results can bear on or be relevant to normative theories in various ways, without amounting to evidence for such theories. For example, as discussed earlier, we can ask of some normative theory whether it does or does not make sense of or provide a normative rationale for features that empirical investigation shows are present in causal cognition. If the normative theory does provide such a rationale, this is a connection between the empirical results and the normative theory even though the former is not construed as providing evidence for the latter. Similarly, as noted in Chapter 1, some normative theories would appear to lack obvious motivation if as an empirical matter people never reasoned in accord with them. At the very least this would be contrary to what the creators of those theories expected and would lead us to think that the theories required strong supporting arguments from some other source. Finding that some people do reason in accord with the normative theory removes this particular concern, but again this does not mean that empirical facts about how people reason are evidence for the normative theory. This is a point that the reader should bear in mind when considering the empirical results reported in subsequent chapters.

perhaps some best systemization of these). This idea was then extended to many other areas by other philosophers. In adopting this conception of justification for inductive reasoning Goodman ignored virtually all of statistics, which does assess inductive judgments by standards that are independent of agreement with other judgments. In classical or frequentist statistics this standard is provided by, for example, calculations of error rates associated with various tests. In Bayesian statistics the standard is agreement with the requirements of the probability calculus (coherence) and conformity to Bayes' theorem. That these statistical standards have turned out to be superior from the point of view of normative justification to Goodman's proposal does not seem controversial. Why think that the situation should be any different with respect to causal reasoning?

A related possible connection between the empirical and the normative, also considered previously, is this: suppose a normative theory suggests that if certain features F were present in people's causal reasoning, this would be normatively appropriate and conducive to the success of that reasoning. As argued in Chapter 1, it would then be natural to investigate empirically whether those features are in fact present in people's reasoning. Indeed, a number of experiments in ERC discussed in subsequent chapters have been undertaken for just this reason. This illustrates another way that philosophical or normative theories of causation can play a positive or constructive role in experimental work: theory can *enable* or *motivate* or provide a rationale for doing certain experiments and a basis for interpreting their results. In a number of cases, it probably would not occur to anyone to do an experiment exploring whether these features are present in the absence of the normative theory.[72] For example, it is unlikely that anyone would have done the experiments described in Chapter 4 exploring whether subjects recognize the difference between conditioning and intervening in the absence of a normative theory telling us that this difference is of central importance in causal reasoning. Note that in such cases the results themselves may be interesting, surprising, and valuable even for those who are skeptical of the overall correctness of the motivating theory. For example, even those who are skeptical about interventionism as either an overarching descriptive or normative theory ought to find experimental results about people's ability to distinguish conditioning and intervening of considerable interest. Note also that this focus on the enabling (rather than the evidential or testing) role of experiments fits naturally with the idea, urged previously, that we think of many experimental results as more like demonstrations of what some class of subjects *can* do or do with some frequency than as tests of hypotheses regarding what they *always* do.

As yet another possibility, again already mentioned in passing, suppose that we find experimentally that people's causal cognition, when successful, exhibits feature G, where G is some feature that is not assigned a role in any current normative theory. This might motivate us to consider the possibility of constructing a new normative theory that explains the role of G in successful causal cognition. I describe several possible instances of this in the accompanying footnote and subsequent chapters.[73]

[72] On several occasions when I have presented this material to audiences of philosophers, they have responded that I have not shown that it is *logically impossible* for anyone to think of the experiment in the absence of the motivating theory. Of course I agree, but so what? It remains relevant that, as a matter of empirical fact, often people do not think of appropriate experiments in the absence of motivating theories.

[73] Some readers may think it implausible that empirical discoveries about human cognition, including causal cognition, can serve as a source of normative ideas. However, this approach has been extensively employed in vision science, with empirical discoveries of strategies employed in human visual processing that are normatively successful being used to suggest normatively good strategies that might be incorporated into machines involved in visual scene processing. Taking discoveries

3.9. Experimental Philosophy

As noted in passing in previous sections, many of the themes I have discussed in this chapter also arise in connection with work in experimental philosophy (X-phi) and the evaluation of its significance for various traditional philosophical problems. I can further clarify the approach that I favor by commenting on some of the similarities and differences between it and X-phi.

In a broad sense, the phrase "experimental (or, more generally, empirical) philosophy" might be taken to characterize any attempt to bring empirical results to bear on a philosophical issue. With this understanding, a great deal of work in philosophy of science (e.g., attempts to use special and general relativity to settle "philosophical" questions about the nature of space and time) and many projects in ethics and political philosophy pursued in a naturalistic spirit (e.g., Kitcher 2011) might qualify as "experimental philosophy." However, this phrase is commonly used much more narrowly in contemporary discussion, to encompass research carried out by philosophers, sometimes but by no means always survey-like in character, with adult humans (rather than, e.g., children) as subjects. An important subset of this research focuses on issues or concepts of long-standing philosophical interest and is organized around concerns, both positive and negative, about the role of "intuitions" or judgments about cases in philosophical argument. There is also some tendency to express research results as claims, either positive or negative, about concepts. (The folk concept of X does or does not agree with the philosopher's concept, or there is no folk concept.) Of course one of the features that distinguishes X-phi from more traditional philosophical approaches is that, like ERC, it makes use of a number of subjects, often without philosophical training, rather than a single one or an unsystematically selected small group (the armchair philosopher and colleagues).

from cognitive psychology and neuroscience about how successful human cognition occurs and then using this as a source of normative ideas that might be incorporated into machine learning is a less familiar idea but is now being actively explored—see, for example, Lake et al. 2017. In connection with human causal cognition, the apparent empirical inadequacies of purely associative models, reviewed in Chapter 4, have been very suggestive for what more normatively adequate models of causal reasoning should look like, including models that might be realized in machines. To spell out one aspect of this, the failures of associative models to adequately describe empirical features of human causal cognition, including features that appear to be normatively successful, suggest that humans succeed in causal reasoning by making use of something more than merely associative processes. We thus look for non-associative models of human causal cognition as a possible source of ideas for what that something more might be. At a minimum, what seems to be required is something more "model-based" than what purely associative processes give us—a position that is advocated recently by writers like Pearl (2018) and Marcus (2018).

On a somewhat different note, let me add, for what it is worth, that the normative ideas about "causal specificity" described in Woodward 2010 also had this sort of "empirical" origin. I first noticed that biologists seemed to care about whether causal relations were specific or not without having any idea about whether there was any normative rationale for their doing so. This in turn suggested a search for such a rationale.

Even conceived in this narrower way, X-phi is (obviously) far from homogenous. One way of carving things up distinguishes three programs:[74] First there is the "negative program" that employs empirical results in a primarily negative or debunking role—to show that many people do not share the judgments about cases to which some traditional philosopher appeals. As argued previously, this is, as far as it goes, a completely reasonable project. On the other hand, to the extent that our interest is in understanding causal cognition or in the various projects (including the normative ones) described earlier in this chapter, negative results of this sort are often of limited usefulness.[75]

Second, we have the "positive program" in which empirical results about people's judgments are used to provide "evidence" for some philosophical theory of X, often with the theory being understood as an account either of the concept of X or perhaps X's underlying nature. Although I have been skeptical about interpreting such results as claims about concepts or natures, I have agreed that such results can sometimes play the "positive" role of providing information about practices connected to causal cognition. I will add that the positive project is also attractive because it is constructive and does not just consist in amassing negative results refuting claims of traditional philosophers. However, as I have also argued, a limitation of this program is that survey results and reports of judgments are just not the sort of evidence that can be used to assess many important claims about causal cognition, at least when such results are taken in themselves and not guided by normative theory and modeling suggested by such theory. In this respect the positive program in X-phi inherits a number of limitations of the old-fashioned "armchair intuitions about concepts" methodology to which it is opposed.

A third strand in X-phi that has been particularly emphasized by Josh Knobe instead thinks of X-phi or at least large portions of it as continuous with cognitive

[74] I take this contrast between the negative and positive programs in X-phi from Alexander, Mallon, and Weinberg 2010. As they note, other writers employ a similar distinction. Examples of the positive program include the use of responses to vignettes to support claims about the folk concept of free will (and whether or not it is compatibilist—e.g., Nichols and Knobe 2007) and the folk concept of intentional action (e.g., Knobe 2003). Examples of the negative program include Alexander and Weinberg 2007 (challenging claims made by analytic epistemologists about the folk concept of knowledge) and Machery et al. 2004 (challenging the universality of Kripkean intuitions about reference). The negative program also includes papers presenting evidence that the intuitive judgments of the folk or philosophers are influenced by such normatively irrelevant factors as order effects (e.g., Swain et al. 2008), context effects, and small variations in wording.

[75] As noted, the negative program in X-phi often appeals to the fact ordinary people disagree with one another and with professional philosophers in their judgments about cases. This is taken to undermine the "reliability" of judgments about cases methods. To the extent that the method is claimed to be a source of information, either about the nature of X or about how others judge, I agree that disagreement can sometimes support skepticism about such claims. On the hand, as I have argued, the existence of such disagreement *need* not in itself have any particular normative significance. (If it is true that many people fail to judge in accord with modus tollens, this does not show that this form of inference is normatively inappropriate.)

science.[76] This strand largely abandons the "intuitions/judgments about cases as evidence for concepts" focus of either the negative or positive programs. Instead, on Knobe's conception, X-phi involves (or at least should involve) the study of effects (generic features of reasoning, judgment, and so on), the representations that underlie these, and the factors that influence them. These results may tell us something about concepts, but they usually do not take the form of claims like "Our concept of X is . . ." To the extent that concepts enter into research of this sort, they are thought of as the upshot of many more specific cognitive processes and effects, with the latter being the focus of empirical research. This third conception is (in comparison with both traditional intuition-based methodologies and both the negative and positive X-phi programs) much more in alignment with the views of cognitive scientists regarding their own research and also reflects the significance I attach to much of the ERC I discuss. Indeed a number of recent papers in this tradition to which philosophers have contributed (e.g., Icard, Kominsky, and Knobe 2017; Henne et al. 2019) look very much like standard cognitive science and are published in cognitive science journals. This research often involves genuine experiments or causal modeling of observational results, as well as computational theories that attempt to explain these results. This is the sort of approach to descriptive issues about causal cognition that I have endorsed here.

There are several other issues that arise in connection with X-phi and its application to causal cognition that are worth comment. Philosophers who wish to defend armchair philosophy against X-phi results that seem to show that the folk don't judge as the armchair philosopher does sometimes appeal to the "expertise" objection. This is the objection that philosophers who make judgments about cases have a kind of expertise (due to their professional training and so on) that makes them more reliable judges than ordinary people presented with the same cases, so that the judgments of the latter should be discounted when these depart from trained philosophical judgments.

It is tempting to dismiss such arguments as question-begging, particularly when they are not accompanied by independent standards for when judgments are reliable and evidence that such standards are more likely to be met by philosophers. However, in the case of causation and perhaps more generally, they raise some additional issues that are worth consideration. First, in the best empirical work on causal cognition, whether conducted by psychologists or philosophers, elaborate care is taken to determine whether the subjects correctly understand the cases or experimental tasks with which they are presented—that they are not confused or misinformed about these. Similarly an effort is made to

[76] See, for example, Knobe 2016.

identify subjects who seem not to be paying attention or who respond randomly. The assumption is that the responses of such confused and inattentive subjects are unlikely to be informative about what the psychologist is trying to learn. Such subjects can be dealt with by, for example, giving them quizzes about the information presented to them and excluding those who lack comprehension. This is not exactly a matter of using only subjects who are "experts" (whatever that might mean), but it does involve a methodology that does not always take subject responses at face value, regardless of how confused they seem to be. To the extent that subjects in X-phi experiments are not filtered by comprehension tests, philosophers who insist that the results of such experiments are hard to interpret seem to me to have a good point. However, as I have emphasized, this is a correctable problem—there is nothing in X-phi methodology that precludes the use of comprehension tests, exclusion of unmotivated subjects, and so on. Indeed, the best work in X-phi does employ such screening.

A second issue concerns what "expertise" might mean in the case of causal cognition or the understanding of causation. It is plausible that the overwhelming majority of people who spend time worrying about Gettier cases and the like are professional philosophers with training in epistemology. Other people rarely think about such matters. So it is arguable that philosophers are the experts about Gettier cases, if anyone is. By contrast there are many groups of people besides philosophers with professional training that seems relevant to expertise in causal reasoning, including statisticians, computer scientists, psychologists, and scientists of many other sorts. For this reason alone arguments that philosophers, in virtue of their training, have special expertise not possessed by others in matters related to causal reasoning or the identification of causal relationships seem even more implausible than they may be in connection with other subject matters. To put the matter differently, philosophers who care about what the experts think in connection with causal judgment should look at the judgments of other possible experts besides philosophers.[77] Note also that while ordinary folk may not ever reflect on whether Gettier cases count as knowledge and typically receive no feedback on whether or not their judgments about such cases are correct, this is not true for causal judgment and causal reasoning. Ordinary people engage in causal reasoning and learning all the time and receive feedback from many sources (the world, other people) on whether their judgments are correct. Moreover, as claimed earlier (and as supported by evidence in subsequent chapters), ordinary people often make causal judgments that are normatively correct, so that in many cases one can't plausibly dismiss such judgments

[77] It would be interesting to compare the judgments of philosophers about standard scenarios involving causation with those of other experts.

on the grounds that they reflect confused and uninformed responses, however plausible this assessment may be for lay judgment about other sorts of topics. People who are professionally involved in causal reasoning outside of philosophy receive even more (and higher quality) feedback. Again, it is arguable that this makes the claim that philosophers have superior expertise less appealing in the case of causation.

4
Some Empirical Results Concerning Causal Learning and Representation

4.1. Introduction

This chapter surveys a number of empirical results relevant to causal learning and representation, both in adult humans and in other subjects, including children and nonhuman animals. I try to connect these results to various normative and descriptive theories of causation, including associationist, counterfactual, interventionist, and process-oriented accounts. Experiments supporting the descriptive plausibility of the interventionist framework are highlighted. I acknowledge upfront that my discussion is highly selective; the empirical literature on causal cognition is huge, and it is impossible to survey all of it. I have tried to focus on results that seem particularly philosophically interesting and that (as best I can judge) reflect careful experimentation and analysis. In some cases when there are other results that are discordant with those that I discuss, I have indicated this in footnotes, but I have not tried to be systematic in this respect.

4.2. What Is It to Have Causal Representations? Some Preliminary Remarks

As a point of departure and at the risk of belaboring the obvious, let me begin with a distinction: it is crucial to distinguish between the claim (i) that some relationship *is* causal and that a subject has learned something about the relationship and the claim (ii) that the subject *represents* that connection *as* causal (i.e., has acquired a representation that represents that relationship as a causal relationship). When a pigeon learns that pecking at a target will produce a food pellet, the pigeon learns a relationship that is causal (let us suppose the pecking causes the appearance of the food pellet), but it is a further question whether the pigeon has acquired a full-fledged causal representation in the sense of (ii).

What then is it to represent a relationship as causal or to acquire a full-fledged capacity for causal representations and causal cognition? Roughly speaking, my paradigm is an adult human being and the sorts of representations these sometimes possess. What such representations involve is something that I will build

Causation with a Human Face. James Woodward, Oxford University Press. © Oxford University Press 2021.
DOI: 10.1093/oso/9780197585412.003.0005

up to and motivate in the course of this chapter, but to anticipate and to provide a guide for what follows, I will take the following considerations to be relevant, although none are strictly necessary and the list that follows is not meant to be exhaustive. It should also be obvious that satisfaction of many of the individual conditions can be a matter of degree, and some conditions can be satisfied even if others are not. So, as will be illustrated in the subsequent discussion, subjects that are not adult humans can have representations that are cause-like in some respects but not others. As will be apparent, many of the criteria are motivated by interventionist ideas, since they have to do with features of representations that facilitate manipulation and control.

4.2.1. Criteria for Causal Representation

(4.2.1a) Sensitivity to the difference between merely correlational and causal relationships as reflected in the subject's inferences and behavior, including, where appropriate, verbal behavior.[1]

(4.2.1b) Sensitivity to the normative significance of the difference between the results of observations versus the results of intervening as reflected in the subject's inferences and behavior. (Recall Section 2.4.) Related to this is the extent to which the subject employs representations that integrate information concerning the results of its own interventions, observations of the interventions of other agents, and observations of covariation that is produced "naturally" rather than by the interventions of any agent into a single representation. In other words, the agent should be able to go back and forth between "seeing" and "doing" in forming causal beliefs. (Cf. Section 4.7.)

(4.2.1c) The extent to which the subject employs representations that integrate geometrical-mechanical aspects of causation in the sense described in Section 2.9 (and the perceptual cues on which these are based) with difference-making aspects. This is also related to the

[1] Although I provide criteria for causal representation, I won't try to provide a general theory of what *representation* (as opposed to some other psychological or neurobiological state) involves. A view broadly similar to mine can be found in Poldrack 2020. I follow Poldrack in assuming that representations in the relevant sense represent items or relations in the external environment (spatial relations, causal relations), that they are structured or model-like, with the structure often being compositional (as illustrated by directed graphs or spatial maps), and that they are typically somewhat decoupled from immediate perceptual stimuli. As for why we should assume that people and other subjects have representations at all, I claim that this has been a successful hypothesis in neuroscience and cognitive psychology. In the particular case of causal cognition, this assumption is supported by empirical evidence described in subsequent chapters.

extent to which the subject exhibits perception/action integration or dissociation in causal understanding. One motivation for this criterion is that geometrical-mechanical relations are important potential sources of information about difference-making information, and the most successful causal representations will make use of this connection.

(4.2.1d) The extent to which the subject's representations decouple (rather than "fuse") the representation of means and ends and incorporate detailed information about how to alter means in face of changing circumstances to achieve the same goal. This is related to whether the subject can recognize irrelevancies in means/ends routines and represent distinctions between more direct and more indirect means/ends relationships and causally intermediate variables—roughly whether the subject has a contrast between direct and indirect causes. A related consideration is the extent to which subjects possess representations that allow them to generalize to new circumstances and situations on the basis of considerations other than perceptual similarity. As noted in what follows, this is closely connected to the role of invariance in causal reasoning.

(4.2.1e) The extent to which the subject's representations of causal information are not encapsulated or available only to specialized systems but are rather available more generally to other systems for reasoning, inference, action, and planning. Arguably, this is also connected to the possibility of "insight" learning as opposed to reliance on extensive trial-and-error learning in the acquisition of causal information. Relatedly, it is connected to capacities for explicit reasoning about hypothetical possibilities ("what would happen if") conducted in the absence of trial-and-error learning.

(4.2.1f) The extent to which causal representation is maplike or model-like in the sense of integrating representations of individual cause-effect relationships into a single, overall, allocentric (as opposed to egocentric) representation, so that not only individual causal links but relationships among these are encoded and, moreover, encoded in the form of variables that are not too egocentric (see the later discussion). This maplike requirement includes the capacity to represent complex causal structures involving several variables, such as structures in which, for example, two effects E_1 and E_2 are represented as effects of the same common cause C and, relatedly, the ability to distinguish such structures from other complex structures such as those in which the same three variables are related by means of a chain structure. Such representation must also encode directional features of causal claims—i.e., the difference between $X \rightarrow Y$ and $Y \rightarrow X$ should be represented.

As I said earlier, my paradigm is adult human causal representation—as a matter of empirical fact, in some cases adult human causal cognition seems to possesses all of these features. Although some psychologists (and philosophers) claim that humans have an "innate" concept of causation that comes "on line" or is "triggered" by the appropriate experiences early in infancy and then remains largely unchanged through adulthood, my view is that the empirical evidence strongly supports the view that the capacities associated with causal thinking change and develop over time. At various points in their development human infants and small children possess representations and abilities with some of the features described here, but not all of them, with more being acquired in the course of development and these being better integrated with one another. Possession of integrated adult causal representations is thus the end product of learning and development rather than something present at the outset.

To illustrate some of the criteria just described, consider an infant who learns that by kicking, it can cause a mobile to which its foot is attached to move. The relationship between the movement of the foot and movement of the mobile is certainly causal and, assuming that what is learned involves the acquisition of a representation of some kind, it is arguable that the infant will have some sort of representation of this relationship, where the relationship is causal. On the other hand, there are many reasons for doubting that the infant has acquired a full-fledged causal representation (a representation of this relationship as causal) in the sense that the infant represents the relationship as having the full suite of features (4.2.1a)–(4.2.1f). For example, the infant need not be representing (as an adult causal representation would) the relationship as one that might also hold between impulses communicated to the mobile by other events besides foot movements and the subsequent movement of the mobile. That is, if P is the impulse, M the movement of the mobile, and F the infant's foot movement, an adult might represent this as in Figure 4.1, where C includes other possible causes of P, such as a breeze or the movement of the adult's hand causing P, which in turn causes M. By contrast, the infant's representation may be more local and egocentric and not contain a representation of the role of P at all; one possibility is that its content may be simply something like "Depending on whether I move my

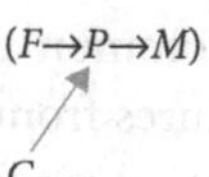

Figure 4.1

foot, the mobile moves." (I discuss this sort of egocentric representation in more detail later—obviously it is an empirical question whether the infant has only this sort of egocentric representation.)

Having distinguished the claim that a subject represents (in some fashion) a relationship that *is* causal from the claim that the subject represents that relationship *as* causal, I must add several further clarificatory remarks. First, it is important to avoid the "opposite" mistake of inferring from the fact that a subject represents a relationship that is causal but not as causal to the conclusion that the subject's representation of the relationship plays no role in the acquisition of the adult capacity for fully causal representation. For example, successful imitation of a tool using routine by a conspecific sometimes (perhaps often) occurs in the absence of detailed causal understanding.[2] It does not follow from this, however, that in humans the capacity to imitate plays no role in the acquisition of the adult capacity to reason causally or to employ full-fledged adult causal representations. Instead, a great deal of empirical evidence suggests that the acquisition of representations of relationships that *are* causal on the basis of imitation but which are not necessarily represented *as* causal plays an important bootstrapping or scaffolding role in the achievement of modes of thinking in which relationships are represented as causal. It is entirely possible that a similar conclusion holds for various abilities involved in associative learning (discussed later)—these may provide part of the basis or scaffolding on which adult human capacities for causal reasoning are acquired without being equivalent to those capacities.

A second point, which will also receive more detailed discussion later, is this: although there are good reasons, both conceptual and empirical, for distinguishing at least some of the abilities and representations that are involved in adult human causal learning from those that are involved in certain other forms of learning, such as associative learning, this should not be taken to imply that there is some single unitary ability that is present in adult causal cognition and that underlies all of (4.2.1a)–(4.2.1f). The abilities described under (4.2.1a)–(4.2.1f) are to at least some extent conceptually distinguishable and can (and do) dissociate empirically in some subjects, including both young humans and nonhuman animals. Their integrated presence in an adult subject is a developmental *achievement* that requires explanation, rather than something that occurs automatically in virtue of the acquisition of some single ability or in virtue of acquisition of (or innate possession of) "the concept" of causation.

A third, related point is that rather than asking the dichotomous question "Do nonhuman animals (or young humans) exhibit a "capacity for causal cognition" or "causal understanding," it is (as we shall see) a better research strategy to ask

2 Woodward 2011.

about whether they possess this or that more specific ability that is related to causal learning and understanding as these exist in human adults and to ask how these abilities connect (or support) or fail to connect with one another in different varieties of subjects. It is also a mistake to suppose that because animals or infants possess some of the skills that go into the complete suite of abilities that make up adult human causal cognition, they must also automatically possess other abilities in that suite. Instead, we should always ask what the specific ability is whose presence or absence is suggested by this or that experimental result, and we should be wary of inferring, absent specific supporting evidence, from the presence or absence of one ability to the presence or absence of others, even if we find these abilities associated in adult humans. When we do find that various abilities associated with human causal cognition co-occur in an integrated way, we should try to understand how and by what processes this integration is achieved, rather than regarding it as inevitable and automatic or not in need of explanation.

As an illustration discussed in more detail later, adult humans are able to use geometrical-mechanical cues to causal relationships (having to do with, e.g., spatiotemporal contact) that may be obtained from passive observation to guide actions aimed at manipulation and control—that is, humans integrate causal representations based on geometrical-mechanical cues with causal representations that are relevant to intervention and action. However, there is evidence that human infants as well as many nonhuman animals fail to do this or at least fail to do it as completely and effectively as adult humans do. Nonhumans and human infants may exhibit sensitivity to geometrical-mechanical cues of a sort that adult humans exploit in causal cognition, and they may also learn various routines for manipulation and control that again would suggest causal understanding in an adult human, but they may fail to put the two together. Understanding how such integration develops or is acquired is thus crucial to understanding adult human causal cognition.[3]

A related point is that it is doubtful that it is useful to look for single dividing line between nonhuman and human causal cognition—a sort of "mental Rubicon" that only adult humans have crossed. Adult human causal cognition probably differs from causal cognition in nonhuman animals (and in human infants) along a number of different dimensions and in the way the abilities displayed along those dimensions are integrated. For similar reasons, it is also probably a mistake to suppose that different species of animals can be arrayed along a single dimension representing degree of causal understanding, with, say,

[3] A corollary is that experiments that probe the extent of such integration and how it is achieved or develops (to the extent it does) can be particularly revealing in understanding the causal competences of different animals. For examples, see Sections 4.4 and 4.5.

primates possessing more of this than other mammals, humans possessing more than other primates, and so on. Instead different species will have specialized skills that reflect the particular ecological niches in which they are located, and these will vary along many different dimensions. A corvid, say, may be superior to many primates in some causal learning tasks and inferior in others.

I see these contentions as echoing and reinforcing the arguments of earlier chapters about not framing results from experiments on causal cognition in terms of claims about "our" or "the" concept of causation. Among other limitations, this framing encourages thinking in terms of a dichotomy—either some ability or reasoning pattern is part of our concept of causation or not, and a subject either has this concept or not. Worse, this framing distracts attention from the important developmental question of how integration among the different abilities that go into adult human causal cognition is achieved (since attention is focused instead on the question of whether the concept is properly attributed to some subject or not). It also encourages the idea that we can explain the full suite of adult abilities simply by talking about whether the subject has "grasped the concept" of causation.

4.3. Associative and Causal Learning

I turn next to a discussion of associative learning—that is, the kind of learning that is involved in classical conditioning of the sort studied by Pavlov[4] and in operant or instrumental conditioning. This is relevant to the topic of causal learning and cognition for a number of different reasons. First, although virtually no one claims that all features of human causal cognition can be fully captured within an associationist framework, a number of investigators maintain research programs that emphasize the continuities between the learning of associative relationships and causal learning—see Section 4.4 for a discussion of some of these claimed continuities. Second, if (as I believe) not all features of causal learning and reasoning can be understood in associationist terms, this requires seeing how far purely associationist accounts can take us in understanding causal learning, reasoning, and judgment. If there are features of causal cognition that cannot be captured within an associationist framework, one can only identify them by seeing what associationist frameworks are capable of explaining. Third, it is uncontroversial that humans and other animals learn through classical and instrumental conditioning.[5] This raises the possibility that some (maybe a substantial part) of

[4] I don't mean Pavlov had the right theory of such conditioning—see the subsequent discussion.

[5] For a striking example that might be interpreted as involving a relationship learned on the basis of classical conditioning and not as learning of a causal relationship, see Le Pelly et al. 2017, 6. In this experiment, subjects were trained to look at a target rather than a distractor. There were two possible distractors, one of which signaled that the subject would receive a high reward and the other a low

what some regard as non-associationist forms of causal learning and reasoning instead reflect the operation of associative mechanisms. Again to assess this claim, we need to understand what associative learning mechanisms are and are not capable of. This is particularly relevant to the assessment of claims about the existence of various forms of causal learning and cognition in nonhumans and young children where associative mechanisms may be particularly important.[6]

Yet another reason for beginning with associationist models of learning is this: associationist learning and particularly classical conditioning are often taken to be highly general and independent of subject matter: many theories of classical conditioning assume that subjects can learn associations between pretty much any two arbitrary stimuli that an animal is able to perceptually detect—colors and shapes, tones and shocks, and so on. In particular, the associated stimuli need not be related as cause and effect—for example, the associated stimuli might be the correlated effects of a common cause.[7] Indeed the stimuli may not be sorts of things that we ordinarily think of as capable of standing in causal relationships—one might learn to associate dogs, cats, and goldfish (all pets), beef, lamb, and chicken (forms of meat), and so on. In other words connections that are due to classificatory relationships or to relations of similarity or spatial or temporal proximity may be learned by associative processes.

Thus to the extent that causal learning is just a matter of (or in some respects highly similar to forms of) associative learning, the implication seems to be that there is nothing special or distinctive about causal learning—it instead involves the application of various domain-general learning strategies that are equally applicable to other sorts of relationships (classificatory etc.) that are non-causal. One can thus think of what follows as an exploration (in part empirical, in part

reward, but the subject received a reward only if the subject did not look at the distractor. Subjects nonetheless looked more often at the high-value distractor, presumably because they learned an association between it and reward value, even if they were informed that doing so caused omission of the reward. In other words the association they learned was insensitive to their explicit causal knowledge, or at least the subjects were unable to use their explicit causal knowledge to avoid being influenced by the association they had learned. This suggests (although it does not strictly imply) that the association between target and distractor may be stored in some format or system that is distinct from causal representation.

[6] It is sometimes claimed that associationist models are "simpler" than more cognitive models and that for this reason the latter should be employed only if the former can be ruled out. I do not endorse this line of thought, which seems problematic on a number of grounds: First, it is dubious that associationist models are generally simpler—for example, such models often employ more free parameters than more cognitive models. Second, even if associationist models are generally simpler, my view is that this is not a reason for believing that they are likely to be true. Instead my reason for considering associationist models is that good methodology requires comparing rival theories and looking for evidence that discriminates among them. I see the evidence as favoring cognitive models for a range of causal reasoning and judgment tasks.

[7] Think of a regularity theory like Mackie's INUS condition account, which arguably (cf. Chapter 2) fails to distinguish such correlations from relationships we think of as causal.

more "theoretical") of the extent to which causal reasoning is "special." In addition, psychological accounts that assimilate causal learning to associative learning of course have philosophical counterparts: these are theories that hold that causal relationships are simply relations of association or correlation that perhaps satisfy various additional domain-independent criteria that are not distinctively causal and thus can be characterized in non-causal terms. For example, these additional criteria might embody spatiotemporal conditions (causes must temporally precede their effects and/or be spatially contiguous with them). Or they may have to do with considerations like simplicity or informativeness, again understood in a domain-independent way, that does not presuppose distinctively causal assumptions: perhaps we think of *C* as causing *E* when *C* and *E* are correlated, and a representation in which they are linked is simpler in some way than alternatives, where this notion of simplicity can be characterized in a way that is not specific to causation. To anticipate, my own view is that causal learning (and reasoning and representation) is "special," at least in adult humans, and hence that purely associationist accounts are inadequate, on both empirical and normative/theoretical grounds.

4.4. Associative Learning: A Slightly More Detailed Look

Before turning to a more detailed look at associative learning, a caveat is in order. Theories of associative learning have become increasing complex and sophisticated over time, with many modifications introduced to accommodate empirical difficulties with earlier versions of such theories. For example, Dickinson and Burke 1996 and Dickinson 2001 describe an associative theory with "within compound associations" that accounts for (or at least is consistent with) the phenomenon of backward blocking (see later discussion), which the standard Rescorla-Wagner theory of associative learning cannot explain.[8] Since there are many different recent versions of associative theories that differ in important ways from one another and there is disagreement about their empirical status, in order to make the subsequent discussion manageable,[9] I focus on the Rescorla-Wagner theory as my paradigm of an associationist theory and compare this with some other treatments of causal learning. This comparison is standard in much of the literature on causal learning (see, e.g., Cheng 1997; Gopnik et al.

[8] Similarly, Dickinson (2012) describes an account of aspects of instrumental conditioning that is much richer (e.g., it involves learning associations between internal associations of considerable complexity) than the description in terms of learning action/outcome relations that I have employed.

[9] Put differently: I acknowledge that a thorough discussion of these issues would require a comparison of the many different associative theories with one another and with non-associative accounts of causal learning, but this just is not possible without making my discussion hugely longer and very unwieldly.

2004; Gopnik and Schulz 2007). However, readers should be aware that it is arguable that some, although by no means all, of the experimental results described subsequently that pose problems for associationist theories such as RW may be accommodated by other associationist theories.[10]

We can think of associative learning at the most general level as a matter of learning an association either between two or more stimuli (classical conditioning) or between a behavior and a stimulus (operant conditioning).[11] In the simplest case of classical conditioning an unconditioned stimulus (US) (e.g., presentation of meat) that evokes a reflexive response (e.g., salivation) is paired with a conditioned stimulus (CS), such as the ringing of a bell, that does not initially evoke the reflexive response on its own. After repeated paring of the US and CS, the CS evokes the reflexive response on its own (the dog salivates at the sound of the bell). The subject thus learns to associate the CS and the US or to "expect" or "predict" the US on the basis of the CS. For this reason the CS is often described as the "cue" and the US as the "outcome," so that the subject is understood as learning which cues predict the outcome. In operant conditioning (or instrumental learning) the subject produces a behavior that is associated with an outcome that is either rewarding or punishing (e.g., a pigeon pecks at a target and a food pellet is released) and, on receipt of the reward or punishment, the subject learns to either produce the behavior or to avoid producing it. Note that in classical conditioning, the subject learns an association between two events that are outside of its control—the learning occurs through passive observation. By contrast, in instrumental conditioning the subject acts on the world (or affects it through behavior) and learns an association between its behavior and an outcome.

In classical conditioning, the relation that is learned is an association or correlation. Focusing on the events CS and US,[12] this correlation may be due to the fact that CS causes US, but the association may arise in some other way as well. For example, in Pavlov's experiment, the tone is correlated with but does not cause the presentation of the meat. Instead both are due to the actions of the

[10] This raises the general question of what makes a theory of learning or representation associationist and when additional elements introduced into an associationist theory make it "non-associationist." This is not an easy question to answer, but I will make some suggestions in what follows.

[11] I follow the usual textbook illustrations of associative learning in terms of conditioning experiments like Pavlov's, but I don't mean to suggest that the correct way to understand such experiments within an associationist framework is in terms of notions like stimulus/response "contiguity" or "pairing." As emphasized in Rescorla (1988), modern associationist accounts understand such experiments in terms of the learning of predictive relationships. I follow this understanding here.

[12] In a theory like RW the associative links that are learned will reflect information about the occurrence of other events besides CS and US since this will affect whether CS increases the predictability of US. However, the important point for present purposes is that what is learned are (predictive) associations.

experimenter. By contrast, in operant conditioning, the ecologically normal case is one in which a causal relationship is present between the animal's action and the outcome—the pigeon's pecking causes the release of the food pellet. Similarly when an animal learns to hit a nut with a rock so that the nut breaks, the breaking usually will be caused by the animal's action, rather than by some third factor. In saying that this is the ecologically normal case, I mean that while it is of course possible that the outcome may be caused in some other way than by the animal's behavior (for example, by some factor that causes the outcome and happens to be correlated with the animal's behavior), in normal environments this is not what usually happens. This point is of some significance for causal learning since it means that relations learned in operant conditioning are typically not confounded.

There are a number of striking similarities and continuities between associative learning and causal learning in humans. For example, both associative learning in rats and other animals and human judgments about whether a causal relationship exists as well as judgments of the causal strength of that relationship (as expressed in verbal reports) exhibit a similar sensitivity to temporal delay—relationships that involve a relatively short temporal delay between cue and outcome are learned more readily, with learning declining as the temporal delay increases. This is true for both classical and operant conditioning and is also reflected in verbal reports of human causal judgment.

Next consider an experimental setup in which both the cue *C* and the outcome *O* are binary events (i.e., they either occur or fail to occur) and these are the only two events of interest. In such cases, both animal learning and human causal judgment are (independently of temporal relations) highly sensitive to the *contingency* Δp between *C* and *O*—that is, to $\Delta p = Pr(O/C) - Pr(O/-C)$. For example, both animal learning (as measured by rates of lever pressing in instrumental learning tasks) and human judgments of causal strength are higher the greater the value of Δp and (with some very important complications in the case of humans, discussed in Chapter 7) tend to decline as Δp approaches zero. Indeed, as we shall see, some prominent theories of human causal judgment model such judgments as tracking Δp.

4.4.1. Forward Blocking

An even more striking similarity between associative learning and causal judgment is the existence of (forward) blocking. Suppose an animal learns an association between a stimulus A and some outcome, which, following the usual convention, we represent as A+ (the + reflects that the outcome occurs more frequently in the presence of A than in its absence). The same outcome is then paired with a compound stimulus containing A and B (AB) in such a way that the outcome does not occur any more frequently in the presence of AB than in the presence of A. In this case the animal will not learn or will learn only a weak association between B and the outcome—a weaker association than if B alone had been paired with the outcome (i.e., if the subject had just been exposed to B+). In other words, the prior exposure to the association between A and the outcome (A+) "blocks" the acquisition of an association between B and the outcome, even if the outcome always occurs in the presence of B (or occurs more frequently in the presence of B than in its absence), since B is always paired with A. This makes sense if we think of the animal as learning predictive relationships or, more specifically, those cues that provide an *improvement* in the capacity to predict the outcome. That is, the animal learns the association between A and the outcome because A is a good predictor of the outcome, but the compound cue AB adds nothing beyond what is provided by A alone to the prediction of the outcome, and hence the predictive relationship learned between B and the outcome is weak or nonexistent.[13] Blocking also (sometimes) occurs in human causal learning and judgment—for example, in an experiment due to Aitken et al. 2000 in which subjects are told that various foods have been ingested that are or are not followed by an allergic reaction, judgments of the strength with which a food causes allergy follow a similar pattern to that described here, with cue/outcome relationships that do not improve the capacity to predict outcomes given lower ratings of causal strength.[14]

4.4.2. Discounting

As yet another example, both rat behavior and human causal judgment are subject to a discounting or signaling effect in which the usual reaction of nonresponse to a non-contingent reward schedule (i.e., one for which $\Delta p = 0$) in an instrumental learning task does not occur when rewards that are not paired with the instrumental action are preceded by a brief visual signal. As Dickinson and Balleine (2000) remark, "The intuitive explanation [of this effect] is that the signal marks the presence of a potential cause of the unpaired outcomes, thereby discounting these outcomes in the evaluation of control exerted by the

[13] In other words, forward blocking is "rational" or normatively appropriate in such contexts to the extent that the goal is just the learning of cue outcome relations that improve the subject's ability to predict the outcome relative to previously learned cue outcome associations.

[14] I'm ignoring a number of complications here. Although humans exhibit blocking (both forward and backward—see subsequent discussion) in some experimental scenarios, they do not do so in others. Relatedly, blocking is not always "rational"—whether it is depends on a number of different factors. In general, blocking phenomena are complex, and there is less than full agreement about what the experimental results show.

instrumental action" (2000, 192). Again this sort of behavior is like that which one would expect from causal learning.

4.4.3. The Rescorla-Wagner Model

These results show that associative learning is in some respects "smarter" than one might naively have supposed. They raise the general question of whether (or to what extent) causal learning in humans (and in other animals, to the extent that it occurs) is just a matter of associative learning and, in connection with this, what the differences, if any, between the two are. To explore these questions, I introduce the Rescorla-Wagner (RW) model, which (although it has been subject to a number of modifications) is still one of the most prominent and influential account of associative learning in the literature. This model will allow us to address questions about what associative models of learning can and cannot account for in a more disciplined way.

The RW model describes how the strength V_{ij} of the association between a cue i and an outcome j (both taken to be binary events) changes in response to successive trials in terms of the following relationship:

$$\Delta V_{ij} = a_i b_j (\lambda - \Sigma V_{ij}).$$

Here ΔV_{ij} is the change in associative strength between i and j as a result of the current trial, a_i and b_j are parameters representing the salience of i and j respectively, and λ represents the occurrence or nonoccurrence of the outcome (commonly taken to be 1 when the outcome occurs, 0 otherwise). ΣV_{ij} is the outcome predicted by the model on the basis of previous trials, so that the expression in parentheses reflects the discrepancy between the actual outcome and the predicted or expected outcome. It is thus a measure of how "surprising" the outcome is, given the previously presented cues. The model is applied iteratively as successive trials occur with V_{ij} being repeatedly updated by ΔV_{ij}. If after repeated trials $\Delta V_{ij} = 0$, the model has reached an equilibrium with no further changes in V_{ij}. The RW model does not have a unique equilibrium for all values of a_i, b_j, but it does have such an equilibrium in special cases (e.g., when a_i, b_j are constant across trials in cases in which the outcome does and does not occur and certain other conditions are met).[15] In the special case in which there is a single cue that may be either present or absent and certain other conditions are met, it is roughly correct that the model converges on Δp as defined earlier. As a consequence, it

15 See Danks 2003.

is common in the psychology and the animal learning literature to think of the claim that the perceived "strength" of the relationship between i and j (as reflected either in behavior or in verbal judgment) will track Δp as a "prediction" of (or as rationalized by) the RW model. In other words, this is taken to be a prediction of an influential associationist model of learning, which can be compared with non-associationist models. Those who favor the RW model, either as descriptive or normative, in turn take this to motivate use of Δp as a measure of causal strength.

Put abstractly, the RW model may be thought of as embodying an "error correction" algorithm that computes the discrepancy between the expected and actual outcome and updates its expectations accordingly. It is thus not surprising that it can account for a number of the phenomena associated with associative learning described previously—it tracks contingency and Δp, given the right additional assumptions, and it also can account for forward blocking: when cue A predicts the outcome and the compound cue AB does not result in any improvement in the predictability of the outcome, then the RW model has the consequence that no associative link of nonzero strength is learned between B and the outcome. There are certain other learning phenomena such as backward blocking (see Section 4.4.4) that the RW model cannot account for, but extensions or modifications of that model that (arguably) remain associative in spirit can predict at least some aspects of such phenomena.

4.4.4. Backward Blocking

In backward blocking, subjects are presented with two cues A and B together that are paired with an outcome (AB+). Subjects then see A alone paired with the outcome (A+) and are asked for a judgment (e.g., of causal strength) about the relationship between B and the outcome (B+). Backward blocking occurs if subjects give a lower strength rating to the relationship between B and the outcome (or are less likely to describe B as a cause) after observing AB+ and A+ or in comparison with some other relationship in which subjects see B paired with the outcome a similar number of times.[16] Adult humans and children exhibit backward

[16] For example, in an experiment described in Sobel et al. 2004 children observed a control condition in which blocks A and B were placed on a blicket detector together twice, and the detector activated both times. Then children observed that A did not activate the detector by itself. In a second condition, the backward blocking condition, the children saw two new blocks, A and B, which activated the detector together twice. Then they observed that block A activated the detector by itself. Note that in both cases B is associated with the outcome two out of three times and that B is absent on the third trial. Children categorized B as a blicket in the control condition 100 percent of the time and B as a blicket 31 percent of the time in the backward blocking condition, thus exhibiting backward blocking. (In these experiments the children were asked to make a categorical judgment about whether the block was a cause.)

blocking in a range of circumstances, and nonhuman animals exhibit behavioral analogues.[17] The RW model cannot account for backward blocking since according to that model changes in associations between cues and outcomes only occur when cues are present, and subjects never see whether B alone is paired with the outcome. On the other hand, as noted earlier, there are other associative models that do predict backward blocking, although at the cost of additional complexity. A natural "rational" causal interpretation of backward blocking is that when subjects see AB followed by the outcome (AB+) and then A alone followed by the outcome, they recognize that in order to explain the AB+ pattern, there is no need to postulate that B causes the outcome; A alone suffices. One possible normative view is that in this sort of case one has no evidence either way about whether B is a cause. Subjects who are guided by this view might express their uncertainty about the causal status of B by means of lower causal strength judgments for B in comparison with their strength judgments when they only see AB+, as assumed in the Bayesian account of Sobel et al. 2004. Alternatively, one might think that, after seeing A+, it is rational to conclude that B is less likely to be a cause (rather than that one has no evidence about the causal status of B).[18] This judgment will also be reflected in lower causal strength ratings. Both non-associationist accounts of causal learning such as the constraint-based methods described in Spirtes et al. 1993/2000 and Gopnik et al. 2004 as well as Bayesian models (Sobel et al. 2004) can account for backward blocking, although they differ somewhat in how they understand this phenomenon.[19]

Are there phenomena associated with learning or causal learning that cannot be accounted for by *any* associative model, including both the RW model and plausible modifications that remain within a broadly associationist framework?[20]

[17] That is, blocking or reduction in the strength of an association, determined by some physiological measure such as extent of salivation.

[18] Yet another possibility is that one might rationally conclude that there is no need to postulate a causal connection between B and the outcome, given what one has observed, and thus decline to do so. Concluding that there is no need to assume that B causes the outcome is not obviously the same thing as concluding that one has evidence against B causing the outcome. Giving subjects additional options (e.g., the option of choosing "There is no evidence either way about whether B causes the outcome") might help to clarify how subjects reason in cases of this sort.

[19] In the Bayesian model of Sobel et al. 2004, seeing that A alone is followed by the outcome establishes that the posterior probability that A causes the outcome is 1. By contrast, on their model the evidence in a backward blocking experiment is uninformative about whether B causes the outcome, so that the subject's posterior probability for this claim is the same as the subject's prior. Thus when subjects' priors that A or B causes the outcome are low, their posterior that A is a cause will be much higher than their posterior that B is a cause, and backward blocking will be strong. By contrast, when subjects' priors that A and B are causes are high, backward blocking will be less strong. (The authors report evidence consisting of manipulating subjects' priors that support this prediction.) Glymour (2001) shows that in Cheng's causal power model, which is a particular parameterization of a Bayesian network as a noisy or-gate (see Chapter 7), a particular interpretation of backward blocking follows from the assumptions of the model, assuming that the causal powers of A and B are independent.

[20] I noted earlier that the general question of what makes a model "associative" is by no means straightforward. One possible suggestion is that what is distinctive about an associative model is

One good candidate is described in an experiment by Waldman and Holyoak (1992, reviewed in Pelly et al. 2017). Associative models, including RW, rely (in addition to contingency information) on information about temporal relationships. It is assumed that information about the cue arrives temporally before information about the outcome and that in cases in which the cue causes the outcome, this temporal information is used as a clue to causal order—that the cue causes the outcome rather than conversely. Waldman and Holyoak used two contrasting experimental scenarios that exploit this feature. In the first ("predictive") scenario, subjects are presented with two cues A and B that are described in such a way that they would be naturally interpreted as potential causes of an effect X—for example, A and B may be the ingestion of two foods, and X an allergic reaction. Note that here the causal order is the same as the temporal order in which information about these events is presented—first the subjects get the information about which foods are ingested, then information about whether the allergic reaction occurs. (Of course, this is the same as the temporal order in which these events actually occur.) In this scenario it is natural to expect "cue competition"—A and B "compete" as causes of X, and if one were to discover that A by itself is capable of producing X, it is a natural thought that it would be rational to lower one's estimate of the strength of the $B \rightarrow X$ relationship. This is sometimes called "explaining away" in the literature on causal reasoning[21] and in fact is what human subjects do with some frequency in this scenario.[22]

In the second experiment, subjects are presented with a cover story in which X is naturally interpreted as the cause of A and B—that is, the causal order between A and X and B and X is reversed in the second scenario. For example, A and B might be symptoms of some disease X. Crucially, though, the temporal

that it only postulates internal structures and representations that represent or are derivative from patterns of statistical correlation and spatiotemporal relationships. Non-associative models of causal cognition make use of structures and representations that cannot be understood in this way, but instead have additional content that one thinks of as distinctively causal. For example, accounts according to which causal relationships are represented by directed graphs, with the directed edges in the graphs not being understood just as representations of statistical associations and temporal priority, but rather as having additional content having to do, for example, with the results of interventions, will be non-associationist. As far as descriptive psychology goes, the question then becomes whether this additional structure is required to explain features of causal judgment and reasoning—a question I answer in the affirmative later.

[21] See Pearl 2001, 17.

[22] Here I have oversimplified certain features of the original Waldman and Holyoak experiment to increase intelligibility. In the original experiment Waldman and Holyoak did not explicitly ask causal strength questions regarding the two scenarios, but rather asked for predictiveness ratings from cues to outcomes vs. outcomes to cues, obtaining the asymmetry described in the text. Predictiveness ratings were employed to make the results of the experiment more convincing to associationists, who employed such ratings in their own work (Michael Waldmann, personal communication). Subsequent research showed that the asymmetry described above also occurred (and indeed was enhanced) when explicitly causal questions were asked or when subjects were encouraged to think in terms of causal structure. The use of causal ratings is clearly more methodologically appropriate.

order in which the information that these events have occurred is presented is contrary to the causal order—the subjects first learn about the symptoms and are then faced with the "diagnostic" task of inferring whether X has caused them. From a rational perspective, although different causes of the same effect compete, different effects of the same cause do not. That is, if one learns that disease X causes nausea, this does not by itself make it more or less likely that X may also cause fever. In this particular experiment, Waldman and Holyoak found that subjects judged accordingly—their judgments of the causal strength of the X → B relationship did not compete with their judgments of the strength of the X → A relationship.[23]

From the perspective of associative models of learning (including the RW model) these two scenarios appear to be the same—in both, one gets information about cues A and B before one gets information about X, and in both, A and B are potentially predictive of X. Because of this, associative models will predict cue competition in both scenarios: to the extent that, for example, A is an excellent predictor of X and B adds little predictive information, the model will arrive at a low estimate of the strength of the B-X relationship, regardless of whether B is a cause or an effect of X. This is contrary to how human subjects actually judge, and it also does not seem to be rationally appropriate behavior. (If you are thinking that the inadequacy of the associative account of the two scenarios is just due to the fact that the associative account tracks the temporal order in which information is learned, and would disappear if the true temporal order in which events occur were tracked, be patient—I will address that suggestion later.)

A second illustration that points to the same conclusion is this: as we noted in Chapter 2, normatively speaking the information that some event is the result of an intervention has a very different significance for causal inference than the information that the event has been produced in some other way—for example, by some cause that may also directly influence the effect of interest. Moreover, as discussed later, as an empirical matter, adult humans are quite sensitive to the significance of this distinction for causal inference. However, associative models, including the RW model, are not sensitive to this distinction: from the perspective of these models all that matters is whether the cue (however produced)

[23] More generally, the experimental evidence strongly supports the claim that many adults are sensitive to considerations having to do with confounding—that is, they recognize that when they want to know whether *C* causes *E* and there are a number of other possible causes of *E* that may affect *C* via routes that do not go through *C*, one needs to control or adjust for the presence of these other causes. Even young children are sensitive to some of the problems posed by confounding and take steps to de-confound possible causal relationships—see, e.g., Kushnir and Gopnik 2005. Confounding, though, involves an asymmetric role between causes and effects; alternative causes of an effect operating through different routes are confounders, but the relationship between *C* and one of its effects E_1 is not "confounded" by the relationship between *C* and a different effect E_2. An appreciation of confounding thus implies an appreciation of the difference between *X* causing *Y* and *Y* causing *X*.

enhances the predictability of the effect expectations—whether or not the cue is due to an intervention makes no difference for the updating process. This is a second fundamental difference between human causal learning (and perhaps some causal learning that occurs in animals—see subsequent discussion—and what associative models (or at least models like RW) predict about what subjects are capable of learning.

One conclusion that is apparently suggested by these and other experiments is that adult humans learn and reason about causal relationships in a way that reflects representations that have additional structure (or incorporate additional information) beyond what can be captured by associative theories. The relationships that associative theories capture (or describe the acquisition of) are relationships (or a subset of relationships) that are useful for prediction or involve surprise reduction. But human causal representation has a structure that goes well beyond this. As noted earlier, from the point of view of classical conditioning, the occurrence of one of the correlated effects E_1 of a common cause C may be an excellent predictor of whether the other correlated effect E_2 will occur (assuming that the occurrence of C has not been previously observed), and the occurrence of E_1 may be a good predictor of C, just as C may be a good predictor of E_1. Despite this, human causal representation distinguishes among the roles played by C, E_1, and E_2. This is reflected in the public devices (graphs, equations, and so on) we have for representing causal relationships and the way in which we take these to be connected to such other notions as intervention and probability. In a directed graph representation of a common-cause structure, we draw arrows from C to E_1 and from C to E_2 but no arrows to represent the other statistical dependence relations present in the example, thus distinguishing causal relationships from non-causal statistical dependence relations. By contrast if we were to (merely) represent the associative links present in this example, we would presumably draw undirected edges connecting each of the three variables with each other. From an interventionist perspective, the presence of an arrow from, for example, C to E_1 but not from E_1 to C or from E_1 to E_2 reflects or encodes the fact there are interventions on C that will change E_1 and not conversely and that there are no interventions on E_1 that will change E_2 or conversely—that is, these connections with interventions are a way of bringing out the additional structure that is associated with the directed graph representation.[24]

[24] In order to forestall some possible misunderstandings, let me remind the reader of some distinctions introduced earlier. First, there is the empirical question of whether associationist models (at least in their current form) can fully account for the observed features of human causal cognition. I claim that as an empirical matter the answer to this question is no and that models with additional structure are needed. Second, I am *not* claiming that associationist models should be rejected as normatively correct accounts of causal reasoning because humans do not fully conform to these models; again, what people do and what they ought to do are different matters. Rather my argument is that there are legitimate jobs that we want causal representation to do (or goals we want to use them to achieve), that those features of human causal representation that go beyond the representation

A similar point holds if we represent this common-cause structure by means of structural equations, writing $E_1 = f(C)$, $E_2 = g(C)$. This encodes the fact there is a locally stable or invariant relationship (invariant under interventions) between C and E_1 and a locally invariant relationship between C and E_2 and that these relationships are different from and independent of one another so that it makes sense to think in terms of changing one of these relationships without changing the other. By contrast, although there is statistical dependence between E_1 and E_2, there is no invariant relationship—again, the absence of such a relationship is additional information that goes beyond just what is captured by the statistical dependence relationships in the example. Again, the empirical evidence strongly suggests that this additional structure is present in human causal representation and, as we have argued, has functional significance—it is connected to the role that causal representation plays in capturing relationships that are relevant to manipulation and control.[25]

My general point, then, is that we seem to need *something* that plays the role of these devices (graphs, structural equations) if we are to capture the structure of human causal representation—both internal representations and their public counterparts. There are reasons why these devices figure both in psychological theories of causal representation and in the public devices we use to represent causal relationships—reasons having to do with the work that they do. We need something that does that work even if we resort to devices other than directed graphs, structural equations, and an interventionist framework.[26]

Let me add that it is of course true that if we consider *conditional* dependence and independence relationships, then, assuming the Causal Markov and Faithfulness conditions, common-cause and common-effect structures will have different statistical signatures. If E_1 and E_2 are effects of C, they will be unconditionally dependent but independent conditional on C. In a common-effect structure with C_1 and C_2 causing E, C_1 and C_2 will be unconditionally independent but dependent conditional on E. This may make it tempting to suppose that we can fully capture the differences among different causal structures in

of associations contribute to the *successful* performance of these jobs, and that the associationist framework does not. In other words, it isn't just that we think in terms of frameworks that are not purely associationist; with some frequency the world cooperates with or supports the use of such non-associationist frameworks in the sense that such frameworks "work" in, for example, allowing successful planning and manipulation. The use of such frameworks is normatively reasonable for this reason, and it thus not so surprising that we find, as an empirical matter, that humans use them.

25 For all that has been said so far, this additional structure may have some other significance not connected to manipulation and control, but it at least has that connection.

26 To put the point differently: I don't take these considerations by themselves to distinctively favor an interventionist framework. My point is rather that human causal representation has distinctive features that go beyond the representation of associations and that an adequate psychological theory needs some way of capturing these.

terms of information about unconditional and conditional and (in) dependence relationships. However, as noted in Chapter 2, this is not correct. A structure in which C is a common cause of E_1 and E_2 is Markov equivalent to (implies the same conditional independence and dependence relations as) a chain structure in which E_1 causes C which causes E_2 and similarly with the roles of E_2 and E_1 reversed. Again, human beings think of these as different structures,[27] which strongly suggests that they are representing something different from (just) patterns of unconditional and conditional relationships among C, E_1, and E_2.

It is worth noting that the features of causal relationships just described (the fact that different structures have different implications for conditional and unconditional (in)dependence relationships and to claims about what would happen under interventions) are closely connected to Waldman and Holyoak's (1992) observation that candidate causes of the same effect compete with one another in a way that effects of the same cause do not. Given a structure in which C_1 and C_2 cause E and in which C_1 does not cause C_2 or conversely and they have no common cause, it follows from the Causal Markov condition that C_1 and C_2 are unconditionally independent, but dependent conditional on E. C_1 and C_2 are dependent conditional on E. Thus in a diagnostic task in which one knows that E has occurred, information about the occurrence of one of the causes provides information provides information about the occurrence of the other. In particular, given that C_1 occurs and is causally related to E, it can be "rational" (given appropriate additional assumptions)[28] to assign a lower probability to the claim that the other cause C_2 is operative. To use Pearl's example (2000), if my car won't start and I learn that the battery is dead, I regard it as less plausible that some other cause of the car not starting is operative. In general, in determining whether a causal relationship exists between C_1 and E, I have to control for or take into account other causes of E. By contrast, in a common-cause structure with no causal relation between the joint effects, given the occurrence of C and satisfaction of the Markov condition, the information that one of the joint effects has occurred provides no additional information about whether the other effect occurs, so that these do not similarly "compete" with one another. In assessing the relationship between C and E_1, I do not need to control for other possible effects of C. The character of human causal reasoning is such that it is sensitive to these sorts of considerations and in particular to the directional features of causation. Empirically adequate accounts need to reflect this.

Although associative accounts of the sort described here thus fail to capture important features of human causal cognition, I should emphasize that this by

[27] This claim is supported by experimental evidence described subsequently.

[28] These include the assumption that C_1 and C_2 do not interact with respect to E. If a fire requires the presence of both a short circuit and oxygen (interaction), observing the fire and the short circuit provides information that oxygen was present (rather than absent).

itself does not show that no account of such cognition that is "Humean" in some broad sense recognized by philosophers is adequate as an account of human causal learning and reasoning. That is, in view of the results described here, it is plausible that human causal representation has additional structure beyond "associative links" of the sort that might be described by the RW model, but for all that has been said so far, the possibility remains open that this additional structure might itself be understood in broadly Humean terms. One natural thought along these lines that has probably already occurred to the reader concerns the relationship of causal direction to time order. We observed earlier that theories of associative learning like the RW model make use of "temporal" information concerning the times at which cues and outcomes are observed, and this creates difficulties when the order in which this information is acquired runs contrary to the causal order, as in diagnostic reasoning. But of course this temporal information does not reflect the time order in which the events in question actually occur, as opposed to when they are learned. Diseases occur before symptoms even if one often learns about the latter before the former. Ordering causal arrows in accordance with (true) time order can thus give us at least some of the additional structure (beyond associative links) that we need—in particular (one might hope) that this gives us the asymmetry of the causal arrow (→). That is, one might think that, from a normative perspective, the true temporal order in which events occur (along with covariational information of course) could be used to order these events causally, relying on the idea that effects do not occur before their causes. Relatedly, one might conjecture, as an empirical matter, that humans use temporal information in the manner just described to construct distinctively causal (rather than purely associationist) representations. In support of this suggestion one might appeal to the following observation: if a set of variables is causally sufficient,[29] such that causes temporally precede their effects, and the variables satisfy the Markov and minimality condition[30] with respect to the accompanying probability distribution, then this picks out a unique causal structure. (Cf. Hitchcock 2018; Pearl 1988; Stern, forthcoming.) On this basis, it might be argued that the problem with associative theories is simply that they don't make use of information about true time order and have to rely instead on an inadequate surrogate: order in which information is acquired. If we instead think of causal representation as built out of (and acquired on the basis of) (in)dependence information, conditional and unconditional, and information about true temporal order, we will arrive at a treatment of human causal learning and representation that is more adequate and yet Humean in spirit.

[29] Roughly speaking, a set V of variables is causally sufficient if there are no omitted common causes that operate in such a way as produce confounding in V—see Spirtes et al. 2000, 22.

[30] Suppose that a graph G and probability distribution P satisfy the Markov condition. G and P satisfy minimality if no proper subgraph of G satisfies the Markov condition with respect to P.

A thorough evaluation of this proposal would take us too far afield, but in my view it is unpromising as a full account of human causal representation and learning. For one thing, human beings can reliably infer causal direction without relying on time order information—this is a good thing given that such information is by no means always available. Indeed, in some cases, the needed time order information does not exist since cause and effect are (at the level of analysis at which we are working), simultaneous,[31] but reliable inference of causal direction may still be possible. One simple case in which such reliable inference to causal direction is possible in the absence of temporal information involves the common-effect structure discussed earlier. When C_1 and C_2 are independent, C_1 and E are dependent, C_2 and E are dependent, and C_1 and C_2 are dependent conditional on E, then (assuming Markov and Faithfulness) a unique causal structure is consistent with this information, one in which C_1 causes E, C_2 causes E, and there is no causal relationship between C_1 and C_2: $C_1 \rightarrow E \leftarrow C_2$.[32]

Here we have used statistical information and additional assumptions to derive the causal order among C_1, C_2, and E without making any use of temporal information. In fact independently of the assumption of Markov and faithfulness, given three variables X, Y, and Z if X and Y are independent, X and Z are dependent, and X and Y are dependent, it is (as an empirical matter) a fairly reliable heuristic that X and Y are causes of Z.[33] There are also other, more sophisticated procedures for inferring causal direction that do not rely on temporal information but rather on generalized ideas about informational independence and connections between causal direction and invariance assumptions. Such procedures can be used to identify causal ordering relationships even when the variables involved occur simultaneously or do not have an exact temporal location.[34]

This is one of several considerations suggesting that there is more content to the way in which we represent and reason about causal order than what one can get out of some combination of straightforward statistical dependence and time order. This thought is reinforced by the observation that causal order claims have connections to probability claims and to claims about what would happen under interventions that appear to be independent of time order considerations. Neither the Markov nor the faithfulness conditions make any explicit reference

[31] This may be true of some economic variables that are defined over temporally extended periods—for example, quarterly GDP and unemployment.

[32] Although the inference to this structure is the normatively correct solution, there is to the best of my knowledge no experimental work bearing on whether adult humans (or other subjects) actually make this inference. In general, there seems to be little empirical work on how ordinary people make inferences about causal direction.

[33] See Woodward 2020b for discussion and for an interventionist account of why this heuristic works.

[34] See Janzing et al. 2012; Woodward 2019; and Woodward 2020b.

to temporal information, but information about causal order is crucial to both. Moreover, consider that within an interventionist framework, $X \rightarrow Y$ if and only if there is some intervention on X that changes Y but it is not true that $X \rightarrow Y$ if and only if there is some intervention on Y that changes X. Instead if $X \rightarrow Y$, no intervention on Y will change X (unless a causal cycle is present in which $Y \rightarrow X$). Again this asymmetry between the roles of X and Y in response to interventions has no obvious connection with temporal order.

I said earlier that one of the issues raised by associative theories and arguably by Humean accounts more generally has to do with the extent to which causal reasoning is "special." Is causal reasoning just a matter of the application of learning and reasoning procedures that are domain general in the sense that they are at work in many other domains besides causation (e.g., classification or categorization), or are those procedures in some respects specific to causation in the sense they do not always yield reliable or sensible result when applied elsewhere? The features of causal reasoning discussed earlier seem to suggest (provide at least some support for the conclusion) that causal reasoning *is* distinctive in some respects. Again consider the Markov and Faithfulness conditions. In general, the arrows in a directed graph might be given many different interpretations—for example, $X \rightarrow Y$ might be interpreted to mean that all instances of Y are instances of X but not conversely. Or $X \rightarrow Y$ might mean that all values of X occur within spatial regions that are spatially contained in regions within which values of Y occur. Presumably, though, there would be no reason to suppose that, assuming there was some accompanying probability distribution, graphs with these interpretations of $\rightarrow$ would satisfy the Markov and Faithfulness conditions. Instead it seems that there is something special about the interpretation of $X \rightarrow Y$ as X is a direct cause of Y that makes these conditions seem plausible when the graph has this interpretation, but not more generally.[35] Similarly for the idea that different causes that influence an effect by different routes compete (one needs to control for such alternative causes) but that this is not true for different effects of the same cause. Even assuming the existence of generalized analogs of "controlling for," this is not something that one expects to be true for all possible binary relationships or for all binary asymmetric relationships—again it seems to have something specifically to do with causation or at least with some more restricted class of which causation is a member.[36] Similarly also for the connection between

[35] I don't mean to claim that only causal interpretations of $\rightarrow$ will satisfy the Markov and Faithfulness conditions—I have no idea whether this is true. I claim only that there are many interpretations of $\rightarrow$ that will not satisfy those conditions.

[36] Some writers have recently argued that there are other "dependence relations" that share at least some of these features of causation. Candidates include "grounding" (Schaffer 2016) and "constitutive relevance" (Gebharter 2017). As I understand these writers, they claim that only some and not all features are shared between such dependence relations and causation. In any case, although I lack the space for detailed discussion, I'm not fully convinced by all of their claims about which features are shared.

causal claims and what happens under interventions. For some possible interpretations of →, the notion of an intervention may not apply at all, and for others there will no reason to suppose that when $X \rightarrow Y$ some intervention on X will change Y or that it makes sense to think of such an intervention as detaching X from its parents while leaving other relationships in the graph unchanged. If it is correct that causal relationships have special distinctive features that show up in their connections with notions of intervention and statistical dependence, this will be a general reason for expecting that associative theories (which do not incorporate these connections) will have a difficult time accounting for distinctively causal learning and representation.

My discussion so far has primarily focused on defending the claim that adult human causal cognition seems to involve representations and reasoning that go beyond the representation of associative relations. As I have acknowledged, it is logically possible that this psychological claim might be true and yet that this additional structure is superfluous or even dysfunctional for successful causal reasoning. However, as I have also argued, this seems unlikely. Instead, the additional features described earlier seem to encode genuine information that it is both possible and desirable to learn about. For example, a creature that can represent the difference between *X* causes *Y* and *Y* causes *X* or between *X* causing *Y* which causes Z and, alternatively, *X* causing *Y* and *Z* (in contrast to one that represents only associations) has an obvious advantage in connection with tasks having to do with manipulation and planning, assuming that these richer features often enough correctly track what happens.[37]

In what follows I will take the considerations just described to motivate the idea that the connections between causal claims and what happens under interventions that are captured by the interventionist framework as well as the connections with probability described by the Markov and other conditions do real work in capturing what is distinctive about causal representations. Perhaps in principle these connections might be fully replaced by something that looks more domain-general and "Humean" but that also serves to pick out distinctively causal representations, but at present no one has shown how to do this. Moreover, as argued earlier, given the functionality of the interventionist

[37] Recall that, as suggested earlier, "correctness" means simply that the creature's causal representations exhibit normatively appropriate relations to interventions, probability, and so on—for example, when the creature represents that *X* causes *Y*, this is correct if interventions on *X* are followed by changes in *Y* and so on. The presence of such additional representational structure (i.e., that goes beyond the representation of associations) and its apparent functionality are thus understood as connected to the minimal metaphysical claims concerning intervention-supporting relations being "out there" in nature, described in the introduction. Put differently, someone who acknowledges the facts about human causal representation just described but who claims that such representations have little or nothing to do with what is actually out there, owes us an alternative explanation of the apparent functionality of these representations. Why do they work to the extent that they do?

framework, one would expect such an alternative "Humean" framework to reproduce many of the "special features" of causation associated with the interventionist account. Given that this is the case and that the interventionist account at least gives us one way of representing the distinctive features associated with causal reasoning, this provides a reason to look at the extent to which, as an empirical matter, various subjects recognize and exploit these distinctive features in their own reasoning—from a functional or "rational mind" viewpoint, it would be surprising if they did not.[38]

Finally, let me note the obvious point that the contrast between associationist models of human causal cognition and those that postulate additional a richer representational structure is just one example of the more general ongoing debate between associationist and more cognitive models of cognition—a debate that encompasses language learning, perceptual processing, spatial navigation, and much else. I make no claims about the superiority of cognitive approaches in general, but I will note that one of the advantages of considering this debate in the context of causal cognition is that we have relatively precisely specified mathematical models of both sorts of approaches as well as a substantial amount of relevant empirical data. This makes the problem of adjudicating between these two approaches somewhat more tractable than it is in other contexts, where the competing alternatives are less well specified and where there are lots of appeals

[38] This is perhaps an appropriate point to address an inevitable question. I have defended theories of human causal cognition that postulate additional structure—Bayes nets, etc.—that go beyond purely associationist links. It is natural to wonder how psychologically or neurobiologically "realistic" such claims are. Am I claiming that we literally have directed graphs, structural equations, and so on in our brains (or minds)? Here is how I would think about this issue. At present we know very little about how, if at all, Bayes nets and many similar structures (involving equations, structured probability distributions such as those postulated in hierarchical Bayesian models), understood as accounts of cognitive processes, might be implemented neurobiologically. (Arguably the situation with respect to perceptual processing is a bit better—see, e.g., Doya 2006.) A similar point holds for most psychological theories that postulate higher-level cognitive representations and computations over these. Nonetheless it seems reasonable to hold that current empirical evidence supports the claim that there must be something or other in the human mind/brain that plays the kind of role or provides the kind of information that is carried by graphical models, hierarchically structured probability distributions, and the like. At present psychological theories constructed in terms of such models are a useful and fruitful way of capturing this fact and exploring its implications. Perhaps at some point in the future, some alternative set of representational structures will be developed that can be mapped more straightforwardly onto the brain, but in my view these alternative structures will need to do much of the same work and capture the same empirical facts currently captured by structures like Bayes nets. The stance I recommend regarding the "realism" of graphical and other sorts of models of higher-level cognition in current psychology involves a commitment of the sort just described and nothing more. Whether this is enough to count as genuine realism I leave to others.

Let me also add (in case it is not obvious) that I do *not* claim that the representational structures involved in causal cognition are innate or genetically specified. As suggested throughout this chapter, the representational and computational capacities involved in adult causal cognition develop over time and involve learning. I make no claims about what must be present at the outset for such learning and development to occur. We should separate the issue of what representational structures are involved in adult causal cognition from the question of how these structures are acquired.

to intuitions about what is or is not "possible in principle." At the very least, causal cognition is an illuminating context for examining this debate, and one that hitherto has received relatively little attention from philosophers.

With this as background I turn to a look at some empirical results bearing on the adequacy of various accounts that differ from or go beyond associationism, assessing these as theories of causal cognition and representation in adult humans and other animals.

4.5. Difference-Making and Counterfactual Theories of Causation Considered Psychologically: Some Clarifications and Preliminary Considerations

I noted in Chapter 2 that interventionist theories are just one species of the more general category of counterfactual theories of causation. These in turn fall into the still more general category of difference-making theories, which embody the idea that causes make a difference to their effects. If some version of a difference-making theory is correct as an empirical account of human causal cognition (or at least captures important aspects of causal judgment and representation), one would expect that causal judgment should be sensitive in some way to contingency information or to information about patterns of dependence between cause and effect. In the special case of causes and effects that are binary variables, one would expect sensitivity to the difference between the probability or frequency in which the candidate effect occurs in the presence of the cause and the probability with which the effect occurs in its absence. (I take "sensitivity" to have to do with whether such contingency information influences judgment and reasoning; it does not mean that some particular measure of contingency such as Δp is the only factor that influences causal judgment.) Many experimental results show that this is the case—both judgments about whether *C* causes *E* and judgments about the "strength" of the causal relation between *C* and *E* reflect the influence of contingency information, although, as we shall see, there is considerable disagreement about exactly how contingency information influences or maps onto such judgment. For this reason, philosophical theories according to which causation has nothing to do with difference-making, such as some versions of causal process theories, seem problematic when construed as descriptive theories of human causal cognition.[39]

[39] As I have been at pains to emphasize, of course the observation that, as a descriptive matter, causal judgment is sensitive to difference-making information is consistent with some non-difference-making theory being the correct normative theory of causal judgment. A straightforward version of this position would presumably hold that people are massively mistaken when their causal judgments are sensitive to difference-making information. A serious defense of this view would need to go on to explain how, despite making this mistake, people can apparently reason successfully about

The sensitivity of causal judgment to contingency information is, as observed in Chapter 2, consistent both with various versions of probabilistic theories of causation that make no use of counterfactuals and with theories that connect causation to counterfactuals, including interventionist counterfactuals. Is there evidence that specifically connects causal judgment with counterfactuals (and to interventionist-like counterfactuals), at least in humans?[40]

Although there are influential philosophical theories such as Lewis 1973a that are built on such a connection, many philosophers continue to regard counterfactuals in general (and a fortiori, their use in a theory of causation) with great skepticism. It is contended that counterfactuals are unclear, untestable, unscientific, and in various ways unnatural and artificial. Perhaps they are mere philosophical inventions that correspond to nothing in the way ordinary people actually think and reason.

One reason for resisting this negative assessment is the existence of normative theories of causal reasoning (not just in philosophy) in which counterfactuals play an important role.[41] These of course include theories (such as Pearl 2000)[42] that make explicit use of interventionist ideas, but are not limited to these—as noted earlier, another important example is the increasingly widely used potential response framework in statistics and econometrics, which explicitly understands causal claims in terms of counterfactuals or claims about the outcomes of hypothetical experiments (again see Rubin 1974; Angrist and Pischke 2009). More generally, although there is disagreement about this, it is common for historians and social scientists to contend that causal claims in their disciplines can be clarified by connecting them to counterfactuals. (See, e.g., Davis 1968; King et al. 1994.) Unless these practices are rejected out of hand as misguided, they suggest that not all uses of counterfactuals or all proposed connections between causal claims and counterfactuals suffer from the deficiencies described earlier.

Of course this leaves open the possibility that while connections between counterfactuals and causal claims can be useful in certain specialized disciplinary contexts, they play little role in ordinary causal reasoning. In fact, however,

causal relationships by relying on difference-making information. I will not speculate about such an explanation might involve.

[40] See my subsequent discussion for the sorts of evidence that might be relevant to establishing that nonverbal subjects connect causal claims and counterfactuals.

[41] See also Wilson and Woodward 2019 for arguments that counterfactuals play an essential role in a great deal of physical reasoning, both descriptively and normatively.

[42] I include Pearl among writers who make explicit use of counterfactual ideas in part because I find it natural to interpret his "do" operator in counterfactual terms—that is, as describing what would happen to *Y* if one were to do *X*. Pearl thinks of counterfactuals as entering into causal reasoning in a somewhat different and more restricted way in accord with the hierarchy he describes in his 2018. I will ignore this complication in what follows since Pearl shares with other writers the assumption that counterfactuals play a central role in causal reasoning.

there is considerable evidence that people employ counterfactuals extensively in various forms of ordinary reasoning and planning and that they connect causal claims and counterfactuals in something like the way that interventionist and Lewis-style[43] counterfactual theories suggest.[44] Moreover, there is also evidence that counterfactual thinking is very important developmentally—children utilize such thinking in pretend play and various imaginative exercises, as emphasized by Gopnik 2009.

Before turning to some of this evidence, however, some additional distinctions and clarifications will be helpful since they bear on an apparent puzzle about how connecting causal claims to counterfactuals can *ever* be a useful or illuminating thing to do. Recall that Lewis's account of the relationship between causation and counterfactuals is reductive. He attempted to reduce causal claims to claims about counterfactual dependence, where the counterfactuals in question did not themselves presuppose causal notions. If this reduction was successful, one might think that there is no puzzle about how counterfactuals could be used to elucidate causal claims: They give us an independent purchase on causal claims that does not presuppose causal notions. By contrast, other philosophers have claimed instead that the connection goes in the "opposite" direction, with causation as the more basic notion and counterfactuals elucidated in terms of this notion rather than vice versa (see, e.g., Edgington 2011). On such accounts, one can't appeal to counterfactuals to elucidate causal claims. The interventionist account does not adopt either of these two alternatives. On the one hand, the interventionist counterfactuals employed by that account presuppose the notion of an intervention that is a causal notion, so that the possibility of reducing causal claims to some notion of counterfactual dependence that does not presuppose any causal commitments is excluded. On the other hand, the interventionist account holds, in common with accounts like potential response theory, that one can help to elucidate causal claims by connecting them to counterfactuals. So on the interventionist account neither causation nor counterfactual is more basic or "fundamental" than the other (at least in the sense that we can reduce one of these to the other). Instead the two are interconnected and mutually constraining in the way that interventionism describes.

This may seem puzzling. On the one hand, I have argued that the interventionist counterfactuals that are connected to causal claims presuppose causal

[43] By a "Lewis-style" theory I mean a theory that connects causal claims to non-backtracking counterfactuals, but does not necessarily involve a commitment to possible worlds.

[44] This evidence is described in more detail subsequently. In the case of adult humans and verbal children, the evidence includes verbal behavior in which counterfactual reasoning is explicitly connected to causal claims. I take such evidence to be *sufficient* to establish that subjects connect causal claims and counterfactuals in a way that is consistent with counterfactual theories of causation. However, I do not assume that such verbal behavior is *necessary* to establish that subjects show such a connection—again see the discussion that follows.

notions, so that we cannot use the former to provide a reductive account of causation. On the other hand, I claim that we can nonetheless use the connection between causal claims and counterfactuals to help elucidate (or provide a purchase on) the former. Indeed, I have claimed that, as a matter of empirical fact, people regularly do this. Of course one might take the view that the sense of illumination people seem to get from connecting causal claims and counterfactuals is an illusion of some kind (people think they are getting illumination, but they can't be, since causation is "prior" to counterfactuals). However, given the functional orientation I have recommended in this book, I suggest instead that it is more promising to ask how it is possible for the connection to be illuminating in the absence of a reduction. I see this as in part a question about the cognitive psychology of causal cognition (what features of causal cognition and representation make it possible for this connection to be useful and illuminating?), but it also has a normative component (how can it be right to regard such a connection as illuminating?)

Part of the explanation for how this is possible (as suggested earlier and described in more detail in Woodward 2003) is that the causal claims that are elucidated in terms of interventionist counterfactuals are different from the causal claims (or causal information) associated with the antecedents of those counterfactuals: when the truth of *C* causes *E* is connected to the truth of counterfactuals about what would happen to *E* if intervention *I* were to be carried out on *C*, determining that an intervention on *C* has occurred does not require that we have already settled whether *C* causes *E*. The relation between the intervention *I* and *C* is a different causal relation from the causal relation, if any, between *C* and *E*, and we can use our grasp of the $I \rightarrow C$ relation to help elucidate the $C \rightarrow E$ relationship.

However, there is more to be said, particularly if our interest is in the psychology of causal reasoning. One consideration is this: when initially formulated or conceived, causal claims that take a form like "*C*s cause *E*s" can be (and not infrequently are) unclear or indeterminate in the sense that it may be less than obvious just what they commit us to or how they might be tested. This is so for many different reasons. One of the simplest possibilities (but by no means the only one—see the later discussion) is that it may be left unspecified whether a claim of the form "*C*s cause *E*s" is to be interpreted as a claim about a direct effect, a total effect or any one of a number of other possibilities. Differences among these causal claims can be connected to different sets of interventionist counterfactuals in the manner described in Chapter 2. We can this use such connections to elucidate the differences among different causal claims and to clarify and make more explicit how a given causal claim should be interpreted and what it commits one to.

There are other ways in which causal claims can be clarified or disambiguated by connecting them to claims about the outcomes of hypothetical experiments. Suppose someone claims that "being a woman causes one to be discriminated against in hiring decisions." It may be unclear how this claim should be understood. One possibility is that being genetically female (in the sense of having XX chromosomes and/or perhaps female external sex characteristics) causes one to be an object of discrimination so that the associated interventionist counterfactual is something like this:

> (4.5a) If an intervention were to be performed that changes an individual's gender from female to male at conception (while leaving everything else unchanged as far as possible), then that individual would be more likely to be hired.

Another (perhaps more plausible possibility) is that what is intended by the original causal claim is something more like this:

> (4.5b) Intervening to change an employer's beliefs about the gender of an applicant (while leaving other information about the applicant unchanged) will change that person's probability of being hired.

Obviously these are different counterfactuals that likely have different truth values. Counterfactual (4.5b) reflects something more like the "direct effect" of employee beliefs about gender on hiring, while counterfactual (4.5a) reflects the cumulative consequences (the total effect, including the way in which gender may have affected the applicant's education and other qualifications) of being one gender rather than another for hiring. We thus can distinguish among different possible interpretations of the original causal claim by distinguishing among different counterfactuals with which these interpretations are associated.

This in turn suggests one way in which the connection between causal claims and counterfactuals of the sort embodied in the interventionist claim (M) from Chapter 2 can be useful or illuminating, despite its non-reductive character: thinking in terms of the connection between causal claims and interventionist counterfactuals allows us to spell out the content of the former more precisely and explicitly in terms of the different counterfactuals to which they commit us. Relatedly, it has implications for how they might be tested (since the evidence relevant to (4.5a) is quite different from the evidence relevant to (4.5b)). Moreover, in some cases, unsuccessfully trying to connect causal claims with associated counterfactuals may persuade us that the causal claims are, as originally formulated, unclear and ambiguous and require reformulation. In fact, at the level of methodology, contentions like these (about clarifying causal claims by connecting them with counterfactuals) are explicitly endorsed by researchers in disciplines as different as statistics, econometrics, and history.[45] If such claims about the potential elucidatory role of connecting causal claims to counterfactuals are correct, this would explain the observation made previously—that researchers in many different disciplines find it helpful to associate causal claims with counterfactuals concerning hypothetical experiments. It also suggests that the same might be true for causal cognition involving laypeople or nonspecialists.

These suggestions/speculations motivate a number of further empirical questions. For example, it would be worthwhile to investigate to what extent it is true (as the preceding proposal suggests) that various subjects (both experts and others) perform better at causal learning and reasoning tasks if they are prompted to associate causal claims with hypothetical experiments in the manner described.[46]

It would also be interesting to learn more about the psychological mechanisms that underlie whatever abilities are at work when we associate causal claims with counterfactuals and get clarification from doing so. In an interesting series of papers (e.g., 2010), Gendler invokes dual-systems theory to explain the apparent fact that we seem capable of learning from thought experiments. Very roughly, dual-systems theory claims that human psychological processing is organized into two systems. System 1 is fast, automatic, "intuitive," and may operate unconsciously, while system 2 is slower, more deliberative, and reliant on explicit, controlled conscious processing. Gendler's idea is that often when we engage in a thought experiment, we "run" or otherwise make use of information that is present in system 1 and plug it into or make it available for processing in a more explicit form by system 2.

One can think, as I do, that the system 1 versus system 2 dichotomy is oversimplified and in crucial respects unclear, and yet still think that there is something right about Gendler's basic idea and that some version of it can also be used to help elucidate how associating causal claims with hypothetical experiments can be illuminating. The thought would be that when one entertains a causal claim like "*X* causes *Y*," not everything that is relevant to reasoning with the claim or testing it is (at least initially) explicit and available for critical assessment. One can entertain the claim that "*X* causes *Y*" without thinking, at least very clearly

[45] For the latter, see Davis 1968.

[46] There is some evidence that experts in international relations who systematically entertain counterfactuals when advancing causal judgments are more reliable than experts who do not do this—cf. Tetlock 2005. In this connection it is also important to note that the answer to this question may be different depending upon the kind of causal claim involved. Mandel 2003 reports evidence that prompting subjects to consider various counterfactuals has little or no effect on *causal selection*. Of course this is compatible with the possibility that prompting subjects to consider various counterfactuals has an influence on other sorts of causal judgment and reasoning tasks.

and explicitly, about just what would be involved in changing or manipulating *X* or how one expects *Y* to change under various possible manipulations of *X*. Associating "*X* causes *Y*" with claims about the outcomes of various hypothetical experiments forces one to be explicit about such matters. Often one does this by drawing on information that one "has" in some sense (perhaps it is present in one's system 1 or in some other format that may not be immediately or spontaneously accessible for explicit reasoning), but which one has not previously integrated explicitly into one's causal judgment.

Put this way, the idea underlying a psychological version of (M) (i.e., (M) construed as a descriptive thesis that captures aspects of what is involved in human causal cognition) is not so much that whenever one entertains a causal claim, one is necessarily thinking of it or explicitly representing it in terms of claims about interventionist counterfactuals, but rather that one can clarify or make precise what one was initially thinking by expanding the causal claim along the lines indicated by (M) and that this tends to result in more successful causal reasoning.

Again, this is a suggestion that might in principle be tested empirically. In fact experiments by Gerstenberg et al. (2017) seem to support this suggestion. Subjects were presented with videos featuring colliding billiard balls and then asked explicitly causal questions about these: for example, did the collision between ball 1 and ball 2 cause ball 2 to go into a slot? Evidence (in the form of both verbal judgments and from eye-tracking) from the experiments shows that subjects appear to answer these questions by running counterfactual simulations—for example, they appear to answer causal questions about the effect of a collision of ball 1 on ball 2 by simulating what the trajectory of ball 2 would have been if it had not collided with ball 1. This is shown both by their verbal judgments about such counterfactuals and their spontaneous looking behavior—they look (accurately) at where ball 2 would have gone if it had not collided with ball 1.[47] Again, this might seem paradoxical. If subjects can answer

[47] The "collisions" viewed by the subjects were generated by a "physics engine" which conformed to the laws of Newtonian mechanics with Gaussian noise added. In saying that the subject's counterfactual judgments were "accurate," I mean that the subjects judged in accord with the trajectories the physics engine would have generated under the various counterfactual scenarios. Since Newtonian mechanics is a good theory of such collisions, these correspond to the trajectories that would have actually occurred under such conditions. The experimenters found that the subjects spontaneously engaged in counterfactual simulations in answering causal questions (in comparison with the absence of such simulation in answering non-causal questions about actual outcomes), that subjects' answers to counterfactual questions and eye-tracking behavior closely tracked their answers to causal questions, and that their degree of confidence in their causal judgments closely tracked their degree of confidence in the corresponding counterfactual judgments. As the authors remark, results such as these seriously undercut the descriptive adequacy of causal process theories such as Dowe's, according to which in making causal judgments subjects only pay attention to what actually happens and do not make counterfactual judgments. Of course it still might be maintained that such process accounts are normatively correct, with the participants in the Gerstenberg et al. experiments making a mistake of some kind in how they go about answering causal questions. However, as noted earlier,

such causal questions by running simulations of the sort described, doesn't this show that they already "have" a causal theory in their possession that provides answers to the questions? Why then do they need to run simulations about counterfactual possibilities? The answer seems to be that such exercises allow the subjects to access (or to access more readily and reliably) information that otherwise would be less accessible or available for reasoning.

It is worth noting these suggestions about the way in which focusing on connections to counterfactuals can aid causal reasoning and judgment do not require adoption of all of the commitments of dual-systems theory. Rather, all that is required is the idea that information relevant to causal judgment and reasoning can be stored in different formats, some of which involve clearer or more accessible connections with counterfactuals than others and that causal judgment can be improved by connecting up the information in these formats.

A second issue raised by experimental explorations of the connection between causal claims and counterfactuals has to do with the evidence that might show that subjects are in fact guided by their understanding of such a connection. One simple strategy is simply to ask subjects to make explicit counterfactual judgments—if they do so in a way that connects to causal judgments in a normatively appropriate ways, this can be taken to show that they have some grasp of the connection between causal claims and counterfactuals. This assessment is strengthened if subjects spontaneously make counterfactual claims and link them to causal judgments or use the former to reason about the latter. This in fact is the sort of evidence that is obtained both in the experiments by Gerstenberg et al. described earlier and in some of the experiments described subsequently, such as those due to Harris and to Lagnado and Sloman.

Although evidence of this sort is (at least often) plausibly regarded as *sufficient* to establish that subjects can reason counterfactually, it is arguable that it is not necessary. (If it is necessary, then we cannot get evidence that preverbal children or nonverbal animals make use of counterfactuals.) One possible additional source of information is eye-tracking, along the lines described in the Gerstenberg et al. experiments.

Another possibility is that something like a grasp of counterfactuals and their connection with causal claims can also be revealed through evidence of planning or some forms of insight learning, even if counterfactual judgments are not expressed verbally. Cases of this sort include those in which the subject seems to immediately recognize that some action is appropriate as a way of producing an outcome without any trial-and-error learning. When there are alternative ways of reaching some goal, each somewhat complex and some better than others and

the subject's counterfactual (and causal) judgments were accurate, so it is not obvious how to provide a non-question-begging characterization of what the "mistake" in question consists in.

the subject makes an intelligent choice of means, it is often natural to think of the subject as undergoing a reasoning or simulation process that has something like the following form: "If I were to do *X*, *Y* would result, which would lead to *G*, which I want, but also to *B*, which I don't want; on the other hand it I were to do *Z*, *W* would result, which would lead to *G* but not *B*," etc. Similarly when faced with an unfamiliar task involving a choice among unfamiliar tools, systematic choice of the appropriate tool without extensive trial-and-error learning might be regarded as evidence that the subject is capable of entertaining or "running" reasoning of the form, "if I use tool t, then . . ." in advance of actually using the tool. Here we are assuming that a capacity to plan is closely bound up with a capacity to represent counterfactuals or something like them. However, while this line of thought seems, as I say, natural, I agree that it is by no means clear when apparent planning is evidence for counterfactual reasoning in nonverbal subjects. A better understanding of what it might mean to attribute representation or simulation of counterfactuals to nonverbal creatures and what evidence would support such attributions would be very welcome, as a contribution both to psychology and to philosophy.[48]

4.6. Counterfactuals and Causal Judgment: Some Empirical Results

Turning now to some additional experimental evidence for a link between counterfactuals and causal judgment, I begin with a charming set of experiments involving young children described in Harris (2000).[49] Harris presented children aged three to four with a number of scenarios that probed the way in which they connected causal and counterfactual judgments. He found, for example, that when children were presented with a causal sequence (Carol walks across the floor in her muddy shoes and makes the floor dirty) and then asked counterfactual questions about what would have happened under different possible

[48] Another possible source of evidence would be the activation of neural structures in nonverbal animals that are homologous to those known to be activated in adult humans when they reason counterfactually. In humans the relevant structures include the dorsolateral prefrontal cortex and orbital frontal cortex.

[49] These claims about the abilities of young children to reason with counterfactuals and to connect these with causal judgments have been challenged by several authors, including Beck et al. 2011. Recent research instead appears to support the views of Harris and indeed assigns a central role in cognitive development to children's capacity for pretense and counterfactual reasoning (Buchsbaum et al. 2012; Kushnir and Gopnik 2005). It may be that the negative results reported by Beck et al. were due to the use of complex scenarios that children had difficulty understanding rather than to any general inability to engage in counterfactual reasoning (Kushnir, personal communication). Some of the psychological literature on this topic also seems to involve confusions about how counterfactuals work. For more detailed discussion see Woodward 2011b.

antecedents (what would have happened if Carol had taken her shoes off), a large majority give correct answers (i.e., answers that respect the intuitive connection between causal and counterfactual claims). They are also able to discriminate correctly between counterfactual alterations in the scenario that would have led to the same and to different outcomes—that is, which alterations in behavior would have avoided mud on the floor and which would not.

Children do not connect causal and counterfactual claims only when explicitly prompted to do so by a question about what would happen under a counterfactual possibility; they also do this when asked why an outcome occurred or how it might have been prevented. For example, in a scenario in which Sally has a choice between drawing with a pen and drawing with a pencil, chooses the pen, and gets ink on her fingers, children who are asked why Sally's fingers got inky motivate the causal role of the pen by appealing to what would have happened if she had instead used the pencil. Indeed, children spontaneously invoke what would have happened under alternative possibilities in arriving at causal judgments even when those alternatives are not explicitly mentioned in or prompted by the scenarios they are given. Harris's conclusion is that "counterfactual thinking comes readily to very young children and is deployed in their causal analysis of an outcome" (2000, 136). Note that here we again have the idea that counterfactual thinking can facilitate causal reasoning.

These claims about the role of counterfactuals in children's reasoning may seem surprising if one is accustomed, as many philosophers are, to understanding counterfactuals in terms of goings on in Lewisian possible worlds that are distinct from the actual world and similarity relationships among these.[50] Clearly small children (and presumably most adults) don't have anything remotely like Lewis's full framework explicitly in mind when they use counterfactual reasoning. This may or may not tell us something about the "psychological reality" of Lewis's theory, but whatever conclusions one draws about this, we should not allow skepticism about possible worlds to lead us to ignore the everyday uses of counterfactual and causal thinking, by both children and adults, in planning and in anticipating what the consequences of various possible courses of action would be (without necessarily performing the actions in question). These are ordinary, natural, practically useful activities and (as we have seen) ones that even small children appear to engage in. If we think of counterfactuals used for these sorts of purposes (notice, by the way, that these are counterfactuals

[50] Philosophers who think that there is no alternative to understanding counterfactuals in terms of possible worlds are reminded that there are many other devices for representing counterfactual claims, including graphs and equations, as described in Chapter 2. We need not interpret these as carrying with them commitments to Lewisian possible worlds. Indeed there are compelling reasons not to interpret them in this way—see Wilson 2017.

in which the antecedents correspond to manipulations), we should be able to see that there is nothing particularly problematic or obscure about them.

4.7. Experimental Evidence Involving Distinctively Interventionist Counterfactuals

Turning now specifically to the notion of an intervention and its connection to counterfactual and causal judgments, a natural worry is that this notion is too complex and cognitively sophisticated to be psychologically realistic. In assessing this worry we need to distinguish two issues:

(4.7a) Do most people consciously or explicitly represent to themselves the full technical definition of a normatively appropriate notion of intervention when they engage in causal reasoning?
(4.7b) Do people connect the results of interventions, interventionist-type counterfactuals, and causal claims in the way that the interventionist account suggests?

I assume that the answer to (4.7a) is almost certainly no for most people without special training. On the other hand, there is considerable evidence that the answer to (4.7b) is yes for many people at least some of the time. (For more on the contrast between (4.7a) and (4.7b) see Section 4.8.)

To begin with, there is evidence that in a substantial range of situations, adults learn causal relationships more reliably and quickly when they are able to perform interventions than when they must rely entirely on passive observations (Sloman and Lagnado 2005; Sobel and Kushnir 2003). This true for infants as well—Sommerville (2007) reports a series of experiments that show that infants who actively intervene to obtain a toy by pulling a cloth on which it rests learn to distinguish relevant causal relationships between the cloth and toy (e.g., whether the toy rests on the cloth or is merely alongside the toy) more readily than those who rely on passive looking.[51] Adult humans are able to design interventions that allow them to distinguish among different causal structures (e.g., between chain and common-cause structures) and when presented with causal structures are able to use them to predict what the results of interventions on those structures

[51] Remember that the question of whether the infant performs an intervention is different from the question of whether the infant conceptualizes what it is doing as an intervention. If you are inclined to doubt that the infant is genuinely intervening in such a case, consider that the pulling on the cloth will often be "spontaneous" and "random" in the sense that it is not correlated with other environmental events that might cause movement of the cloth independently of the pulling. The case, discussed earlier, in which an infant causes a mobile to move by kicking it and learning the relationships between kicking and the movement is even clearer.

would be—that is, they recognize that in a common-cause structure in which C causes E_1 and E_2, intervening on E_1 will not change E_2, while in a chain structure $E_1 \rightarrow C \rightarrow E_2$, an appropriate intervention on E_1 will change E_2.[52] Indeed a significant number (although by no means all) adults intervene optimally in learning causal structures when given a choice among which interventions to perform, choosing those interventions that are maximally informative about this structure. For example, when presented with a scenario in which there are several possible candidates for the correct causal structure, one of which is a chain structure in which *X* causes *Y* which causes *Z*, people choose to intervene on the more diagnostic intermediate variable *Y*, rather than on *X* or *Z*. (Steyvers et al. 2003). This suggests some appreciation of the connection between interventions and causal structure.[53]

A similar conclusion is suggested by a series of experiments by Lagnado and Sloman (Lagnado and Sloman 2004; Sloman and Lagnado 2005). They report the following: in an experiment in which adults are told that billiard ball 1 causes ball 2 to move, which causes ball 3 to move, almost all judge that if ball 2 were unable to move, ball 1 would still have moved, and that ball 3 would not have. On other hand, when presented with a parallel scenario involving conditionals that lack an obvious causal interpretation and are of the form if p then q, if q then r, subjects' responses are far more variable, with a considerable number willing to infer not-p from the information that not-q. In another words, most subjects endorse the non-backtracking counterfactuals associated with interventionist accounts in the causal scenario but respond differently to non-causal conditionals, where a considerable number do endorse a backtracking, noninterventionist interpretation.[54]

In a second series of experiments, subjects were presented with a chain structure in which they are told that *A* causes *B* which causes *C*. They were then told either (a) "Someone intervened directly on *B*, preventing it from happening" or (b) we "observe" that *B* didn't happen. Again consistently with the interventionist account, subjects treat the "intervention" condition (a) very differently from the "observation" condition (b). For example, they judge that probability of *A* is higher in the intervention condition than in the observation condition—that is, they don't backtrack in the former, and are more likely to in the latter.[55]

52 Assume here that the variables are binary.

53 Another set of experiments that point to a similar conclusion is Waldman and Hagmayer 2005. They presented subjects with a fictitious common-cause model and a fictitious causal chain model involving the same variables. Subjects were able to predict what would happen under interventions on the variables figuring in these models and to use information from interventions to discriminate among them.

54 Note that this is another illustration of the specialness of causation—people treat conditionals describing a cause-to-effect connection very differently from conditionals that do not have such an interpretation.

55 I should emphasize, though, that these are aggregate results. Some subjects are sensitive to the difference between conditioning and intervening and some are not.

In other words, the subjects behave as though they were aware of the normative significance of the difference between observing and intervening and make causal inferences in accord with this normative requirement. This result may be compared with purely associative theories of causal learning, which, as we noted earlier, do not distinguish between observing and intervening.[56]

Taken together, these and other experiments involving more complex causal structures suggest that adult humans do indeed connect counterfactuals that have an interventionist interpretation with causal claims in the way that normative theorizing suggests and that they recognize a difference between, on the one hand, such interventionist counterfactuals and, on the other hand, other sorts of backtracking, noninterventionist conditionals. I emphasize that the results show *both* that adult humans can infer from causal structure to predictions of the outcomes of interventions on that structure *and* that they can use information generated by their own interventions to distinguish among causal structures—that is, they can infer in both directions. I will also note that these results seem inconsistent with claims (e.g., Bennett 1984) in the philosophical literature that people do not preferentially employ non-backtracking counterfactuals in contexts involving causal reasoning.[57]

[56] Again note that the subjects do not seem to be employing "generic" counterfactuals that cover both backtracking and non-backtracking interventionist interpretations. Instead subjects treat interventionist counterfactuals quite differently than other counterfactuals, assuming that backtracking is inappropriate for the former, even if backtracking is appropriate for other counterfactuals, such as those involving inferring backwards from effects to causes as in "diagnostic" reasoning. This goes to the question of whether it makes sense to try to construct a general theory that applies to all counterfactuals rather than a theory that applies distinctively to interventionist counterfactuals.

[57] In my opinion a substantial amount of the discussion of backtracking in the philosophical literature has been maladroit (or at least has lost track of the main issues). Lewis (e.g., 1973a, 1973b) claimed (or at least is often interpreted as claiming) that humans rarely backtrack in evaluating counterfactuals or at least that the "standard" interpretation of counterfactuals is one that does not involve backtracking. This provoked critics to respond that backtracking counterfactuals are common and, in at least some cases, normatively appropriate. From my point of view, there was no reason for Lewis to tie his views to these claims about how "common" or frequent or "standard" backtracking counterfactuals are. What is important for Lewis's account is that backtracking and non-backtracking counterfactuals behave differently (they are evaluated by different criteria) and that it is counterfactuals of the non-backtracking (more specifically, interventionist) sort that are closely connected to causal claims of the Cs causes *E*s sort (that is, cause to effect claims). The experimental results described here support both these claims, even if it is true that, contrary to what Lewis thought, backtracking is common in other contexts. One also encounters in the philosophical literature the anti-Lewisian idea that backtracking is somehow *required* in assessing counterfactuals—that non-backtracking is (always) objectionable, perhaps because it requires the postulation of "miracles" (see Dorr 2016). Putting aside the issues of whether non-backtracking requires miracles and whether miracles are objectionable, the fact remains that people commonly interpret counterfactuals in a non-backtracking way without falling into any obvious incoherence, as the experimental results described earlier illustrate. If people operate with such non-backtracking counterfactuals and these serve their purposes, one could be meant by the claim that all counterfactuals "must" be interpreted in backtracking manner? It seems to me that a more plausible view is this: it is appropriate to use backtracking counterfactuals to answer certain questions such as those having to do with how the past would have had to have been different had the present been different. As noted previously, such backtracking counterfactuals are important in diagnostic reasoning. However, this does not mean that it is misguided to use non-backtracking counterfactuals to answer other sorts of questions such

What about human children? Schulz, Gopnik, and Glymour (2007) investigated the abilities of young children (three to five years old) in a series of experiments, finding that when presented with simple causal structures, their subjects were able both to predict the effects of interventions from knowledge of causal structure and to infer causal structure from knowledge of interventions.[58] Moreover, these authors also find that children are able to employ what they call the *conditional intervention principle* to infer causal structure. To explain this, recall the definition of direct cause from Chapter 2: *X* is a direct cause of *Y* in causal structure G if, holding fixed all other variables besides *X* and *Y* in variable set V at some value via interventions, there is an intervention on *X* that is associated with a change in the value of *Y*. Basically the conditional intervention principle is, as illustrated later, an application of this characterization, describing how one can learn about direct causal relationships from *combinations* of interventions (rather than just single interventions).

In the experiments conducted by these authors, children (three and a half to five years old) were presented with a gear toy that consisted of two interlocking gears A and B and a switch S. When the switch was on, both gears revolved, and when it was off, neither revolved. Each gear was removable but only when the switch was off, so that what the children saw when the toy was intact was either both gears moving together or neither moving. In principle there were four possible causal structures compatible with this behavior: (a) when S is on, this caused A to revolve, which then caused B to revolve; (b) when S was on, this caused B to revolve, which then caused A to revolve; (c) S being on was a common cause of both A and B with neither A nor B causing the other to move; or (d) the switch being on and A being present caused B to spin and the switch being on and B being present caused A to spin.

Note that this is not a case in which spatiotemporal information or information about force transmission/connecting processes or previously acquired mechanism knowledge can be used to figure out the correct causal structure—the available information about this is compatible with each of the possible structures (a)–(d). Similarly for information involving observed correlations with the intact toy.

Nonetheless, the structures may be distinguished on the basis of intervention information. A simple way of determining which structure is correct involving just single interventions would be to, for example, remove gear A while S was on and see if B continued to move and do the same for B. Recall, however, that the

as those having to do with whether *C*s cause *E*s. The two kinds of counterfactuals are just different, with different truth conditions.

[58] The results reported here from Harris 2000 support a similar assessment.

device was set up in such a way that the gears could only be removed when S was off. In this case, a more complicated strategy could be followed, which uses the conditional intervention principle: First, remove gear A but keep B in place while the switch is off, then turn the switch on, and observe whether gear B moves—this tells you whether the switch's being on is a direct cause of B moving. Then turn the switch off, replace A and remove B, flip the switch, and observe whether A moves. Here one is using combinations of interventions on both the switch and one of the gears to determine the causal structure. If, for example, one finds that when A is removed and the switch is on, B does not revolve but that when B is removed and the switch is on, A does revolve, this shows that the correct causal structure is S → A → B—that is, the switch influences B only through A. Remarkably, the children were able to follow this strategy to determine the correct causal structure of the device, thus showing sensitivity to such structural considerations as the difference between direct and indirect causes.[59]

4.8. Interventions and Voluntary Actions

I noted earlier that in many situations subjects (including adult humans and children) make more reliable causal inferences and learn more quickly when they are able to intervene. From a functional or design viewpoint, one thus might expect that subjects will have more confidence in causal inferences and judgments that are directly associated with their interventions and perhaps that some of these inferences will be fairly automatic.[60] This suggests the following hypothesis: human beings (and perhaps some animals) have (i) a default tendency to behave or reason as though they take their own voluntary actions to have the characteristics of interventions (at least when these are not obviously confounded—see the later discussion), and, associated with this, (ii) a strong tendency to take changes that temporally follow those actions with a short delay as caused by them.[61] If (iii) the default tendency in (i) is often correct (or if we are fairly good at recognizing when it is likely to be correct), then, on interventionist principles, (iv) the tendency in (ii) will also often be correct. "Voluntary" here is not meant to suggest anything metaphysically portentous but rather just the common-sense distinction between, for example, deliberately pouring the milk into one's coffee and spilling it in accidentally. I assume that humans and some

[59] However, somewhat discordant results for young children have been reported—for example in Frosch et al. 2012. These authors also speculate about how differences in their experimental design from that employed in Schulz et al. may account for their different results. They do endorse the empirical results reporting a close connection between causal judgment and interventionist counterfactuals, including counterfactuals involving conditional interventions, in adults.

[60] For evidence that this is the case for young children, see Kushnir et al. 2009.

[61] For some suggestions among broadly similar lines, see Glymour 2004.

animals can recognize this distinction,[62] both in their own behavior and the behavior of conspecifics.

I suggested previously that it is not psychologically realistic to suppose that most adults (and still less small children or nonhuman animals) operate with an explicit representation of the full technical definition of the notion of an intervention. Nor is it plausible that most people appeal to this in justifying causal inferences. Thus the idea that people have a default tendency to treat their voluntary actions as interventions should be understood in the following way: they make the kinds of inferences about the causal relationship between those actions and close-by outcomes that temporally follow them that would be normatively justified if those actions had the characteristics of interventions.[63] The existence of causal illusions in which we judge or even "experience" salient changes that follow our voluntary actions as caused by them suggests that such a heuristic is at work.[64]

I have argued that subjects may make inferences of the sort just described without explicitly thinking of their voluntary actions as intervention in the technical sense and without explicitly appealing to an interventionist framework to justify these inferences.[65] For this to be possible subjects must have some way of determining (some signal that tells them) when they have performed a voluntary action. This seems to be true for human subjects—they often have a characteristic phenomenology that is associated with voluntary action, and they typically have a sense of agency or ownership of their behavior that is not present when they act involuntarily. Moreover, although not infallible this signal is fairly reliable. One may conjecture that something like this (or at least the presence of some functionally equivalent signal) is true of some other animals as well.[66] This is not surprising: presumably it is very important for humans and other animals

[62] Oliver Wendell Holmes observes that a dog can recognize the difference between being kicked deliberately and being kicked accidentally.

[63] Consider a perceptual analogy: people, including young children, and animals have a default tendency to trust their senses in typical cases. Obviously this need not require explicit belief about the conditions under which their senses are reliable or anything similar. Operating in accord with this default tendency, these subjects eventually learn things that, in some cases, may provide them with information about reliability and justification in connection with sensory processing under various conditions, but this information is not required for the process to get underway. I start out trusting what my visual experience suggests more or less unconditionally, learn things by proceeding on this basis, and then learn eventually that my visual experiences can be misleading—partially submerged sticks are not really bent. The earlier learning is genuine even if the information about conditions making for unreliablity comes later—indeed the earlier learning may be required for the latter.

[64] An example reporting an actual experience: very shortly after inserting the key to unlock my car door, a car alarm went off in a neighboring car, leaving me with the very strong impression that my action caused the alarm to go off.

[65] Nonetheless there is evidence that adult humans often think of their choices when they deliberate as having some of the characteristics of interventions—see Hagemeyer and Sloman 2009.

[66] See Wegner 2002. Of course, as Wegner documents, there are illusions of agency but their existence does not show (and Wegner does not claim) that the feeling of agency is generally an unreliable clue to voluntariness.

to have some way of distinguishing those cases in which a change occurs in their environments or in their bodies that results from their voluntary actions from those cases in which the change comes about in some other way—not as a result of a movement of their bodies at all, or as a result of a movement that is non-voluntary. It is plausible that one role for the feeling of ownership of one's action is to provide information that helps organisms to monitor this distinction. Once this feeling is available, it may be used for many purposes, including causal inference (cf. Glymour 2004).

For the heuristic under discussion to be normatively justifiable, it is not enough that we infer from voluntary actions to causation. It must also be the case that (as claimed by (iii) in the preceding discussion) our voluntary actions often have the characteristics of interventions. Of course not all voluntary actions are interventions. In a badly designed clinical trial, an experimenter might be subconsciously influenced, in his decisions to give a drug to some patients and withhold it from others, by the health of the patients—his decisions are (in some relevant sense) "voluntary" and yet correlated with an independent cause of recovery in a way that means that the conditions for an intervention are not satisfied. Nonetheless, it seems plausible that many voluntary actions do, as a matter of empirical fact, satisfy the conditions for an intervention of some kind (either soft or hard) or come close to doing so. If I come upon a wall switch in an unfamiliar house and find that there is a regular association between my apparently random flippings of the switch and whether a certain overhead light is on or off, then often enough my flippings will satisfy the conditions for an intervention on the position of the switch with respect to the state of the light. Of course it is logically possible that, for example, a mad scientist has wired up my brain so that he can control when I flip the switch (although I am unaware of this) and at the same time control whether the light is on, thus accounting for the correlation (there being no causal connection between the switch and the light), but, needless to say, this is not an ecologically normal or frequently occurring circumstance. Similarly, if I wonder whether ingesting a certain food is causing my indigestion, I can eat or refrain from eating the food in a variety of different circumstances and see whether indigestion follows—a suitably chosen variety of circumstances can make it unlikely that some confounding influence rather than the food is the cause of my indigestion. Moreover, deliberate planned variation and conscious inference are not necessary for this heuristic to be reliable. As noted earlier, a baby whose leg is attached to a rattle will readily learn to use the rattle to make a noise by kicking. Assuming (as is likely) the baby's initial kicks are unsystematic and random in the sense of being uncorrelated with other causes that might activate the rattle, the kicks will count as interventions, and the baby is warranted in behaving as if its kicking causes the sound—that is, in kicking when it wants the noise to occur.[67]

Going further, it might be conjectured that involuntary behavior is less likely to meet the conditions for an intervention.[68] If this is so, and subjects are guided by the heuristic described earlier, one might expect that the impression of causal efficacy for outcomes following such behavior should be attenuated. Premack and Premack (2003) report this is the case. More recently, Kushnir and Gopnik (2005) found that children are less willing to infer causal relationships involving events that follow their behavior if their behavior is obviously confounded or involuntary—for example, if their hand is moved by another person.

The heuristic just described is one of many that can play a role in "jump-starting" causal learning and also helps to address some worries about "circularity" directed at interventionism as a psychological thesis. If subjects have to possess the full normatively correct notion of intervention to learn about causal relationships from interventions, then (I agree) it is hard to see how they could acquire a full-fledged adult concept or representation of causation via this route since the full notion of an intervention presupposes possession of such a concept. Suppose, however, as argued previously, that one can learn relationships that are in fact causal by performing voluntary actions (where these are in fact interventions) even if not recognized as such. This then provides one starting point for the acquisition of causal knowledge. More generally one can learn representations that have some of the characteristics of the adult causal representations without acquiring representations that have all of these characteristics, acquiring the latter only as the result of a prolonged developmental process.

4.9. Primate Causal Cognition

Nonhuman animals display an impressive range of abilities in various learning tasks, some of which probably reflect associative procedures and some of which may have other bases.[69] Despite this, it is a striking fact that nonhuman animals, including primates, are greatly inferior to humans, including small children, at

[67] Gopnik (e.g., 2012) suggests that a similar observation holds for children's apparently "random" play with an unfamiliar mechanical toy. Bonawitz and Schulz 2007 find that children engage in more exploratory play when evidence is confounded, so that the random or spontaneous nature of the play serves as a strategy for de-confounding.

[68] Whether this is correct is of course an empirical matter. I don't claim that it is obviously so, merely that the conjecture is worth exploring.

[69] The literature on animal causal cognition (and other forms of cognition in the same ballpark) is huge. Needless to say, what follows is not intended as a systematic survey but is rather an attempt to describe some observations that have a bearing on some of the distinctive features of human causal cognition.

many tasks involving causal learning, especially those involving tool use, object manipulation, and an understanding of "folk physics." This is so despite the fact that nonhuman primates and many other mammals not only have capacities for associative learning but also abilities on object permanence tasks and sensitivity to nonphysical behavior (as when one solid object appears to pass through another and so on) that are apparently not so very different from those possessed by human children and adults.[70] This suggests that while these various abilities may well be necessary for the acquisition of the causal learning abilities and understanding possessed by human beings, they are not sufficient. Can we say something about what more might be involved?

In approaching this question, let me begin by briefly describing some representative experimental results involving nonhuman primates. (These results underscore my earlier point that although humans make plenty of mistakes in causal learning and judgment, they are nonetheless strikingly more successful at these tasks than other animals. Again I stress that this is something that needs explanation.) Most of the experiments I will describe involve tool use because these are the abilities on which primatologists have tended to focus. There have been fewer experimental investigations of other sorts of learning and reasoning tasks such as those involving contingency information. Obviously this focus on tool use privileges sensitivity to "mechanical" properties that are important to successful tool use and has other limitations as well.[71]

In experiments conducted by Kohler and subsequently repeated by others, apes (including chimps, orangutans, and gorillas) were presented with problems that required stacking several boxes on top of each other in order to reach a food reward. In comparison with humans, including relatively young children, the apes had great difficulty. They behaved as though they had no understanding of the physical principles underlying the balancing of the boxes and the achievement of structures capable of providing stable support—as Kohler put it, they had "practically no statics" (1927, 149, quoted in Povinelli 2000, 79). The structures they succeeded in building, after considerable trial and error, were

[70] The fact that nonhuman animals have many of the same abilities that human infants and young children demonstrate in tasks having to do with object permanence and sensitivity to nonphysical behavior as revealed in looking-time experiments and yet nonhuman animals perform badly on other tests of causal learning is one reason why, contrary to what a number of psychologists have claimed, the former tasks probably do not reveal anything like a mastery of the full set of capacities that go with adult human causal cognition. In other words, looking-time experiments per se are likely over-interpreted if they are taken to show the presence of adult human causal concepts.

[71] These tool use experiments obviously differ in important ways from many of the experiments discussed previously, which involve inferences based on contingency information, including contingencies produced by interventions. The literature on tool use experiments, at this point, seems largely disjoint from experimental results concerning learning by animals from contingency information. For arguments that the two need to be brought closer together and that reliance on tool use experiments alone has important limitations, see Edwards et al. 2011.

highly unstable, and completely neglected center-of-gravity considerations, with boxes at an upper level extending in a haphazard way far over the edges of lower-level boxes. Subjects even on occasion removed lower-level boxes from beneath boxes they supported. Errors of this sort were made repeatedly, suggesting what from a human perspective would be described as complete lack of insight into the causal principles governing the construction of stable structures. When stable structures were achieved, this appeared to be the result of trial-and-error learning (rather than "insight"). There was little evidence that the apes were able to reason hypothetically or counterfactually about what would happen if they were to create this or that structure, without actually creating the structures in question, and then use this counterfactual reasoning to guide their actions in the way that, for example, the children in Harris's experiments were able to reason.[72] So this is a case in which the apes seem to lack (or are unable to access) certain kinds of "what would happen if" representations that young children use in "insight" learning—or at least the apes are unable to exploit such representations without repeated experience.

In another series of experiments, conducted by Visalberghi and Trinca (1989), a desirable food item was placed in a transparent hollow tube and the animals were given various tools that might be used to push it out. Both apes and monkeys were able to solve some variants of this problem. For example, when given a bundle of sticks that was too thick to fit into the tube, they unbundled the sticks and used appropriately sized sticks to dislodge the food item. On the other hand, they also frequently behaved as though the lacked what from a human perspective would count as a real understanding of the causal structure of the task. For example, they inserted sticks that were too short to reach the reward when a stick of appropriate length was available. They attempted to use sticks with crosspieces that blocked insertion into the tube. They also inserted non-rigid objects like tape that were incapable of displacing the food. In still other experiments, the animals failed to choose implements with a hook at the end, which would have been effective in retrieving desired objects instead of straight sticks, and they chose nonrigid implements when rigid tools were required.

Povinelli's summary is that the animals "appear to understand very little about why their successful actions are effective" (2000, 104). In particular, according to Povinelli, the animals appeared to not understand the significance of the mechanical properties of the systems they were dealing with—properties such as weight, rigidity, shape, center of mass, and so on. Instead, as both Povinelli

[72] For what would count as counterfactual reasoning on the part of nonverbal animals, recall the discussion in Section 4.5. It is unclear what accounts for the difference between apes and children. Does the fact that the children can use language (even if their linguistic abilities are inferior to those of an adult) facilitate their use of counterfactual or "What would happen if..." reasoning? Can other systems of representation besides natural language facilitate counterfactual reasoning?

(2000) and Tomasello and Call (1997) remark, they often acted as though (any) spatiotemporal contact between the target object they wished to manipulate and the means employed was sufficient to achieve the desired manipulation.[73]

Both Povinelli (2000) and (Tomasello and Call) 1997 go on to suggest a more general characterization of the deficits exhibited in the experiments: they claim that these stem from the animals' lack of various abstract concepts having to do with "unobservables" (Povinelli 2000, 300, mentions gravity, force, shape, and mass, among others) that humans think of as mediating causal relationships. According to these writers, in contrast to humans, apes operate entirely within a framework of properties that can be readily perceived, and this underlies their lack of causal understanding.

Philosophers of science are likely to find this invocation of "unobservables" a bit puzzling, and not just because of general skepticism about the observable/unobservable distinction. If we think of a property as "observable" for a subject as long as the subject can reliably discriminate whether or not it is present (or among different values if the property is quantitative) by sensory or perceptual means, then (one might think) it is not clear that properties like weight or rigidity are literally unobservable by apes—surely an ape can recognize the difference between a light or heavy object when the animal lifts it. There is, however,

[73] Note that if the apes move the tool employed as a means, there will be "force transmission" of some kind (or energy/momentum transmission) between the tool and target. This illustrates the great difference between, on the one hand, information about whether there is a causal (or some sort of connecting) process linking tool and target, and, on the other, an understanding of how to produce desired changes in the target by manipulating the tool. The latter is arguably bound up with some sort of grasp of interventionist counterfactuals and is a central component of human causal cognition. To expand on this point: suppose that we interpret causal process theories of the sort advocated by Salmon/Dowe as (in part) theories about how subjects (human and otherwise) represent and reason about causal relationships. I suggest that the experimental results about primate cognition just described show an important limitation in causal process theories construed in this way. In particular, it is striking that the limitations on the apes' understanding described in these experiments does not seem to be *just* a matter of their failure to grasp the abstract notion of a causal process (as a process that transmits force, energy etc.) or an inability to recognize particular instances of such a process in these system of interest. As I noted in Chapter 2, as a conceptual matter, grasp of the notion of a causal process and recognition of its presence is *not* sufficient for the sort of detailed knowledge of dependency relationships that is required for successful manipulation in tasks like balancing boxes or extracting food from a tube. (This was the point of the transmission of chalk / pool cue example in Chapter 2.) What needs to be explained is the apes' lack of this latter sort of knowledge (dependency information required for successful manipulation). Whenever a primate moves a food source with a stick—whether it is pushed in an appropriate or inappropriate direction or with an appropriate instrument—there will be transmission of "force" and energy, the presence of a mechanism, and so on. A creature that could recognize when force/energy was being transmitted and whose heuristic was "To cause a desired outcome, transmit force to the outcome (or the object associated with the outcome)," would not get useful guidance from this heuristic about exactly what it should do to balance boxes in the stacking task or to expel food from the tube. To accomplish this, far more specific information about how the outcome that the agent wishes to affect depends on variation in other factors (including factors that are not linked too closely to egocentric sensory experience) that the agent is able to control is required. This looks more like information of the sort represented by (TC) and (DC) (Chapter 2) than information about force transmission.

an alternative way of understanding this claim that makes it more plausible.[74] Suppose that when an ape learns to discriminate among objects according to (what we would call) "weight," the discrimination is made on the basis of sensory feedback and bodily sensations associated with differential effort in lifting. If apes' "concept" of weight is very closely linked to these bodily sensations, then it becomes more understandable why they are apparently unable to make use of information about weight in other sorts of contexts requiring causal reasoning—why, for example, they are unable to recognize the relevance of weight to support relationships. To recognize the relevance of weight to these contexts requires possession of a more abstract way of thinking about weight that is not so closely tied to sensory and motor experience. Similarly for properties like rigidity. Thus the sense in which these properties are not "observable" by apes may be that the apes' representation or categorizations of these properties is such that those representations do not allow for certain kinds of learning and reasoning or exportation to new circumstances in which the relevant sensory/motor feedback may not be available.

On this way of thinking about the matter, the apes (in comparison with humans) operate with the "wrong variables" to enable the sophisticated causal learning required for the tasks described earlier. Instead their variables are too closely linked to egocentric sensory experience. From the perspective of the interventionist account, we might describe this as a situation in which certain intervention-supporting relations cannot be learned by the apes because the variables in terms of which those relations are framed are unavailable to the apes.[75] Whether or not this analysis is accepted, it seems clear, as a more general point, that whatever the apes' grasp of notions like weight and rigidity, they do not understand their causal relevance to the tasks with which they are dealing and cannot integrate these notions into causal representations that successfully guide action in connection with those tasks.

The idea that the apes lack the right variables (and hence fail to learn dependency relationships based on those variables) gives us one way of explaining at least some of their deficits in causal understanding. Here is an alternative line of argument, which I see as complimentary to and not in competition with the "wrong variables" analysis, and which also fits naturally into an interventionist framework (as well as making use of some additional ideas from Tomasello and Call). This focuses on three different sources of difference-making information and, associated with these, three different possibilities for learning representations that

[74] Thanks to Daniel Povinelli for a very helpful conversation that corrected a serious misunderstanding of his views.

[75] Recall the remarks on variable choice in Chapter 1, and for related discussion, see Woodward 2016.

have cause-like aspects even if not they are not necessarily fully causal (where, as explained above, adult humans provide a standard for this).

(4.9a) Let us say that S is an *egocentric proto-causal learner*[76] if S is capable of learning contingencies between S's own actions and outcomes caused by those actions, as when a baby learns that kicking its foot will move a mobile to which the foot is attached. One might think of such an egocentric learner as learning that (or as having representations to the effect that) if I do X, goal G follows, and if I don't do X, G does not follow. Ordinary operant or instrumental conditioning (or learning) falls into this category, but I leave it as an open question whether there are egocentric forms of learning action/outcome associations that do not involve operant conditioning.

(4.9b) S is an *agent-causal learner* if S can learn about causal (or cause-like) relationships both from action/outcome contingencies involving its own actions/interventions and by observing action/outcome contingencies involving the interventions of other agents and if, in addition, S is able to integrate the information from these two sources in the sense that S is able to appreciate that the outcomes of others' interventions have implications for what would result from S's own interventions. In other words, S is able to learn, by observing that some other subject S*'s action A produces outcome O, that S itself could produce O by doing A.

(4.9c) S is an *observation/action* (OA) *causal learner* if S learns action/outcome contingencies involving its own actions; action/outcome contingencies involving the actions of other agents; and contingencies deriving from patterns of covariation in nature that may occur "naturally," rather than being produced by any agent and, moreover, suitably integrates these, regarding each as a source of information about the other. In other words, S is both an agent-causal learner and regards the results of observational learning not involving agents as relevant to its own action-outcome learning.

I assume that adult humans are, at least in many cases, OA learners. This claim is supported by empirical evidence (some reviewed previously) that adult humans are able to learn about causal relationships both from their and others' interventions and from observations not involving interventions and to integrate information from both sources into unified representations. Relatedly, it

[76] If you don't like the label "proto-causal," perhaps thinking there is nothing "causal" about what is learned by the learner described under (4.9a), just substitute a label like "action/outcome" learner. Similarly, substitute whatever label you want for "agent causal learner."

also seems uncontroversial that adult humans think of causal relationships in terms of the assumption that the very same relationship can be present between their own actions and outcomes, between others' actions and outcomes, and between naturally occurring events not involving the actions of any agents. Thus when I put water on a plant and it grows, when I observe you put water on a plant and it grows, and when I observe rain falling on a plant and it grows, I assume that the very same sort of causal relationship between water and plant growth is present in all three cases and that observation of any one of these cases can furnish information about the others.

However, it seems logically or conceptually possible for a creature to be an egocentric proto-causal learner only and not an agent-causal or OA learner. Such a creature would be able to learn contingencies linking its own actions to the outcomes they produce, but either (i) would not be able to learn about causal relationships from the interventions of others or from covariational information not involving other agents or (ii) would not be able to put together what it learns from its own interventions with what it learns from covariational information from other sources. Thus such a creature would not, for example, infer from covariational information from other sources to how such information might be relevant to producing desired outcomes from its own interventions.

It also seems conceptually or logically possible that a creature might be only an agent-causal learner and not an OA learner. Such a creature could learn from its own interventions and from observing the outcomes of the interventions (or manipulations) of other agents and could represent that the very same relationship that is present between A and O when it does A can also be present when another agent does A, but does not learn from observations not involving other agents and does not represent that the same relationship between A and O that is present when another agent does A can also be present in nature, independently of the activities of any agent.[77]

Indeed, these are not just logical possibilities. Tomasello and Call (1997) suggest that, as a matter of empirical fact, apes are not OA causal learners in the sense described here, even though they are presumably (at least) egocentric proto-causal learners. That is, although they learn from the results of their own

[77] To avoid confusion let me emphasize that what characterizes an agent-causal learner is the sources of information from which such a learner can learn. The notion of an agent-causal learner is not meant to suggest that the adult human notion of causation is somehow reducible to or acquired just from the experience of agency, as is advocated by agency theories of causation. There are many reasons, discussed in Woodward 2003 and 2009 as well as in what follows for rejecting such a view. Indeed, the point of introducing the notion of an agent-causal learner is to make it clear that such a learner possesses something less than the full adult human representation of causation. However, I take it to be fully consistent with this that learning from one's own interventions and by observing the results of the interventions of others plays an important role in the acquisition of the full adult capacity for causal reasoning—as argued throughout this chapter, there is a great deal of empirical evidence that this is the case.

interventions, and also track naturally occurring covariation, they do not move back and forth between these, applying the results of observations of naturally occurring covariation to the design of their own interventions. They illustrate by means of the following thought experiment:

> We are not convinced that apes need to be using a concept of causality in the experimental tasks purporting to illustrate its use, at least not in the human-like sense of one independent event forcing another to occur. More convincing would be a situation in which an individual observes a contiguity of two events, infers a cause as intermediary, and then finds a novel way to manipulate that cause. For example, suppose that an individual ape, who has never before observed such an event, for the first time observes the wind blowing a tree such that the fruit falls to the ground. If it understands the causal relations involved, that the movement of the limb is what caused the fruit to fall, it should be able to devise other ways to make the limb move and so make the fruit fall. . . . we believe that most primatologists would be astounded to see the ape, *just on the basis of having observed the wind make fruit fall*, proceed to shake a limb, or pull an attached vine, to create the same movement of the limb. Again, the problem is that the wind is completely independent of the observing individual and so causal analysis would have to proceed without references to the organism's own behavior and the feedback it might receive from that (thus, it might be able to learn to shake the limb if its own movements had previously led to a limb shaking and the fruit falling as a result). Moreover, performing some novel behavior to make the fruit fall would involve an even deeper causal analysis of the web of possible ways that the cause could be repeated so as to reinstate the desired effect.

Tomasello and Call (1997, 389) make use of the diagrams in Figure 4.2 to clarify what they are claiming about the apes' understanding.[78]

If the ape were able to infer on the basis of observing the wind shake the fruit loose that it could intervene (or manipulate) to shake fruit loose, it would, to this extent, exhibit an ability characteristic of an OA learner. Tomasello and Call suggest that apes will not be able to perform this task—thus that they are not OA learners but instead either agent or egocentric causal learners. To the best of my knowledge experiments that would test these claims have not been done with nonhuman primates.[79] They would be worth doing.

[78] The second diagram in Figure 4.2 represents an example in which an animal ("self") is able to recognize that a second animal can be caused to be afraid by the fall of a rock, a predator, or a noise made by self.

[79] Some years ago I attended a workshop at which both Povinelli and Call were present. My recollection (which may well be faulty) is that I suggested to Povinelli that it would be worth doing an

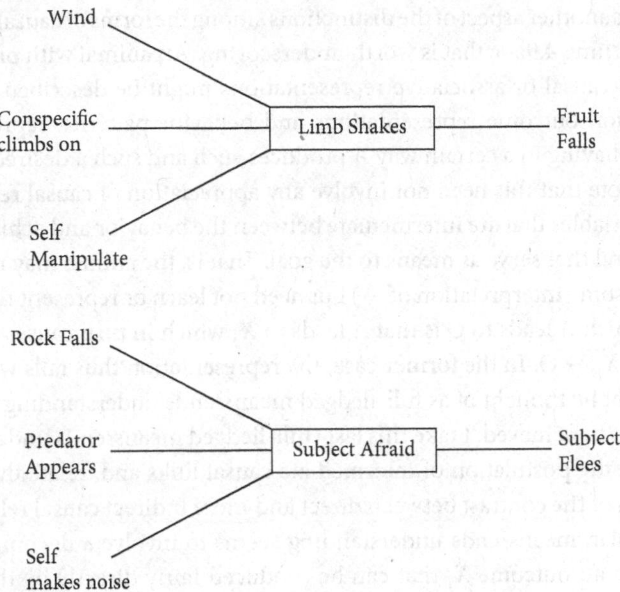

Figure 4.2

The distinction between these three forms of learning and representation suggests that susceptibility to operant/instrumental conditioning by itself shows at best that the subject is an egocentric proto-causal learner; perhaps with some sort of egocentric proto-causal representations or alternatively, representations of associations. Susceptibility to instrumental conditioning does not by itself show that the subject is capable of OA learning or forms of representation associated with this.[80]

actual experiment of the sort described here to see what the result would be. He responded that it was so obvious that apes would succeed at the task (i.e., would behave as OA learners) that it was not worthwhile to actually do the experiment. I then repeated this to Call, who responded that it was so obvious that apes would not succeed at the task that it was not worthwhile to actually do the experiment. So as far as I know, the experiment remains a thought experiment.

[80] Some readers may be aware of a form of learning that occurs in humans and other animals called Pavlovian to Instrumental Transfer (PIT) and wonder whether this is relevant to the issues under discussion. In a typical PIT experiment an animal is trained to associate a stimulus *S* with a reward via classical conditioning and also trained to produce some action *A* (e.g., lever pressing) that leads to the same reward via instrumental conditioning. The presentation of *S* is found to increase the action *A*, so that to that extent there is transfer between the two kinds of conditioning. The kind of transference discussed in the preceding text is different from this: imported into an associative learning framework, the question is whether, when the animal is capable of producing some outcome *X*, and *X* is associated with a reward R via classical but not instrumental conditioning (i.e., the animal is not subjected to instrumental conditioning in connection with the *X-R* relationship), the animal produces *X* in order to get *R*.

There is another aspect of the distinctions among the forms of causal and proto-causal learning 4.9a–c that is worth underscoring. An animal with only egocentric proto-causal or associative representations might be described possessing fused action-outcome representations and behavior patterns: representations that its behaving in a certain way *A* produces such and such a desired outcome/goal *O*. Note that this need not involve any appreciation of causal relationships among variables that are intermediate between the behavior and achievement of the goal and that serve as means to the goal. That is, the animal may represent $A \rightarrow O$ (for some interpretation of $\rightarrow$) but need not learn or represent that, say, the way in which *A* leads to *O* is that *A* leads to X_1 which in turn causes *O*—that is, that $A \rightarrow X_1 \rightarrow O$. In the former case, the representation thus falls well short of what might be thought of as full-fledged means/ends understanding of how the goal might be achieved. I take this last (full-fledged means/ends understanding) to involve the postulation of intermediate causal links and, relatedly, some appreciation of the contrast between direct and more indirect causal relationships. In particular, means/ends understanding seems to involve a decomposition of a task into an outcome X_1 that can be produced fairly directly by the subject's action *A* and a further outcome or goal *O* that is more directly caused by X_1 and less directly by *A*. Note that this last link ($X_1 \rightarrow O$) and similar links involving other intermediate variables) are not action-outcome links; instead they are links between events or variables that are not actions or manipulations. Thus to learn such relationships the subject must have representations of dependency relationships that exist in the world independently of its own actions and in this respect must have some of the capacities of an OA learner.

In Tomasello and Call's diagram, the intermediate variable X_1 is described by the variable "limb shakes," and this in turn causes the outcome described by "fruit falls." As is apparent from their diagram, it is the introduction of this intermediate variable that makes possible (or corresponds to) the recognition that there are different ways (involving both actions and events occurring in nature) in which the same goal (fruit falls) might be brought about, all of which have in common the fact that they operate through the intermediate variable "limb shaking."[81] Consider, by contrast, an animal that only has a representation connecting its own action on the limb to the production of the desired goal state but does not decompose this causal sequence into intermediate steps in the manner described earlier. In other words, the animal has only a representation of the form, "If I do *A*, desired outcome *O* results" with no representation

[81] Note also that there is nothing "unobservable" (in any ordinary sense) about this intermediate variable—that is, if the apes fail at the task under discussion, it is not because they fail to postulate *unobservable* intermediate variables but rather because they fail to recognize the relevance of an observable intermediate variable.

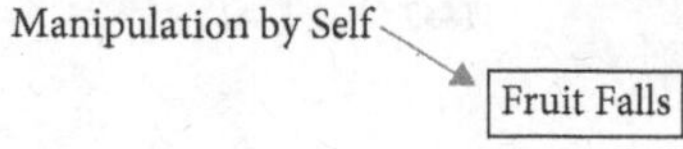

Figure 4.3

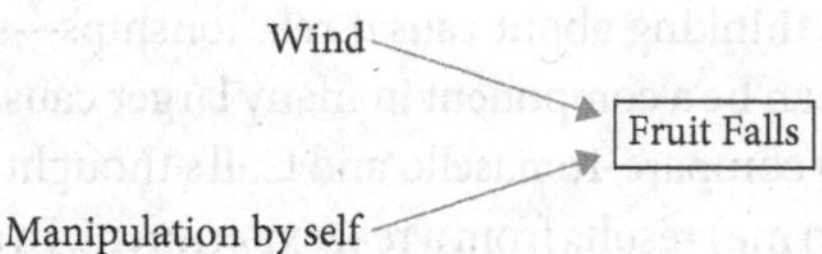

Figure 4.4

of intermediate causal links or means leading from *A* to *O*. In Tomasello and Call's example, this animal's representation would look something like Figure 4.3 (where the directed arrow needn't mean "causes" but may rather represent something like "brings about," having only some of the features of a fully causal representation). or perhaps Figure 4.4.

In general, the postulation of intermediate links (and with it an appreciation that causal relationships can be more or less direct) as in Figure 4.2 goes hand in hand with a decoupling of the final outcome goal from the means (intermediate variables) used to achieve it as well as a move away from a purely egocentric representation of dependency relationships. Note also that this requires representation in terms of variables that are not tied too directly to the animal's own egocentric sensory experiences: a variable the content of which corresponds to something like "I have sensory experiences associated with difficulties in lifting" will not be a good candidate for a variable that can figure in the characterization of a dependency relation that is appropriately self and agent independent. (So in this respect the diagnosis of the apes' failures in terms of their having thc wrong variables and the present diagnosis are mutually reinforcing.) Note also that once the role of the $X_1 \rightarrow O$ link is recognized (that is, the intermediate link X_1 between *A* and *O*), as in Figure 4.2, this results in a representation that is more generalizable (and which consists in links that are likely to be more *invariant* in the sense described in Chapters 2 and 5) than those representations that look like Figures 4.3 and 4.4.[82] With a representation that incorporates the information in Figure 4.2, one is able to see that any prior event, whether an action or natural occurrence, that leads to limb shakes will produce the same result. Thus means/ ends decoupling tends to be associated with identification of relationships that

[82] More distal and indirect causal relations are not always less invariant than the intermediate links that compose them, but they often are—there are more ways for the latter to break down, as noted in Chapter 5.

$I \to T \quad L \to F$

Figure 4.5

are more stable or invariant. Means/ends understanding also goes hand in hand with compositional thinking about causal relationships—seeing how the same causal relationship can be a component in many larger causal structures.

It is interesting to compare Tomasello and Call's thought experiment with the surprising (at least to me) results from a real experiment. In a study conducted by Blaisdell et al. (2006) rats were first exposed to an observational learning phase (i.e., learning that did not involve interventions) in which, it was claimed, the rats acquired a common-cause model in which a light (*L*), was represented as a common cause of both a tone (*T*) and whether or not food was present (*F*): $T \leftarrow L \rightarrow F$.[83] In subsequent tests, when the rats were presented with the tone, they behaved (according to the experimenters' interpretation)[84] as though they believed that food was present (they increased their search for food, as measured by nose poking), which is of course consistent with their adoption of the common-cause model. In the next, crucial "intervention" phase of the experiment, a lever was introduced, the pressing of which by the rats caused the tone to be presented. In this case, the rats were less inclined to search for food after tones caused by the lever press, despite the fact that tone and food were associated in the observational phase of the experiment. This of course is the normatively appropriate behavior if the rats grasped the causal structure of the situation they were dealing with: when the rats intervened on the tone, this "broke" the connection between the tone and the light and rendered the tone statistically independent of the food presentation (Figure 4.5).[85]

If these experimental results are taken at face value, they do show, as the authors claim, that rats can, in some respects, "distinguish between causal and spurious correlations" and "that they are capable of deriving predictions for novel actions after purely observational learning" (Waldmann et al. 2008, 469), although the "prediction" in this case concerns the absence of a correlation between their lever presses and the presentation of food. Note, however, this experiment does *not* provide evidence that rats can design interventions to achieve novel goal objects on the basis of purely observational information—that is, that they can do what

[83] Obviously the attribution of representations that are in any sense "causal" to rats is controversial. I will not enter into discussion of the appropriateness of this here, since my focus is more on intervening/conditioning contrast in connection with rat behavior. However, rat learning is surprisingly sophisticated and does exhibit some features often associated with causal cognition—see Pineno 2010.

[84] This interpretation has not gone unchallenged. Bowers and Timberlake (2018) contend that nose-poking is not a good measure of whether the rats expect food.

[85] This relies on the "arrow breaking" interpretation of interventions described in Chapter 2.

Tomasello and Call describe in their thought experiment, inferring, for example, on the basis of observations of the wind shaking the tree branches and the fruit fall that if they were to shake the branches, that would make the fruit fall. It is this latter sort of integration that is required for full OA learning.[86] Indeed, Blaisdell et al. also trained rats on a "chain" model in which (according to the authors) the rats acquire a representation in which *T* causes *L* which in turn causes *F*. They do *not* report that after observing the association between *T*, *L*, and *F*, the rats spontaneously intervene on *T* in order to get *F*, which they presumably want. It is this sort of observation that would provide evidence that the rats produce interventions to achieve goal objects on the basis of purely observational information. Thus, even given the results of Blaisdell's experiments, the possibility remains that the ability to integrate (or at least fully or extensively integrate) causal information based on relationships associated with interventions with causal information based on purely observational patterns of information into a common system of representations is one of the features of adult human causal cognition that separates us from most or all other animals.

But what about young human children? The conceptual distinctions among egocentric, agent, and full causal learners suggests the possibility that there may be a stage in human development in which these abilities dissociate, with children behaving as egocentric or agent but not full causal learners. An experimental paper on which I was a coauthor (Bonawitz et al. 2010) reports results addressing this question. Compressing greatly (and omitting a number of other questions explored in the paper), the relevant results were as follows: we employed an experimental paradigm in which children in two different age groups (toddlers vs. preschoolers: mean age 47.2 vs. 24.4 months) observed a block slide toward a toy airplane that activated when the block touched the base of the airplane. In one condition (the so-called ghost condition), no agent was involved in the movement of the block—it appeared to move spontaneously, activating the plane on touching. In another condition (the "agent" condition), an experimenter moved the block, demonstrating that it would activate the plane on touching. The children were then asked to make the airplane go by themselves, after being given experience with moving the block so that it was clear that they knew how to do this. A very large majority of preschoolers were able to use their previous observations of the association between the movement of the block to

[86] Also relevant to this question are the results in Fawcett et al. 2002. Starlings who were presented with a trained conspecific that performed one of two actions in manipulating a plug on a bottle in order to obtain a food reward were more likely to reproduce that action themselves in order to obtain the reward. However, when presented with a "ghost" condition in which the plug moved in similar ways spontaneously (i.e., not as the result of the action of a conspecific), the starlings did not act so as to produce the plug motion they had seen. The starlings thus behaved like agent-causal but not OA learners. Thanks to Alison Gopnik for drawing my attention to this paper.

design an intervention of their own to activate the airplane by moving the block. By contrast, none of the toddlers were able to do this, despite the fact that we had independent evidence that they had learned (from the ghost condition) that there was a predictive association between movement of the block and plane activation. We knew that they had learned such a predictive association because the children engaged in predictive looking when exposed to the association between the movement of the block in the ghost condition and activation of the plane—seeing the movement of the block, they looked to the plane as though they expected it to activate. But despite having learned this predictive association, they were unable to use it to design successful interventions. By contrast, many of the toddlers were able to design an intervention of their own to use the block to activate the plane when they had previously observed the experimenter activate it. That is, they were able to learn from another agent but not from the information in the ghost condition.

Although the matter needs more detailed investigation, this is prima facie evidence that there is a stage in the development of human causal cognition in which children (toddlers in this case) are able to learn to design their own interventions from observations of the interventions of other agents but not from the observation of otherwise similar associations not involving interventions. In other words, at this point the toddlers appear to be in something like an agent-causal stage of learning. This is then followed by a stage in which fully causal learning is achieved, as evidenced by the behavior of the older children. This interpretation is consistent with a great deal of other developmental evidence showing that even very young children seem to be sensitive to the difference between agents and non-agents in their environments and are primed for or adept at learning from the behavior of other agents (particularly the intentional or goal-directed actions of other agents) as opposed to other sources of information. One possibility (cf. Meltzoff 2007) is that as soon as children represent their own actions, they also represent the actions of others in a common amodal code, and hence they are never in a purely egocentric stage of causal learning.

This investigation also provides another example of an experiment that is motivated or suggested by a normative theory: unless one has the philosophical/computational ideas that there are important connections between causation and intervention, that some human actions are paradigmatic interventions, and that causal learning from both interventions and observations is possible, one is unlikely to undertake the sort of investigation just described. Given these ideas, it becomes natural to ask how and when in development information from interventions is integrated with other sources into more unified causal representations. Note, though, as suggested earlier, that this is not a matter of the experimental results providing "evidence" for interventionism, construed as a normative theory—instead the normative theory plays something more like a motivating or enabling role and provides a framework for interpreting the experimental results. Again, normative theory can be relevant to the design of experiments and interpreting experimental results without those results "confirming" the normative theory (whatever that might mean).

Several of the experiments described here may also be used to illustrate how influence can flow in the opposite direction—how empirical results can have implications for philosophical theorizing. Consider an "agency" theory of causation like that defended by Menzies and Price (1993), according to which our concept of causation is derived from our subjective experience of agency, which is then "projected" onto the world. When this is interpreted as an empirical claim, experimental results about learning from interventions suggest there is something in the neighborhood of this claim that is correct, in the sense that our capacities to act as agents and to manipulate plays an important role in the development of our capacities for causal learning and cognition. On the other hand, these experimental results also suggest the need for a somewhat more complex story than the one told by Menzies and Price. To see this, recall a standard philosophical criticism of agency theories: they face problems explaining how the notion of causation comes to be extended to unmanipulable causes such as earthquakes, where no one has the relevant experience of agency—a criticism that Menzies and Price address. However, the experimental results discussed previously suggest that a related problem arises, so to speak, much earlier—there is already a problem about how causal learner X moves from her own subjective experience of agency to a recognition that causal relations are present when other agents act and produce effects, since in such cases X presumably does not have the experience of (her own) agency. Moreover, there is a parallel problem about cases in which the subject infers the existence of a causal relationship involving causes that are straightforwardly manipulable, but which are not in fact manipulated by any agent, as when wind shakes fruit from a tree. Similarly for cases involving intermediate variables that are not actions as when "limb shakes" is conceptualized as causing "fruit falls"—this too cannot be directly understood in terms of the experience of agency. "Projection" can function as a label for our willingness to extend the notion of causation from our own manipulations to such cases, but as best I can see, it provides no insight into how we develop the capacity to do this or the factors influencing such development. (Nor does it help us to understand why it is normatively justified to make this extension.) Moreover, if Meltzoff is correct, it may be an empirical mistake to suppose that young humans begin the process of developing the full suite of abilities associated with human causal cognition in a purely "egocentric" stage centered around their own experience of agency—instead we may be agent-causal from the outset. If so, although agency may be crucial to the development of causal cognition, it may be that what is central is not just the *experience* of agency, but also other features associated with the capacity to act effectively as an agent, such as the ability to represent means/ends structures of actions exhibited both by self and by others. If so, this should inform philosophical theories that connect causation and agency.

5
Invariance

5.1. Introduction

This chapter is perhaps the most "philosophical" or "theoretical" one in this book. It is an exploration of the notion of invariance and the role it plays in causal analysis. I focus on how this notion is to be understood, the different forms it can take, and why it is important from a normative perspective. The following two chapters (6 and 7) then look at more specific applications of these ideas and empirical research relevant to the role that invariance-based considerations play in causal reasoning. This chapter thus sets up a framework for understanding and evaluating the more specific empirical models discussed in the following chapters. As before, I follow a "rational model" approach, assuming that because the search for invariant relationships of various sorts makes normative sense, it is often reasonable to expect that, as an empirical matter, the identification of such relationships and reasoning in terms of them will play an important role in people's causal cognition.

My focus on invariance is motivated by the claim that we—ordinary people and scientists alike—value invariant relationships (in the various forms and versions described in what follows) and are justified in doing so. Other things being equal, to the extent that we are interested in causal analysis, we try to discover invariant relationships (and relationships that are more rather than less invariant), we choose variables and functional forms (see subsequent discussion) and other conceptualizations that best enable the formulation of relevantly invariant relationships, we judge in accord with measures of causal strength that reflect the extent to which relationships are invariant, and in general our causal thinking is structured by and tracks a concern with invariance. Moreover, these practices are rational in the sense that they are effective means to goals that we have, including goals having to do with prediction, control, and generalization of causal relationships to new contexts. Since, other things being equal, it is good or desirable to discover relationships that are relatively invariant along various dimensions, such relationships are often regarded as "good" or "paradigmatic" causal relationships in comparison with causal relationships that satisfy minimal conditions for causation such as (M) (Chapter 2) but are comparatively less

Causation with a Human Face. James Woodward, Oxford University Press. © Oxford University Press 2021.
DOI: 10.1093/oso/9780197585412.003.0006

invariant.[1] As we shall see, it is the more invariant relationships that, other things being equal, tend to receive higher "causal strength" ratings in empirical studies.

5.2. Some Preliminaries

The generic idea of invariance is the notion of a relationship that remains the same and "continues to hold" or "operate"[2] across various sorts of changes. It is thus akin to (certain understandings of) notions like stability, robustness, and insensitivity that have figured in recent philosophical discussion.[3] Recall from Chapter 2 that I take invariance under at least some interventions to be a necessary condition for a relationship to qualify as causal or nomic at all. If, say, C is a candidate cause variable that can take at least two values c_1 and c_2 and E the candidate effect variable with at least two values e_1 and e_2, (and, for simplicity, confining ourselves for the moment to deterministic relationships that are representable by some function), to qualify as causal at all the relationship $E = f(C)$ must be invariant under at least some interventions that change the value of C from c_1 to c_2, where this means that f correctly describes how the value of E changes in response to this change in C. If f does not describe how E changes in response to changes in C for any interventions that change values of C, as is the case for the relationship between barometric pressure and the occurrence or nonoccurrence of a storm, then there is no causal relationship between C and E. These are the minimal conditions for causation (and its absence) described by (M) (Chapter 2). Given a relationship that satisfies this minimal condition, we can then go on to ask about exactly which changes of various sorts are such that the relationship is invariant under those changes. For example, if $E = f(C)$ is invariant under some interventions that change the value of C—say interventions that change C from c_1 to c_2—then (assuming that C and E are multi-valued rather than binary) we can ask whether f is also invariant under interventions that change C to various values other than c_1 and c_2, we can ask whether f is invariant under various changes in background variables (conceived here as roughly variables that are distinct from both C and E and not causally between them),

[1] Because of an editing error the reference to (M) was removed from Chapter 2. The reader should think of (M) as just (TC) in Chapter 2 or as the minimal commitments of an interventionist account of causation as described in Chapter 2 or in Woodward, 2003.

[2] I refine what this means in what follows. In previous work I have often described invariance in terms of whether a relationship "holds" across certain sorts of changes. This now strikes me as insufficiently precise, since it fails to indicate whether condition (i) alone holds: (i) the relationship fails to be violated under those changes or whether instead both (i) and (ii) hold, where (ii) is that the relationship continues to be operative across those changes. I intend the latter interpretation—both (i) and (ii). (ii) is elucidated later in this Section.

[3] I've tended to use "invariance," "stability," and (sometimes) "insensitivity" somewhat interchangeably in past books and papers. In what follows I will use the word "invariance" (and sometimes "stability") for the general notion and "insensitivity" for the special case of invariance under changes in background conditions.

and so on. As explained later, in some but not all cases, it will be possible to assign some sort of measure or, alternatively, a basis for ordinal comparisons that allows us to talk of the size or importance of the range of changes over which *f* is invariant: *f* may be invariant under a relatively "large" or "important" range of changes or under a "larger" or "more important" range of changes than some other function *g* relating *C* to *E*. Alternatively *f* may be invariant (again in the sense of satisfying (M) for some interventions) but under only a very narrow range of changes/interventions, in which case it is "fragile" or "relatively nonrobust" (see Chapter 6 for more discussion).

The picture of invariance I advocate thus has both a threshold and, above this, the possibility of further variations or "degrees." Some relationships are not invariant under any interventions and thus are below the threshold—they are not invariant at all. Above this threshold, relationships will be invariant, but they also can differ in the range of circumstances or changes under which they are invariant. Differences in the extent to which relationships satisfying (M) are invariant thus capture one way in which relationships that qualify as minimally causal can nonetheless differ along important dimensions relevant to their role in causal analysis. This framework thus allows us to capture two commonly accepted ideas about causal relationships. On the one hand, we think of the question of whether a causal relationship is present between two variables as requiring a yes- or no-type answer—either a causal relation is present or it is not. On the other hand, when causal relations are present, we recognize further distinctions among them, and this is reflected in judgments that some causal relations are "stronger," more informative, or more paradigmatically causal than others—here we make judgments that admit of degrees. (Specific illustrations of what this involves are provided later.) Invariance is one but not the only dimension along which such variation occurs.

In what follows, I will typically assume (unless I indicate otherwise) that we are dealing with relationships that satisfy the minimal condition for invariance (invariance under some interventions) and thus are "causal." I will explore the various ways those causal relationships can nonetheless differ with respect to the sorts of changes over which they are invariant.

Slightly more precisely, the basic idea of differences in invariance that we will be working with is the notion of causal relationships that differ in the extent to which they remain "operative" across different circumstances or contexts. A causal relationship is relatively more invariant to the extent that the candidate cause acts in the same way with respect to the effect of interest across different circumstances, where "act in the same way" means that the candidate cause continues to cause that effect. This is a more subtle notion than might initially appear. First, (i) it of course requires that the candidate relationship not be violated or break down across the circumstances over which it is invariant. Second (and more subtly), (ii) it also requires that the candidate relationship continue to apply—that is, to characterize what is going on, what causal relationship is operative—across the changes in question. "Continue to apply" should be understood as implying that the causal factor operates in the unified way described by the relationship across the changes in question.[4] A relationship is relatively more invariant to the extent that the causal factors figuring in the relationship "carry" their ability to produce the outcome described by the relationship across different circumstances.

To illustrate the difference between (i) and (ii), consider the following example, which is discussed in more detail later. The gravitational inverse square law,

$$F = Gm_1m_2 / r^2, \tag{5.1}$$

satisfies the minimal condition of being invariant under some interventions—for example, it will continue to hold under some interventions that alter the distance between the gravitating masses and the values of those masses. Some philosophers (e.g., Cartwright 1983; Giere 1999) contend that (5.1) should be interpreted as describing the relationship between the total force between two masses and the values of the variables on the right hand side of (5.1)—that is, that F in (5.1) represents the total force. Under this interpretation—call it (5.1a)—(5.1) will be false whenever there is some other force between the two masses, such as an electromagnetic force, that is nonzero. Under this (5.1a) interpretation, there is a violation of requirement (i) in these circumstances—(5.1) is violated or falsified and hence fails to be invariant when other forces are present. Another possible interpretation of (5.1)—call it (5.1b)—also advocated by some philosophers, takes (5.1) to describe the relationship between the variables on its right-hand side and the total force when no other forces except gravitational forces are operative between the two masses. Thus (5.1) is interpreted to mean something like "When no other forces besides the gravitational force are operative between two masses, the force between the masses is given by $F = Gm_1m_2 / r^2$." (Cartwright and Giere sometimes write as though they adopt this interpretation.) Under this (5.1b) interpretation, (5.1) is not false when other forces are present, but it fails to apply or to characterize what is going on (or even part of what is going on) in such circumstances. Here we have a failure of invariance in the sense of (ii)—the story about what is going on when no other forces besides gravity are present does not describe the causal or explanatory factors that are operative when other forces are present.

Finally, we might interpret (5.1) as (5.1c), describing the relationship between the *gravitational component* F_g of the total force due to the masses m_1 and m_2 and the values of the right-hand side variables in the equation, and we might also

[4] As an illustration (in addition to the one immediately following) there is a sense of "apply" according to which it might be argued that the conjunction of, say, Hooke's law and Boyle's law "applies" to a wider set of systems (e.g., springs and gases) than either conjunct does alone. But this conjunction does not describe a relationship that operates in the same way across these two kinds of systems.

interpret (5.1), so understood, as continuing to apply even when other forces are present. When such other forces are present, the gravitational component of the total force continues to be given by $F_g = Gm_1m_2 / r^2$, with the total force given by the vector sum of all of the forces, gravitational and non-gravitational, between the two objects, but with (5.1) no longer understood as describing this total force. Under this last interpretation (5.1c), the force law is invariant across changes in whether other forces are present, but under the first two interpretations, (5.1a) and (5.1b), it is not invariant under such changes. In other words, under interpretation (5.1c) the gravitational force law "continues to apply" when other forces are present. Thus (5.1) interpreted as (5.1c) is (in this respect) invariant under a larger range of circumstances than (5.1) interpreted as (5.1a) or (5.1b)—larger in the straightforward sense that the circumstances under which (5.1a) and (5.1b) are invariant are a proper subset of the circumstances under which (5.1c) is invariant.

I will suggest later that this last interpretation (5.1c) is the most appropriate interpretation of the inverse square law. Notice also (to anticipate discussion that follows) that this example illustrates one respect (of many) in which the claim that a relationship is invariant across some set of changes is *not* equivalent to the claim that it is exceptionless—either in general or with respect to those changes. When interpreted as (5.1b), (5.1) does not have exceptions (is not violated or falsified) when other forces are present since under interpretation (5.1b), (5.1) is explicitly restricted to cases in which no other forces are present. Nonetheless when interpreted as (5.1b), (5.1) is not invariant across changes in whether (and which) other forces are present since it does not apply to such circumstances or describe what is going on in them.[5] For the sake of brevity, I will sometimes describe invariance as a matter of a generalization's "continuing to hold" across various changes, but this (now) should be understood as meaning both that the generalization is not violated in such circumstances *and* that it applies (describes the causal or explanatory relations that are operative) in such circumstances.

The Newtonian inverse square law is not invariant (it fails to hold) under certain conditions such as very strong gravitational fields, when general relativistic effects become important. Some philosophers will go further, saying that the inverse square law is not literally true anywhere since general relativistic effects are always present even if tiny. I am going to use the notion of invariance in such a

[5] Here is a slightly different way of thinking about the distinction just described. Say that generalization G is violated (or breaks down) in circumstances C if the antecedent of the generalization is satisfied by some system in circumstances C and its consequent is false or violated in those circumstances—thus (5.1) interpreted as (5.1a) is violated when other forces are present. Say that a generalization G fails to apply in circumstances when its antecedent fails to be satisfied in circumstances C—(5.1) interpreted as (5.1b) fails to apply when other forces are present. Invariance of G in circumstances C requires *both* that G not be violated in those circumstance and that G apply in those circumstances.

way that a generalization counts as invariant ("continues to hold") within a certain domain or regime as long as it is what scientists would describe as a "valid" or "effective" generalization within that regime, as the Newtonian inverse square law is for many applications.[6] I will also sometimes also express this by saying that generalizations like the Newtonian inverse square law are "correct " within that regime. Those who want to interpret "correct," "effective," and "valid" in terms of approximate truth are welcome to do so.

Now consider another example. Suppose relationship R between *C* and *E* [R: ($C \rightarrow E$)] is intervention-supporting and thus describes a causal relationship between *C* and *E* (i.e., R holds) in background circumstances B_1 but that in background circumstances B_2 this relationship is violated or breaks down—*C*s are followed by *E*s in B_1 but not in B_2. Suppose also that B_1 and B_2 are mutually exclusive and exhaustive—they cover all possible background circumstances. Now consider the relationship R*: $C.B_1 \rightarrow E$. R* "applies" in the sense of correctly describing what causal relations hold or are operative in exactly the same circumstances as R does—both apply in B_1, but not in B_2. However, the claim associated with R* is (we are supposing) correct and the claim made by R is false since R is "violated" in B_2 (R falsely claims that *E*s are always followed by *C*s). So in this sense $C \rightarrow E$ is deficient in comparison with $C. B_1 \rightarrow E$, in virtue of consideration (i), presented earlier.[7]

Next let me emphasize that as I will understand invariance, the notion has counterfactual or modal import. What matters for the invariance of some relationship R is whether *if* certain changes *were* to occur or circumstances were to be different in some way, R *would* continue to hold, where this includes changes that do not actually occur, as well as changes that do occur. We will see later (especially Chapter 7) that, as an empirical matter, people seem to operate with such a modal conception of invariance when they use invariance-based considerations to guide their causal judgments.[8] Thus generic invariance claims of form

Relationship R is invariant under changes (or across circumstances) *X*

have two placeholders that need to be filled in: we need to specify (1) what relationship R we are talking about and (2) the sorts of changes in *X* over which (it is

[6] One motivation for this usage is that, as discussed in subsequent sections, all known laws and causal generalizations are merely effective.

[7] Perhaps it would not be natural to say that $C \rightarrow E$ is "less invariant" than $C.B_1 \rightarrow E$, but $C \rightarrow E$ is still deficient in the way described—it fails one of the two conditions relevant to invariance.

[8] This is not to say that people pay no attention to (or give no special weight to) considerations having to do with invariance under actually occurring variation but rather that they also consider what would happen under variations that do not actually occur. Moreover, these factors seem to be weighted differently for different kinds of causal claims. For example, actual variation seems to count for more in connection with actual cause claims. See subsequent sections.

claimed) that relationship would continue to hold.[9] There are a number of possibilities for both (1) and (2), all of which I see as different but legitimate aspects (or varieties or components or dimensions of) of invariance. We can spell out the content of different invariance claims by being more specific and precise about (1) and (2).

Note that on this conception invariance claims are (even when the relationship R is specified) always relative to some specified set of changes *X*: relationship R might be invariant with respect to one set of changes or circumstances *X* but not with respect to some different set of changes *X**. Indeed for known invariant relationships, including current candidates for fundamental laws, this is always or virtually always the case—such candidate laws are merely "effective." I emphasize also that in the cases with which I will be concerned, unless explicitly indicated otherwise, it will always be an empirical matter whether some relationship is or is not invariant under some specified set of changes or circumstances.

5.3. Which Relationships Are Invariant?

With respect to (1) in Section 5.2 (which relationships are invariant?), one possibility (the possibility that I have been assuming so far) is that the relevant relationship is (or is described by) a particular *function*. For example, in the case of the gravitational inverse square law, it is the functional relationship $F_g = Gm_1m_2/r^2$ that is invariant under many changes (in the values of the gravitating masses or the distances between them, as well as changes in background conditions—see the subsequent discussion). This is understood to mean that this functional relationship continues to hold in the sense of correctly describing the gravitational force between two masses across such changes.

Another possibility is that the relationship in question has to do with the quantitative values of the *conditional probabilities* associated with a cause *C* and an effect *E*: for example, when *C* is a binary variable, the conditional probabilities *Pr*(*E*/*C*) and *Pr*(*E*/*not C*), might continue to hold under various sorts of changes such as changes in *Pr*(*C*)—the unconditional probability of the cause—or across various changes in background conditions. Note, though, as moment's thought will show, in the common situation in which there are other causes X_i that affect the probability of *E*, *Pr*(*E*/*C*) and *Pr*(*E*/*not C*) will not be invariant under changes in these X_i. It might be the case, however, that the conditional probabilities *Pr*(*E*/

[9] Here I assume that in specifying relationship R we don't need to build into R all of the circumstances in which it holds or fails to hold—the latter information if available at all is specified separately as the range or domain of invariance of R. This is defended in more detail in Section 5.7.

$C.X_i$) etc. are invariant under changes in the value of C when C is a direct cause of E and the X_i are all the other direct causes (parents) of E that are distinct from C.

Yet another possibility is that the relationship that is invariant is more qualitative. For example, in the case of binary variables C and E with values {present, absent}, we might find that interventions that change the value of C from absent to present increases the probability that E = present in circumstances B_i in comparison with the probability of E when C = absent in B_i: $Pr(E/C = \text{present}.B_i) > Pr(E/C = \text{absent}.B_i)$ over a variety of changes of various sorts including changes in B_i, even though the precise amount of this increase varies across such changes. Here what is invariant is the qualitative fact of probability increase across different background circumstances. Still another possibility is that what is invariant is simply that Y is some nontrivial function or other of X (in the sense that at least two different values of X are mapped into different values of Y, so that in this sense Y depends on X, in accord with (M) but where the function linking X to Y itself changes (while remaining nontrivial) under some set of changes (either in the values of X or in background circumstances). Here what is invariant is simply the fact that Y continues to depend on X under those changes. In a graphical representation in which X and Y are the only variables, this would correspond to the case in which the graphical structure $X \rightarrow Y$ is invariant under those changes, even though the precise function (the form of the function or its parameterization) relating these variables is not invariant. More generally, some more complex graphical structure might be invariant under a set of changes in the sense that the structure would continue to hold even though the functions corresponding to the arrows in the graph change: we might have the structure $X \leftarrow Z \rightarrow Y$ holding invariantly under various changes (e.g., changes in the values of X, Y and Z and certain structural changes elsewhere in the graph) even though the exact functions linking Z to X and to Y fail to be invariant across these changes.

Another possibility (discussed more extensively in Chapter 6) is that what is invariant is some relationship of counterfactual dependence between C and E (e.g., in the simplest case that $C \Box\!\!\rightarrow E$ and *not* $C \Box\!\!\rightarrow$ *not* E, where $\Box\!\!\rightarrow$ is the counterfactual conditional). Yet another possibility (developed in detail in Cheng's causal power theory described in Chapter 7) is that what is invariant is the probability that C if present would *cause* E.[10]

As we shall see, as a general rule, the formulation of more invariant relationships typically involves movement away from relationships that relate more directly observable quantities to relationships that are more abstract and less tied to immediate observation.[11] Thus, to anticipate an example discussed in more detail later (but already implicit in the gravitational force law example),

[10] Note that this is different from $Pr(E/C)$ since C and E can be present and yet C can fail to cause E.

[11] We also noted this in connection with the discussion of primate causal learning in Chapter 4.

the relationship between a cause and its total effect (i.e., the overall effect that the cause may produce by multiple routes) is typically (but not always)[12] less invariant than the relationship between the cause and the specific components or contributions along a path (assuming these exist) the cause contributes to the effect. The latter will also often be more abstract or removed from direct observation as illustrated by the contrast between F_g and the total force experienced by some object. Similarly relationships that are fully describable in terms of frequency information about the co-occurrence of cause and effect such as $\Delta p = Pr(E/C) - Pr(E/not\ C)$ are, if invariant at all, often less so than invariant relationships that are not so describable, such as Cheng's notion of causal power discussed in Chapter 7. A concern with invariance (and finding more invariant relationships) is thus one route leading to abstraction in science.

5.4. Under Which Changes Is the Relationship Invariant?

When we turn to question (2)—under which changes is the relationship invariant?—in Section 5.2, there are also many different possibilities. Here I lay out some elements of a basic taxonomy before turning to a more extended discussion in subsequent sections. Needless to say, the list that follows is not meant to be exhaustive; it rather describes some possibilities that arise both in science and in more ordinary contexts. These possibilities encode more specific expectations about the kinds of invariances we sometimes find in nature—as explained subsequently, we can think of these as default expectations or as inductive biases. The following example (or perhaps more accurately analogy) may help to illuminate what I have in mind: laws of nature satisfy (and are expected by physicists to satisfy) various invariance conditions—time translation invariance, invariance under spatial translations, Lorentz invariance, among others. It is of course an empirical question whether any particular candidate for a law satisfies such invariance conditions. Nonetheless all fundamental laws do satisfy many of these conditions, and the conditions in question are also used to guide the search for new laws—there is a kind of default expectation that new laws will satisfy such conditions as well, although this default is defeasible.[13]

[12] "Not always" because it seems possible that when *C* affects *E* via different routes, the overall effect remains the same, even if contributions along the different routes change. Some claim that distal causes of diseases sometimes have this property.

[13] This illustrates another respect in which a concern with invariance differs from a general concern with finding relationships that are as nearly exceptionless as possible—expectations about invariance are more specific expectations of various sorts about particular ways in which relationships will not have exceptions, at least in certain domains. For example, when a system conforms in certain circumstances to a generalization that is a candidate for a law, we do not expect to produce a violation of the generalization simply by spatially translating the system, even if there are other circumstances in which the generalization is violated.

I see many of the invariance conditions described later as playing (albeit more loosely) a similar role in causal reasoning in sciences outside of physics and in ordinary life. Nothing guarantees that as an empirical matter it will always be possible to formulate generalizations and causal relationships in such a way that these are invariant under the various sorts of changes I will discuss. Nonetheless, it turns out that in many cases it *is* possible to formulate generalizations meeting some or many of these requirements. Moreover, the empirical evidence suggests that people (both ordinary folk and scientists) often expect that this will be possible: that is, in formulating generalizations and causal relationships, people often first try out or explore possibilities that meet certain invariance requirements—these are treated as "defaults"—and only retreat from these when the empirical evidence requires this.[14] In this way, expectations about invariance help to guide the discovery of new causal relationships (including new variables that can be used to formulate such relationships) and help to structure the ones we do discover, just as invariance requirements in physics play a role in the discovery of new laws.[15]

5.4.1. Invariance under Interventions

I suggested earlier that to qualify as invariant at all a relationship must be invariant under some interventions on variables figuring in the relationship. Assuming that this minimal condition is met, one can then ask whether that relationship is invariant under some specified *range* of interventions on those variables—for example, under interventions that set those variables to different values within that range. (i.e., noting that $Y = f(X)$ is invariant under interventions that set $X = x_1$, we can ask whether this relationship is invariant under interventions that set X to values other than x_1.) One typical kind of case involves a relationship R holding in some particular set of background circumstances B_i, where these B_i are only loosely specified—for example, B_i may describe a human body in normal physiological condition and R may relate injection S with a saline solution with a certain concentration (the independent variable) to some measure of thirst T (the dependent variable). The question of interest is whether R (which may be a particular functional relationship $F(S) = T$ between S and T or something vaguer

[14] This is a theme emphasized by Patricia Cheng—see Cheng and Liu 2017.

[15] There is, however, a difference in degree between, so to speak, the stringency with which the invariance requirements are enforced in the two cases. Candidates for laws in physics that violate standard invariance requirements in physics may be rejected as bona fide laws on those grounds alone, unless there is overwhelming evidence for them. Except for the requirement of invariance under interventions, candidate causal relations in common sense and elsewhere in science are much less likely to be rejected just because they violate one of the particular invariance requirements discussed in what follows.

such as whether T increases with increasing S) is (at least) invariant under some range of interventions setting S to different values in B_i—of course it may be invariant under other conditions as well.

As another illustration, the inverse square law is invariant under a considerable range of interventions on the variables r, m_1, and m_2 figuring in that law—that is, the law continues to hold for a range of interventions that change the distance between the masses m_1 and m_2, under a range of interventions that add mass to one of the gravitating objects, and so on. Of course, as noted earlier, this law is invariant under many other changes as well.

I emphasize that as I propose to understand the notion of invariance, there is no requirement that to count as invariant a relationship must be invariant under *all* interventions on all possible values of the variables figuring in the relationship. It is enough if there is invariance under *some* interventions setting those variables to some values. For example, the relationship between the extension X and the restoring force F described by Hooke's law,

$$F = -kX, \tag{5.2}$$

for a suitably behaved string is certainly causal, and this is reflected in the fact that for a range of interventions on X, the restoring force will behave in accord with (5.2). On the other hand, if the spring is extended too much it will break: (5.2) is not invariant under such extreme interventions on X. Even if it does not break, it will respond to some interventions in a nonlinear way that does not conform to (5.2). In a similar vein, although the gravitational inverse square law is invariant under a large range of interventions on the variables r, m_1, and m_2, it is not invariant under all such interventions—as noted earlier, for suitably large gravitational fields (among other conditions), the Newtonian gravitational force law fails to hold. On my view, this is consistent with this generalization describing a genuine nomic relationship (and arguably a relationship that is causal in some suitably broad sense). Indeed, as remarked earlier, it is plausible that virtually all laws that are presently known have this feature of failing to be invariant under some possible changes. For example, the field equations of general relativity are widely believed to be invariant under changes or circumstances that are above the length scales at which quantum gravitational effects are unimportant ("above the Planck length") but to break down below these length scales. Similarly the fundamental equations in current formulations of quantum field theory are widely thought to break down at energy scales that are currently experimentally inaccessible. Talk of generalizations being invariant under some (perhaps very wide) range of circumstances but not under all circumstances is a natural way of capturing this. Conceptions of laws as exceptionless generalizations is not.

5.4.2. Invariance under Changes in Variables Figuring in a Relationship When These Changes Are Not the Result of Interventions

Consider a relationship R between X and Y, $Y = f(X)$, that is invariant under some range of interventions that set X to values $x_1, \ldots x_n$ and hence counts as causal. One can also ask whether R is invariant under those same values of X when these are results of causal processes that do not qualify as interventions. Of course one would like the answer to this question to be yes—the impact of X taking the value $X = x_i$ on Y should just depend on the value of X, and not on whether that value is produced by an intervention or in some other way. One may think of this as a kind of invariance condition on R, with the condition having to do with whether R is invariant across changes in how the values of X are produced, whether by an intervention or not. Call this *value invariance.* It is certainly logically possible for a relationship to be invariant under some interventions but fail to exhibit value invariance. However, when this appears to be the case, my view is that this failure is generally taken to provide a reason to try to reformulate the relationship and/or the variables figuring in it so that value invariance is satisfied. That is, we expect to be able to formulate relationships in such a way that if they are invariant under interventions at all, they will also satisfy value invariance for the values for which invariance under interventions hold. (This illustrates a general theme to which we will recur: the role of invariance considerations in leading us to formulate relationships or to choose variables in such a way that they satisfy invariance requirements.)

As an illustration of the significance (and interpretation) of value invariance, consider the following set of linear relationships,

$$Z = aX + bY, \tag{5.3}$$

$$Y = cX \tag{5.4}$$

and the corresponding graph (Figure 5.1).

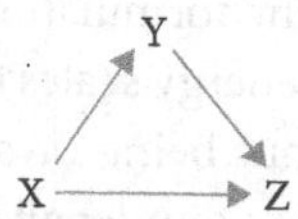

Figure 5.1

Suppose that an intervention on Y with respect to Z occurs that changes Y's value by Δy and that (5.3) is invariant under this intervention, so that the resulting change in Z is $b\Delta y$. Now suppose that the same change Δy in the value of Y occurs as a result of a change of magnitude $1/c\Delta y$ in the value of X. Such a change in Y will not count as an intervention on Y with respect to Z because the change also affects Z through a route (the direct route from X to Z, also represented by (5.3)) that does not go through Y. Moreover, the total change in Z that results from this change in X will reflect both the direct influence of X on Z and the indirect influence of X on Z that goes through Y. The total change in Z will be $(b + a/c)\Delta y$, which of course is different from the change in Z resulting just from the intervention on Y, which is $b\Delta y$. Does this mean that Z has a different effect on Y depending on whether the value of Z is due to an intervention or instead produced by a change in X, so that value invariance is violated? Not according to what our equations say if they are interpreted correctly. The effect of Y on Z is the same in both of the cases described earlier and is captured by (5.3) (and by asking what the change in Z would be under an intervention on Y). In adopting the representation (5.3)–(5.4) we decomposed the total effect of X on Z into components in a way that allows us to think of the effect of Y on Z as invariant and independent of how the value of Y is caused. If we did not do this—if instead we held that the effect of a change of Y on Z was whatever total change in Z was associated with some change in Y, regardless of how the change in Y was caused—we would have defined the effect of Y on Z in such a way that it was highly non-invariant: the magnitude of this effect would vary depending on how the change in Y was caused and on the larger causal structure in which the $Y \rightarrow Z$ link was embedded. Among other considerations, this would make it difficult to generalize or extrapolate the $Y \rightarrow Z$ link to new situations. Thus the decomposition into components (5.3)–(5.4) allows for the possibility of a more invariant representation—one that satisfies value invariance, at least in some range of circumstances. Examples like this reflect the powerful impact of invariance-related considerations on how we think about causal representation.

There is more to be said about value invariance. Suppose it is not true that the effect of X taking the value x on Y is the same regardless of whether $X = x$ was set by an intervention or a nonintervention or that we adopt a representation that does not satisfy this requirement. Another disadvantage of permitting such violations of value invariance is that this fails to capture the connections we naturally expect between causal relationships that hold in experiments (when interventions occur) and causal relations holding in contexts in which interventions are not present—it fails to capture standard expectations about the way in which results from each of these contexts transfer to the other. In other words, when failures of value invariance are present, discovering that Y responds according to some relationship $Y = f(X)$ under interventions setting X to various values would warrant no inferences at all about what the causal relationship between *Y* and those same values of *X* when these are not the results of interventions.[16]

Although, as I said earlier, I do not claim that this is impossible, we certainly have a strong default presumption that relations in our world generally do not behave this way and a corresponding preference for representations that do not exhibit this sort of failure of value invariance. A world in which value invariance systematically fails would be a world (and presumably an accompanying conception of causation) in which, when we intervene experimentally, we inevitably change or distort the targets of our interventions in a way that leads them to behave quite differently than they do in non-experimental contexts, so that we learn nothing from experimental results about causal relationships in non-experimental contexts. This seems contrary to how modern science's conception of how the world and causal relationships behave and certainly contrary to the presumption that one can learn about causal relations in non-experimental contexts from experiments (and vice versa).[17]

Indeed, as noted in Chapter 4, as a matter of descriptive psychology, it is a highly nontrivial feature of how adult humans think about causation that this involves the idea that the very same relationship can be present between two variables both when we intervene and when these variables take values "naturally" as a result of nonintervention processes and that we can use information gathered from one of these contexts to make inferences about causal relationships in the other. In fact, as we have seen (Chapter 4), this may be a feature that is distinctive of the adult human capacity for causal thinking—there is evidence that the causal cognition of very young children and nonhuman primates may not possess this feature.

[16] Illustration: suppose we are investigating a structure in which *C* causes *X* and *C* causes *Y*, with *X* and *Y* correlated because of the common cause *C*, but we don't yet know whether there is also a causal relationship from *X* to *Y* or from *Y* to *X* (or neither). We intervene on *X* (*Y*) and discover that under all such interventions there is no change in *Y* (*X*). Ordinarily we would infer that *X* does not cause *Y* and *Y* does not cause *X*. But in the absence of some invariance principle connecting the effects of *X* and *Y* when they are intervened on to their behavior in nonintervention contexts, it would be possible to hold that, for example, when *C* causes *X* and *C* causes *Y* (i.e., when the common-cause structure is intact), *X* causes *Y* but the link from *X* to *Y* always disappears whenever one intervenes on *X*, disrupting the link between *C* and *X*. This is obviously relevant to some of the issues surrounding the status of the Causal Markov condition in common-cause structures.

[17] There are some researchers who might be interpreted as claiming that there is no systematic relationship between causal relationships that hold under experimental manipulations and relations that hold when there is no such manipulation. For example, Rubin (2013) objects to adopting a causal interpretation of mediating variables (i.e., variables that are causally between the cause *C* that is subject to experimental manipulation and its eventual effect *E*) even in experimental contexts in which *C* is manipulated on the grounds that the mediating variables themselves are not directly manipulated but only observed. On one way of interpreting his objection, because the mediating variables are not themselves directly manipulated, we have no basis for claims about the causal relationships in which they stand since observational evidence provides no secure source of information about the behavior of these variables under manipulation. Again, I claim that this is contrary to how we usually think about causation.

5.4.3. Invariance under Background Conditions

For present purposes, let us think of background conditions simply as factors that (i) are not explicitly represented in the relationship $X \rightarrow Y$ whose invariance we are interested in assessing but where (ii) we do *not* include in background conditions, variables, or conditions that are causally between *X* and *Y*. That is, "mediating" variables that are effects of *X* and causes of *Y* are not treated as background conditions.[18] For example, the colors of the gravitating masses in equation (5.1) ($F = Gm_1m_2 / r^2$) are not explicitly represented in this relationship and thus are background conditions. Obviously (5.1) is invariant under changes in this set of background conditions.

As another illustration, consider the relationship

Striking a match causes it to ignite (5.5)

(interpreted here in accordance with (M) as "ignition is counterfactually dependent on striking, when counterfactual dependence is interpreted in terms of interventionist counterfactuals"). Here the presence of oxygen is a background condition, and (5.5) is not invariant under changes in whether or not oxygen is present. As this example illustrates, many causal generalizations employed in ordinary life omit information about relevant background conditions—relevant in the sense that the generalizations fail to be invariant under at least some variations in those conditions. Invariance considerations figure in the normative assessment of such omissions. Other things being equal, a generalization of form *Cs* cause *Es* will be preferred (judged more normatively acceptable) to the extent that *Cs* cause *Es* in a wide range of (omitted) background conditions B_i even if not all of them. Similarly if those B_i over which the relationship holds are much more common (or perhaps more "normal" in other ways) than those B_i for which it fails to be invariant, the causal relation is more likely to be preferred, perhaps especially in ordinary life contexts.[19] For example, although (5.5) fails to hold when oxygen is absent, such circumstances are relatively rare in ordinary life,

18. This is a rough-and-ready characterization designed to capture the intuitive idea; there are subtleties that I lack the space to explore. I will add that we should *not* include in background conditions "lower level" variables the values of which "realize" *X* and *Y*. For example, a given value of temperature for a gas may be realized by a large number of different combinations K_i of values of the kinetic energies of the individual molecules making up the gas. These K_i are not background conditions for the relation between temperature and other thermodynamic variables. The appropriate way to think about such realizing variables is discussed in Chapter 8 and in more detail in Woodward, forthcoming b.

19. As noted later, in other sorts of contexts (e.g., physical theorizing) this consideration typically matters less.

and (5.5) is invariant or more nearly so across many common circumstances in which oxygen is also present.

As noted previously, we might address the failure of invariance present in (5.5) by building the presence of oxygen into the relationship of interest, replacing (5.5) with

Striking a match in the presence of oxygen above level L causes ignition (5.5*)[19]

We will see, however, that this move of dealing with failures of invariance by incorporating additional background conditions is not always illuminating or desirable.

The contrast between asking whether a relationship is invariant under some interventions on variables figuring in an relationship and asking whether, in addition, that relationship is also invariant under changes in background conditions (or under interventions performed under different background conditions) is related to (I don't claim it is identical with) the commonly drawn distinction between *internal* and *external* validity in connection with causal claims, especially when internal validity is established on the basis of experimental manipulations. Suppose that an experiment is conducted to determine whether, for some set of experimental subjects S, all of whom have headaches, treatment with drug D relieves headaches. Let us make the unrealistic simplifying assumption that the experimental subjects are homogeneous. That is, all respond in the same way to the drug so that if there is a positive "average causal effect,"[21] this describes the response of each individual given the drug. The internal validity of the experiment has to do with whether it establishes the correct conclusion about the causal efficacy of the drug for the subjects in this particular experiment. This is a matter of establishing that if an increase in headache relief occurs, this is due to the treatment and not due to chance or to some confounding factor. Although this is not completely uncontroversial, appropriate experimental manipulations corresponding to interventions (e.g., involving randomized assignment) are generally thought to provide reliable evidence for claims about internal validity in this sort of case. This is also an implication of the interventionist principle (M). External

[20] Recall that (5.5*) is preferable to (5.5) because of consideration (i) in Section 5.2: the former is true and the latter false.

[21] Average causal effect (ACE) is a commonly used measure of causal efficacy in such contexts. In an experiment with no confounding it can be estimated by the difference between the mean response to the drug in the treatment group and the mean in the control group. As noted in what follows, in the more realistic case in which there is some heterogeneity in subject response to the drug, the numerical value of the ACE will of course fail to generalize to populations with a different distribution of subjects.

validity has to do with whether this result about the causal efficacy of the drug among the experimental subjects generalizes to other subjects, including those with different characteristics (e.g., subjects who differ in age, gender, or ethnicity) from those in the experimental group as well as subjects in different environments or background conditions.[22] External validity (or one important aspect of it) thus has to do whether the relation between D and headache relief persists across changes in these characteristics and background conditions.

Suppose one establishes via experimentation or in some other way that D does cause headache relief among the original subjects. Obviously it is a different question—and not one that is answerable just from the experimental result—whether the drug also causes relief for subjects with different characteristics or under different background conditions. The distinction between invariance under at least some interventions (which establishes that a causal relationship exists but possibly only in some very specific context) and invariance under various additional background conditions is meant to map onto the difference just described.

So far I have been talking, in a deliberately vague way, about whether "the causal relationship" between drug and recovery holds for the experimental subjects and for other subjects. Even if we assume that all subjects in the experimental group respond in the same homogeneous way (so the ACE is the individual effect), there are many different possible "causal relationships" whose invariance we can assess. Suppose, to take just one possibility, the drug reduces the incidence of headache via several different pathways or routes—it has a direct biological effect by causing blood vessels in the head to become less constricted, but it also causes subjects to exercise more, with exercise also reducing the incidence of headache. One might be interested in whether the magnitude of total effect of the drug on headache (as measured by, e.g., the probability of no headache, given ingestion of the drug, in comparison with the probability of no headache given no ingestion) along all of these different routes is invariant across different populations of subjects, but one might also be interested in whether the effects along different routes are invariant, assuming that the situation is such that we can make sense of and get information about this last question.[23] As suggested earlier, it is often reasonable to expect that total effects will be less stable or invariant than at least some effects along routes when generalizing to new background circumstances. This might happen, for example, if the relation between drug

[22] I assume that variation along such demographic characteristics should be distinguished from variation in background circumstances—very roughly background conditions are "external" to the individual units or subjects in a way that, for example, gender is not. I leave open the question of how we should think about this distinction. Of course we are interested in invariance across variation in characteristics of units as well as invariance across changes in background conditions.

[23] See Pearl 2001 for a discussion of when various path-specific effects are identifiable.

ingestion and exercise and/or exercise and headache relief changes in the new background circumstances, while the relationship associated with the direct biological effect of the drug on headache remains stable in the new circumstances.

To take another possibility concerning what is meant by "the causal relationship," it is, as noted earlier, common in many experiments in biomedical contexts (as well as in many other disciplines) to focus just on the "average effect" of the manipulated cause. In the case under discussion this corresponds to whether the incidence of headache is lower in the treatment group that receives the drug than in the control group of headache sufferers who do not receive the drug. Obviously, the drug can reduce the incidence of headache on average even if it is inefficacious or headache promoting for some subjects. Supposing that, on average, the drug reduces headache; one can then ask whether the drug has a similar average effect (either in magnitude or in sign) on other populations of subjects. Of course if subjects are heterogeneous in their response to the drug, with, say, the drug acting to reduce headaches among most of those in the original experiment but promoting headaches among a minority, the average effect of the drug will not generalize well to new populations that have a different distribution of subjects in terms of drug response. That is, in the presence of heterogeneity, average effect may be a less invariant measure of causal efficacy than alternative measures.

So far I have focused on invariance under at least some interventions, value invariance and invariance under background conditions. However, there are a number of other respects in which a relationship can be (or fail to be) invariant. I turn next to a discussion of several of these.

5.4.4. Invariance and Interaction

Recall the example from 5.2 in which the relation between the gravitational component *Fg* of the total force on an object m_1 due to a second mass m_2 at a distance r is invariant under changes in whether other forces such as a Coulomb force F_e are present. We may think of such other forces as one kind of background condition but note how they contrast with the role played by the presence or absence of oxygen in match ignition, which was also treated as a background condition. Intuitively, this contrast has to do with the fact that the Coulomb force makes an independent contribution to the total force experienced by the object, as reflected in the fact that we calculate the total force by adding the separate vectorial contributions of the gravitational and Coulomb forces. This addition operation reflects the fact that there is no interaction between the two forces: as Novick and Cheng (2004) argue and as is discussed in more detail in Chapter 7, when variables are continuous and have no ceiling or upper bound, some form of additivity is the natural way of capturing the idea that causal factors operate independently with respect to some outcome of interest and that each makes an independent invariant contribution to that effect.[24]

By contrast, the presence or absence of oxygen does not make an independent contribution to whether the match ignites. Instead ignition involves an *interaction* between the striking and the presence of oxygen. This is why I suggested that although the presence of the Coulomb force qualifies as background condition (according to the definition given earlier) for the gravitational inverse square law and the presence of oxygen is a background condition for the match ignition example, they play distinguishable roles in thinking about causation and invariance.

Additivity is, as I have said, the natural way of representing the idea that a set of causal factors makes the same independent (invariant) contribution to an effect regardless of what other independent sets of causal factors are operative (i.e., that no interaction is present) when we dealing with continuous variables that do not have a ceiling or upper bound. This provides one illustration of the general idea that certain sorts of invariance claims are connected to claims about the functional forms describing causal relationships—that is, additivity captures one aspect of invariance that is connected to the absence of interaction. However, as Liljeholm and Cheng (2007) have also noted, additivity is inappropriate for capturing this aspect of invariance (invariance as independent contribution) when we are dealing with binary variables. Suppose that C_1 and C_2 are possible deterministic causes of E; when $C_1 = 0$ (1), $E = 0$ (1), and when $C_2 = 0$ (1), $E = 0$ (1). Since when $C_1 = 1$, E is already at its ceiling of 1, C_2 cannot "add" anything to the value of E, even when C_2 also = 1. As Liljeholm and Cheng argue (see the more detailed discussion in their 2007 and also Woodward 1990, as well as Chapter 7 for a similar argument), it does not follow merely from this observation about failure of additivity that C_1 and C_2 interact or act non-independently on E, assuming that by interaction we mean a physical or causal interaction of some kind. Instead, what follows is that additivity is an inappropriate condition for whether there is causal interaction in cases involving binary variables. The same core idea of a cause making the same invariant contribution to an effect regardless of whether other causes are present can also be retained for the binary case, but we need a mathematical signature for this that is different from additivity.

To briefly develop this thought and to provide a further illustration of how the notion of invariance as independent contribution (i.e., invariance of the relation between cause and effect under changes in the presence of other factors that causally influence the effect) should be understood when variables are binary, let

[24] Of course whether a relationship is invariant in the way described is a fact about nature. What I mean is that if the relationship is so invariant, additivity is a natural way of representing this.

us vary the example slightly. We now suppose that both C_1 and C_2 are probabilistic causes of E, and we understand this to mean that there is a fixed probability $0 < p_1 < 1$ that C_1 causes E when C_1 is present (when $C_1 = 1$) and a fixed probability $0 < p_2 < 1$ that C_2 causes E when $C_2 = 1$. (Of course neither C_1 nor C_2 causes E when they both are absent [both = 0].) Assume also that the behavior of both C_1 and C_2 is invariant in the sense that each behaves in the same way with respect to the causation of E, regardless of whether the other is present: that is, the probability p_1 that C_1 causes E when $C_1 = 1$ is the same whether or not C_2 is present and causes E and similarly for the behavior of C_2: C_2 causes E with probability p_2 in the presence or absence of C_1. Intuitively this corresponds to the condition that C_1 and C_2 do not interact with respect to their tendency to cause E. Finally, assume that whether C_1 or C_2 causes E when C_1 or C_2 is present is independent of (invariant with respect to changes in) the frequency or probability with which C_1 and C_2 occur: that is both p_1 and p_2 are stable under changes in $Pr(C_1)$ and $Pr(C_2)$. This condition is an example of what I call invariance under changes in the probability or frequency of initial conditions, discussed in Section 5.4.5.

As argued in Woodward 1990,[25] given these assumptions, and the additional assumption that C_1 and C_2 are independent in probability, the probability $Pr(E)$ with which E occurs is

$$Pr(E) = p_1 Pr(C_1 = 1) + p_2 Pr(C_2 = 1) - p_1 Pr(C_1 = 1) p_2 Pr(C_2 = 1). \quad (5.6)$$

The first term on the right gives the probability that E is caused to occur by C_1 and the second term the probability that E is caused to occur by C_2. However, these alternatives are not mutually exclusive—on some occasions E will be caused to occur by both C_1 and C_2. Subtracting the third term on the right corrects for this. In this situation, the probability of E is *not* given just by adding p_1 and p_2 or p_1 and p_2 weighted by the probability of occurrence of C_1 and C_2—that is, by the sum of the first two terms in (5.6). Thus the non-interactive, invariant behavior of C_1 and C_2 is not captured by an additivity condition in the way that it is in the force law example. Indeed, simply adding the first two terms on the right to derive the probability of E might easily yield, depending on the relevant parameter values, the nonsensical result that $Pr(E) > 1$. Subtracting the third term avoids this possibility. Thus, in this case, the absence of interaction between C_1 and C_2 with respect to E is captured by the holding of relation (5.6)—departures from this relation reflect the presence of an interaction between C_1 and C_2 and the fact that the effect of C_1 (C_2) on E is not invariant under changes in C_2 (C_1). So in this

[25] See also Cheng 1997 for a similar argument, discussed in Chapter 7. Cheng also presents empirical evidence that people think about interaction and its absence in the way described.

case, too, assumptions about invariance dictate a particular form for the function (namely (5.6)) that expresses an absence of interaction, although a function different from the additive function that is appropriate for continuous unbounded variables.

Here is a slightly different way of thinking about the example, which again illustrates the power of invariance-based ideas. Suppose that when $C_1 = 1$ and $C_2 = 0$ and no other causes of E are present, we find that $Pr(E = 1 / C_1 = 1) = p_1$ where $0 < p_1 < 1$. Suppose also that whenever $C_1 = 1$ and $E = 1$, we adopt the practice of concluding that $C_1 = 1$ causes $E = 1$—that is, we always attribute causation of E to C_1 in these circumstances regardless of whether $C_2 = 0$ or 1. Since when $C_1 = 1$ and $C_2 = 1$, the probability of occurrence of E will go up in comparison with what that probability is when $C_1 = 1$ and $C_2 = 0$, if we adopt this pattern of attribution of causation, it will follow that the behavior of C_1 with respect to the probability with which it causes E changes (is non-invariant) depending upon whether C_2 is present, despite that fact that (we are assuming) there is no interaction between C_1 and C_2. This contrasts with the way most of us find it natural to think about the example, according to which the increase in $Pr(E)$ when both $C_1 = 1$ and $C_2 = 1$ (in comparison to $C_1 = 1$, $C_2 = 0$) is attributable to the additional occurrences of $E = 1$ that are caused by $C_2 = 1$. The latter way of thinking relies on invariance assumptions of the sort described in this chapter. Interestingly, several recent theories of token or actual causation imply that in a case like that under discussion, when C_1 is a probabilistic cause of E, C_1 causes E whenever both C_1 and E occur—a conclusion that is inconsistent with invariance-based assumptions.[26]

I will argue in Chapter 6 that the general notion of a cause operating in the same invariant way or making the same contribution to an effect regardless of whatever other causes are present (as long as these operate "independently" or without interaction) is crucial to the way in which we think about complicated scenarios regarding overdetermination, preemption, and the like. We could not coherently think about these scenarios in the way that we do without also thinking of causes as exhibiting some degree of invariance in their behavior, including invariance of operation of some kind in the presence of other causes.

"Interaction" or its absence can take other forms as well. Consider a collection or population of units u_i some portion of which will receive a treatment, as in a randomized experiment. One possibility (i) is that the response of each unit to the treatment (with respect to some effect variable) depends only on the value of the treatment that unit receives and nothing else. Another (inconsistent) possibility (ii) is that the response of an individual unit depends not only

[26] See, e.g., Fenton-Glynn 2016. Lewis 1986 holds a similar view, but with the requirement that C_1 raise the probability of E substantially.

on the treatment it receives but on the treatments received by other units in the population and/or that it depends on the distribution of treatments across the populations—for example, on the number of units that receive the treatment or whether certain correlations are present in this distribution. In the latter case (ii), there are interaction or spillover effects among the units. An example is provided by a vaccination program in which the response of a given individual (measured in terms of whether she develops the disease) depends not just on whether she is vaccinated but on how many others in the population have been vaccinated.

The former assumption (i), that there are no such interaction effects among units, is well known in the statistical literature as the Stable Unit Treatment Value Assumption (SUTVA). Rubin (1986) describes this assumption as follows (quoted in Morgan and Winship 2015, 48):

> SUTVA is simply the apriori assumption that the value of Y for unit u when exposed to treatment i will be the same no matter what mechanism is used to assign treatment i to unit u and no matter what treatments the other units receive.

When Rubin describes this as an a priori assumption, he means merely that this is an assumption that the analyst may be justified in making prior to analysis of the data and which is then used in the analysis—not that it is a non-empirical assumption or that it is guaranteed to be a priori true in the philosopher's sense of that word. When this assumption is satisfied, causal analysis is considerably simplified since one does not have to worry about the possibility that the observed response to treatment in the population will vary depending on the pattern of assignment of treatment to the various units. SUTVA may be usefully viewed as a particular sort of invariance assumption. It is the assumption that the response of each individual to the treatment is invariant under changes in which other individuals receive the treatment, how many individuals receive the treatment and so on, as well as under changes in the assignment mechanism. (In other words, the responsiveness to treatment is a fixed and stable feature of the individual, rather than a relational feature dependent on the properties of other individuals or the way in which the treatment is assigned.) In some cases this may be a very reasonable assumption. For example, if we are interested in the effectiveness of a drug in relieving headaches among subjects with headache, it may be reasonable to assume that the tendency of each individual to experience relief when given or not given the drug is independent of the tendency of other individuals to experience relief in similar circumstances—headaches are not contagious, the suppression of my headache does not by itself make it more likely that yours will go away, and so on. Of course in other circumstances, a similar assumption will not be justified.

5.4.5. Invariance under Changes in the Distribution of Initial Conditions

We considered earlier notions of invariance having to do with whether a relationship continues to hold under changes in the values of variables figuring in that relationship, whether these are due to intervention-like causes or not. When the variables in question are random variables governed by some joint probability distribution, we have a special case of the notion of invariance under changes in values of variables figuring in a relationship: given some relationship between random variables $X_1, \ldots X_n$ and E (e.g., the relationship captured by $Pr(E / X_1, \ldots X_n)$ although this is just one possibility), we can ask whether and to what extent this relationship remains invariant under changes in the probability distribution $Pr(X_1, \ldots X_n)$.[27] This is a natural invariance condition to associate with contexts in which causes operate probabilistically—we relied on a particular case of it in the example involving probabilistic causation in Section 5.4.4, when we required that $Pr(E/C_1)$ be invariant under changes in $Pr(C_1)$ and similarly that $Pr(E/C_2)$ be invariant under changes in $Pr(C_2)$.

As before, this invariance condition (invariance under changes in distribution) tracks or reflects important aspects of causal relationships. As a very simple illustration, consider two correlated variables X and Y. The joint distribution $Pr(X,Y)$ can be factored in two different ways:

$$Pr(X, Y) = Pr(Y / X) Pr(X) \tag{5.7}$$

$$Pr(X, Y) = Pr(X / Y) Pr(Y). \tag{5.8}$$

Suppose one is interested in whether there is a causal relationship between X and Y—whether X causes Y, Y causes X, or neither. As I have emphasized, the existence of a causal relationship need not manifest itself in any particular invariance condition (other than invariance under some interventions). Causation requires that some relation or other be invariant under some interventions, and we often find other invariance conditions satisfied as well when there is causation, but which such conditions hold will depend on the details of the case. Nonetheless, there are natural invariance conditions having to do with distributional changes that can be connected to (5.7)–(5.8).

[27] As argued earlier, invariance claims are always relative to a class of changes, so it is entirely possible that the relationship of interest is invariant under some changes in $Pr(X_1, \ldots Xn)$ and not under others. In order to avoid needlessly complicating my exposition, I will not always make this qualification explicit, but the reader should think of it as always implicit.

Suppose one finds that (i) under changes in $Pr(X)$, including changes due to interventions, $Pr(Y/X)$ is stable or invariant.[28] Given such changes in $Pr(X)$ and other natural conditions required to ensure nontriviality, $Pr(Y)$ will also change. It is easy to show that under these assumptions $Pr(X/Y)$ will not be invariant under such changes in $Pr(Y)$. Given (i), the natural conclusion is that X causes Y—in this sort of context we expect that if X causes Y and $Pr(Y/X)$ captures that causal relationship, then $Pr(Y/X)$ will be invariant under changes in the probability distribution of the cause, $Pr(X)$.[29] We have no corresponding expectation that relationships running from effects to causes will be invariant under changes in the probability distribution of the effects. In other words, under these conditions, the causally correct factorization—the factorization that reflects causal structure (assuming that some such factorization exists)—will be the one in which the conditional probability is invariant under changes in the marginal probability or more generally, the factorization (if any) in which the terms in the factorization are each invariant under changes in the others. Thus (5.7) is the factorization that reflects causal structure if the conditional probability occurring in (5.7) is invariant under changes in the other term $Pr(X)$.[30]

This idea can be generalized. Recall the Causal Markov condition (**CM**) from Chapter 2: a directed graph G satisfies the causal Markov condition if and only if

$$Pr(X_i \perp \text{ND}(X_i) \; / \; \text{Par}(X_i) \text{ for all } X_i.$$

This condition turns out to be equivalent to the following factorization condition:

$$(\mathbf{FC})\text{Pr}(X_1 \ldots X_n) = \prod \text{Pr}(X_i \; / \; \text{Par}(X_i)),$$

where $Pr(X_i / Par(X_i)$ is just $Pr(X_i)$ if X_i has no parents (is exogenous).

A joint probability distribution $Pr(X_1 \ldots X_n)$ of course can be factored in many different ways. A natural thought[31] is that the causally correct factorization (the factorization that correctly represents the causal structure of the system

[28] The conditional probability of *Y*'s taking some particular value conditional on *X's* taking some particular value will of course depend on the value of *X*, but when we describe *Pr (Y/X)* as invariant, the idea is that the conditional distribution of *Y*, given *X* does not change depending on the probability with which values of *X* occur and that this is so for all values of *X*.

[29] For details, see Woodward, forthcoming b.

[30] Note that this provides one possible basis for inferring causal direction that is independent of time order. This strategy is exploited by Hoover (2001) in connection with the problem of discovering the causal direction between money and prices. A somewhat similar strategy is among those employed in Janzing et al. 2012 to infer causal direction. Again see Woodward, forthcoming b, for additional discussion.

[31] See Janzing and Schollkopf 2010 for a somewhat similar suggestion.

of interest) is that factorization in which each of the terms in the factorization (FC) can be changed independently of the others—that is, the factorization in which each of the terms and the relationships they describe is invariant under interventions that change the other terms and relations. Suppose, for example, that we find that with three variables *X*, *Y*, and *Z*, the joint distribution *Pr*(*X*, *Y*, *Z*) = *Pr*(*X*/*Z*) *Pr*(*Y*/*Z*) *Pr*(*Z*) and that furthermore each of the terms in this factorization can be changed independently of the others: *Pr*(*X*/*Z*) can be changed independently of (is invariant under changes in) *Pr*(*Y*/*Z*) and *Pr*(*Z*) and so on. Then (according to this line of thinking) the correct causal structure is one in which *Z* is a parent of both *X* and *Y*—that is, *Z* is a common cause of *X* and *Y*. The underlying intuition is that in this case there are two distinct causal relationships, one running from *Z* to *X* ($Z \rightarrow X$) and captured by the term *Pr*(*X*/*Z*) and the other running from *Z* to *Y* ($Z \rightarrow Y$) and captured by the term *Pr*(*Y*/*Z*). Because these relationships are distinct, it should be possible in principle to change or interfere with one of them independently of the other. *Pr*(*Z*) describes the probability distribution of the cause *Z*, and again one expects that it is possible to change this distribution without automatically changing *Pr*(*X*/*Z*) or *Pr*(*Y*/*Z*)—this is what we earlier called invariance under changes in the distribution of initial conditions.

We might call the assumption just described the Modal Markov assumption (MM) to distinguish it from CM. CM is a claim that relates causal structure to the factorization of the joint distribution. MM says that the correct representation of the causal structure is the one in which the terms in the factorization of the joint distribution are changeable independently of one another. Obviously MM embodies invariance assumptions and thus provides an additional illustration of connections between causation and invariance.

5.4.6. Modularity as an Invariance Condition

There is another kind of case that, although it is also subsumable under the general category of invariance under background conditions, is also distinctive enough to deserve separate discussion. Previously we considered whether the gravitational inverse square law was invariant changes in the magnitude of other forces that might or might not be present but we did not consider possibilities under which the laws governing those other forces were themselves different. Such "change in laws" scenarios may or may not make sense, but there are many other cases in which changes or modifications to more local, garden-variety causal relationships are readily empirically realizable. Suppose that we have a structure that realizes a set of such modifiable causal relationships, One can ask, for each of those causal relationships, whether and to what extent it would continue to hold across changes in the other causal relationships. Questions of this sort arise naturally in connections with mechanical devices (or "mechanisms") of various sorts. Suppose one has a Rube Goldberg–type system of interconnected springs, pulleys, levers, weights, balls rolling down planes, and so on, which connects an input event to an output.

A natural thought is that it ought to be possible ("in principle" and quite possibly in practice) to alter the behavior of one of these components without altering the behavior of the others. That is, there are some alterations of each component that will not change the causal relationships characterizing the other components. For example, one might stretch one of the springs in a way that changes the spring constant governing its behavior, or one might alter the angle of an inclined plane without changing the behavior of other components in the system. Of course such changes will alter the *inputs* from those components to other components in the system and hence may change its overall behavior, but this is consistent with the causal relationships governing the behavior of the other components (e.g., the equations governing that behavior) being left unchanged. For example, a change in the behavior of one of the springs (spring 1) that changes its spring constant from k_1 to $k_1{}^*$ may leave the relationship governing the behavior of a second spring (spring 2) intact in the sense that this spring continues to be governed by the relationship $F_2 = k_2X$ regardless of whether the first spring has spring constant k_1 or k^*_1 and similarly for other components in the system. To the extent this is so, we can describe the relationship governing the second spring as invariant under such changes in the first spring. One generally supposes that machines and many other systems (including, particularly those that we think of as operating mechanistically) exhibit some degree of invariance or independent changeability of this sort among the relationships governing its components. To the extent this is so, such systems can be described as *modular* in one meaning of that protean term.[32]

More generally, suppose one has a system of causal relationships that are described by a system of equations involving different dependent variables, such as the system described in Section 5.4.2:

[32] Researchers in many different disciplines have appealed to notions of modularity and have had somewhat different things in mind in invoking this notion. Space precludes detailed discussion, but one thing to keep in mind is that modularity in the sense described earlier (independent changeability) is different from modularity in the sense of informational encapsulation discussed by Fodor (1983). Assumptions of modularity understood as claims of independent changeability are common in biological contexts—for example, in discussions of genetic regulatory networks, see Davidson et al. 2002. Woodward 2013 takes some degree of modularity to be a feature of many of the structures that we think of as "mechanisms."

$$Y = aX \tag{5.9}$$

$$Z = bY + cX. \tag{5.10}$$

One can then ask, for each of those causal relationships (or the equations describing them), whether and to what extent they would continue to hold across changes in the other causal relationships. For example, for what sorts of changes, if any, that disrupt or modify the relationship described by equation (5.9) would equation (5.10) continue to hold? To the extent that some equation characterizing some aspect of the behavior of the system would continue to hold under changes in relationships described by other equations, we might describe this as involving *equation invariance* or *independent changeability* of equations and the relations they represent, regarding this as a modularity condition appropriate for systems of equations. (Note that this is different from the requirement that an individual equation remain invariant under interventions on its independent variables.) Again it is natural to expect some degree of invariance of this sort whenever one is dealing with a system of equations in which each equation describes a separate and distinct causal relationship—indeed one might take this to specify part of what it means to say that the relationships in question are separate and distinct. Note that the "equation wipeout" representation of interventions briefly described in Chapter 2 assumes some degree of equation invariance: when we replace equation (5.9) with $Y = k$ (which is what corresponds to intervening to set $Y = k$) and substitute the result into equation (5.10), using it to calculate the result of this intervention on Y, we assume that (5.10) (i.e., the relationship described by this equation) will remain unchanged under this modification of (5.9). Of course whether this assumption is correct is an empirical matter.

Similar points apply to the representation of causal relationships in terms of directed graphs. As we noted in Chapter 2, it is common to graphically represent the effects of an intervention on a variable X by "breaking" all of the other arrows directed into X, so that X is entirely under the control of the intervention variable I. It is assumed that this intervention will leave intact all of the other arrows in the graph, directed into other variables. This assumption—that the other arrows into and out of other variables besides X remain the same under interventions on X—amounts to an invariance assumption, in particular that the other causal relationships represented by the graph will be invariant under this intervention on X. When the variables in a graph are random variables, Hausman and Woodward (1999; see also Woodward 2003) suggest, one can capture this (particular) invariance assumption by means of the following condition, which they label (**MOD**)—for (a probabilistic version of) modularity.

> MOD $Pr(X/\text{Par}(X)) = Pr(X/\text{Par}(X).\text{Set } Z)$ where Z is any set of variables distinct from X and where Set Z involves setting Z to some value by means of an intervention.

In other words, the probability of X is determined by the full set of parents of X, and setting any other variable Z to any value leaves this probability unchanged or invariant. As formulated, satisfaction of MOD is obviously an all-or-nothing matter. But one can formulate more graded notions of the condition—$Pr(X/\text{Par}(X))$ might remain invariant or approximately so for some substantial range of values of some other variable Z even if this is not the case for all values of Z. Or $Pr(X/\text{Par}(X))$ might be invariant under changes in some variables Z that are of interest but not under all such variables. Graded notions of modularity holding with respect to systems of equations or other structures might be understood similarly.

As characterized earlier, modularity (or its absence) is a feature of representations. The same physical system might have a relatively modular representation at some scale or grain, while some other representation at a different scale might be non-modular. However, if we wish, we can extend the notion of modularity to physical systems or sets of causal relations by saying that these are modular as long as they have some modular representation.

Woodward (1999, 2013) describes a number of advantages (including generalizability to new situations) possessed by systems of causal relationships and their accompanying representations that are relatively modular or equation-invariant and disadvantages to systems lacking these features. I will not repeat these here but will merely note that it is unmysterious why discovering relationships and representations having these features is desirable. It is not surprising that as an empirical matter scientists look for these and value their discovery. There is to my knowledge little discussion in the psychological literature of the extent to which laypeople tend to assume as defaults that systems they are trying to understand are modular or tend to value the discovery of modular causal representations, but this would certainly be something worth exploring.[33]

[33] To underscore a point that I hope is obvious, I am *not* claiming that as an empirical matter the causal representations that laypeople or scientists employ are always or usually modular or that as a normative matter they always ought to employ such representations. Relatedly, I am not claiming that all systems of causal relations have highly modular representations. The assumption that a modular representation can be discovered for a particular system and that such a representation is desirable is (at best) a default assumption—it may turn out that nature cannot be represented in this way. It is certainly worthwhile to investigate the conditions under which modularity assumptions for various systems fail, but the existence of such "counterexamples" does not mean that modular representations when empirically supported are not useful.

5.4.7. Invariance under Different Realizations of the Cause

Suppose that we have an "upper level" variable *X* (e.g., temperature, intention to reach in a certain direction) the values of which can be "realized" by a number of different combinations of values of lower-level variables (different kinetic energies for different molecules making up a gas, different neural states corresponding to the same intention). The upper-level variable thus can be thought of as a kind of coarsening of the more fine-grained, lower-level variable(s). Suppose we are interested in whether in this sort of situation, a causal relationship holds between such an upper-level *X* and some candidate effect *Y*. A natural invariance condition to impose is that for *X* to cause *Y*, different interventions on *X*, which will have different lower-level realizations, will have the same effect on *Y*. In other words, the effect of *X* on *Y* should be insensitive to (invariant under changes in) how *X* is realized.[34] This condition rules out so-called ambiguous interventions (Spirtes and Scheines 2004), which have different effects depending on how *X* is realized. To use Spirtes and Scheines's illustration, suppose that total cholesterol *TC* is the sum of high-density cholesterol and low-density cholesterol and that *HDL* promotes heart health (*H*) while *LDL* has a detrimental effect on *H*. Then the effect of a manipulation of *TC* on *H* will depend on the particular combination of values of *HDL* and *LDL* that realize *TC*, so that this manipulation will be ambiguous and the invariance condition under discussion will fail. A condition of this sort on upper-level causes has recently been imposed by several different researchers, including Ellis (2016) and Chalupa et al. (2015) and is also defended in Woodward 2008, 2020b. This condition is closely connected to the conditional independence condition concerning upper-level variables described in Chapter 8.[35]

5.4.8. The Role of Subject Matter-Specific Considerations

The invariance conditions described so far have involved general templates that can be applied to more specific examples independently of subject matter considerations. For example, one can ask, concerning candidate generalizations from many different subject areas, whether they are invariant under interventions,

[34] This condition is called "realization independence" in Woodward 2008. Note that it is *not* the same concept as what philosophers call multiple realizability: invariance under different realizations of the cause is not satisfied merely because there is more than one way of realizing the cause; instead it has to do with whether *all* such realizations have a uniform effect on *Y*.

[35] Like other invariance conditions, this can be relativized and relaxed in various ways—for example, one might require that "almost all" realizations of the cause have a uniform effect on *Y*, that this be true for realizations meeting certain further conditions, and so on. See Woodward 2020b for additional discussion.

under changes in the probability distribution of initial conditions, and so on. However, subject matter-specific considerations can matter to the assessment of invariance as well. What I have in mind is that in different areas of investigation with different subject matters, it may be that generalizations exhibiting invariance under certain kinds of changes specified by the subject matter are particularly valued—that is, invariance under certain changes may matter normatively a lot or relatively little, where the changes that matter are dictated by subject specific considerations. As noted in Woodward 2000 and 2003, many generalizations in economics would be disrupted by surgical or pharmacological changes that alter fundamental neurological processes in economic agents—the generalizations are not invariant under such changes. However, this sort of failure of invariance is usually not regarded as interesting or important by economists. By contrast, the failure of an economics generalization to be invariant under changes in the information available to (psychologically normal) economic agents or under changes in relative costs or incentives is typically regarded in economics as much more significant and is often taken to indicate that the generalization cannot play a fundamental or foundational role in economic explanation. This is connected to the role of "abnormality" considerations in the assessment of invariance, which is discussed in more detail in Section 5.5.2.

5.5. More on Comparisons of Degree or Extent of Invariance

The picture I have presented so far is that generalizations and relationships have to reach a threshold (be invariant under at least some interventions on variables figuring in them) to qualify as causal at all. But beyond this, there are a number of other invariance-linked features they may possess. By spelling these out we help to clarify the content of the claims in question. This spelling out is a matter of specifying (i) exactly which relationships are claimed to be invariant and (insofar as this is possible) (ii) the range of circumstances or changes under which the relationships are invariant. I say insofar as possible because it is common, indeed typical, not to have full or complete information about (ii)—see Section 5.6.

Given that relationships will differ with respect to the circumstances or changes over which they are invariant, a natural question is whether it is possible to compare "degrees" of invariance or to order relationships with respect to how invariant they are and, if this is possible, on what basis we might do so. Or, at least, one might wonder, given a relationship that is invariant under a certain set of changes, whether there is some basis for judging the extent to which it would be an improvement with respect to invariance if it were also invariant with respect to other changes.

I should say at the outset that I do not have anything like a complete or fully satisfying answer to these questions, but I do have some suggestions that are applicable to specific sorts of cases.

5.5.1. Comparisons Based on Proper Subset Relations

First, one straightforward possibility, to which I have appealed earlier, is that the range of variations over which relationship R is invariant is a proper subset of the range over which a second relationship R* is invariant. In this case it seems unproblematic to say that R* is more invariant than R. Allowing (as suggested earlier) for comparisons among relationships that may hold only to a high degree of approximation, it might be argued that this fits a number of familiar patterns of theory replacement or succession in science. For example, the range of circumstances under which the Newtonian gravitational inverse square law is invariant is, to a good approximation, a proper subset of the range of circumstances under which the field equations of general relativity are invariant. A similar analysis applies to a number of other examples discussed earlier. Experimental work (e.g., by Lombrozo and coauthors) described in Chapter 7 shows that lay subjects seem to make comparisons of invariance (as reflected in their causal strength judgments) on the basis of such subset relations, among other considerations.

On the other hand, there are certainly many cases that don't lend themselves to assessments of invariance on this basis. For example, some sorts of failures of invariance seem to matter more than others, in part (but not entirely) for subject matter-specific reasons (as noted in Section 5.4.8) and it also sometimes seems to make sense to describe a generalization that is invariant under some interventions as nonetheless relatively "fragile" in the sense that it would fail to hold across "many" possible changes, but where this does not appear to involve a comparison with any specific alternative. (The generalization strikes us as "fragile" rather than "more fragile than.") Claims of this sort seem to involve assessments that are not based on proper subset relations. I turn now to some remarks about possible bases for these.

5.5.2. Invariance Assessments Based on Normality or Closeness to Actuality

I said earlier that in assessing the invariance of a relationship, it matters whether the relationship would continue to hold under non-actual as well as actual possibilities. But while this is a consideration that appropriately influences invariance assessments, there are also contexts in which more weight is given to whether a relationship is invariant under certain classes of actual occurrences, including those seen as typical or normal, or those that, in addition to being possible, don't seem highly unlikely or farfetched. For example, in most ordinary contexts, the fact that some causal generalization concerning an automobile fails to be invariant under the highly "abnormal" circumstance that the car is struck by a large meteor will matter less than failures of invariance under more ordinary circumstances—for example, changes in temperature within some normal range or in whether the roads are wet from rain. Given an automotive generalization that is invariant under changes in circumstances of the latter sort but not under meteor impacts and one with the opposite profile, it will be natural to regard the former as more invariant than the latter or at least as invariant under changes that matter more.

Presumably one consideration that guides such judgments is simply whether the circumstance is improbable or statistically unlikely, as it clearly is in the case of the meteor striking the car, in comparison with a temperature within some normal range. (It is unsurprising, given our interests in manipulation and control, that laypeople commonly care more about invariance under such likely changes than invariance under highly unlikely ones and that this in turn influences causal judgment, as we shall see in subsequent chapters.) But there is also considerable evidence that, as an empirical matter, causal judgment in ordinary nonscientific contexts is influenced by notions of abnormality that are not purely statistical but instead bound up with norms and conventions, where these are understood as having some sort of prescriptive force that is somewhat independent of how often they are in fact conformed with.[36]

We noted earlier that subject matter-specific considerations can influence which failures of invariance are important. Often we can think of these as operating by affecting judgments of normality. Thus, as noted previously, certain changes in human psychology will be "abnormal" or involve "large departures from actuality" from the point of view of the economist, while other changes (e.g., in the incentives people face) will not. It also seems plausible that different scientific disciplines differ not just in which circumstances are regarded as abnormal but also in the degree to which they care about such abnormality at all. One might think that in principle physics looks for generalizations that are invariant as possible, even in circumstances that rarely occur or are otherwise unusual, as long as those circumstances are "physically possible" or even

[36] As noted earlier, the influence of such norms as well as considerations having to do with what is statistically likely is arguably particularly strong in the case of some actual cause judgments—see, for example, Hitchcock and Knobe 2009. Of course issues about invariance of generalizations under normal vs. abnormal circumstances are different from issues having to do with the role of abnormality in causal selection.

if, although impossible, they bear some relation to circumstances that are possible.[37] Biologists, by contrast, may care more about what happens (which generalizations are invariant) in "biologically normal" circumstances than in those that are not, where biologically normal circumstances include those that occur in living organisms.[38] For example, whether or not there are invariant generalizations about the behavior of DNA in circumstances that are very different from those obtaining in the interiors of most cells may be of little interest to many biologists (although they may be of interest to chemists). However, even in biology, it would be a mistake to say that the only circumstances that are relevant to the assessment of invariance are those that occur naturally or normally.[39] Experimental manipulations that produce genetically engineered organisms that would never occur naturally as well as other exercises in bioengineering can have considerable biological significance, and we want assessments of invariance to extend to such cases. Nonetheless the more general point is that the kinds of circumstances that will be of interest to the biologist for the assessment of invariance will be different from those of the physicist.

5.6. Invariance Needs to Be Understood against the Background of Our Epistemic and Calculational Limitations

I suggested in the introduction as well as in Chapter 1 that in thinking about causation and related notions such as invariance and proportionality we need to consider how these interact with the fact that we—the users of these notions—are beings with epistemic and calculational limitations. In connection with invariance, the focus thus should be on understanding how this notion works in the presence of such limitations. To illustrate, consider—to engage in a standard metaphysics-of-science fantasy—a Laplacian intelligence (hereafter Larry the Laplacian, ∇^2) who knows "everything." Larry has a complete grasp of a final theory, including all of the most fundamental laws of nature, full and arbitrarily accurate information about all initial and boundary conditions, and no calculational or other cognitive limitations, so that the construction of derivations (and explanations) of everything that happens, including macro-events of arbitrary complexity such as the state of the US stock market on January 2, 2020, from such fundamental laws and initial and boundary conditions is always possible. These fundamental laws, we can safely assume, will be highly invariant and, virtually by

37 As with two-dimensional models of three-dimensional processes.

38 I won't try to specify more exactly what the term "biologically normal circumstances" includes—perhaps it should also include those circumstances that "could" occur, consistently with the characteristics of current or past living organisms.

39 Here I disagree with, for example, Waters 2009.

definition, as "invariant as possible" in the sense that no alternative candidates for laws that are empirically adequate will be more invariant. Although invariance notions will thus apply to these fundamental laws and although Larry can make invariance-based judgments about other, non-fundamental generalizations, such judgments will do little or no work for Larry, in comparison to the role that invariance-based considerations play for investigators like us. There will be no need for Larry to make judgments of comparative invariance involving non-fundamental generalizations; Larry does not need such generalizations, being able to derive everything from first principles. Similarly invariance considerations will play no role for Larry in the discovery of new generalizations—he already knows everything. Also, Larry will never be in the position of trading off invariance against other desiderata such as precision—a possibility that I discuss later.

By contrast, notions of invariance come into their own and have their most fruitful and useful application in contexts in which users have partial knowledge and face other sorts of limitations—that is, in virtually all circumstances actual human beings face. These limitations mean that the kinds of claims about which we make invariance judgments are typically incomplete in various ways: such claims may describe some but not all factors that are causally relevant to some outcome or omit some factors that matter under some conditions but not others. An invariance-based framework captures this in a natural way—it gives us a way of thinking about causation that fits with and reflects our limitations. For example, we can know whether a generalization or relationship is invariant in some range of circumstances even though there may be other circumstances unknown to us in which it breaks down (or in which we may suspect that it breaks down, even if we don't know this for certain). Relatedly we can make comparisons involving invariance while not being in possession of information about exceptionless laws and the like. Of course it might be responded that, metaphysically speaking, underlying (i) every generalization with a limited range of invariance is (ii) an exceptionless fundamental law, which serves as a truth-maker for (i) or from which (i) somehow inherits its invariance. But even if the details of such a view could be worked out (no easy task), this doesn't give us much insight into how humans learn and reason about invariant relationships, given that the laws of form (ii) are unknown. Invoking the in-principle possibility of Larry doesn't help with this. In general invariance-based notions seem to fit better with how we learn about and represent causal relations than accounts that invoke exceptionless fundamental laws—we start with only locally invariant relationships and then, in some cases, generalize and refine them into more and more invariant relationships, rather than thinking in terms of exceptionless laws from the outset. Exceptionless laws, if discoverable at all, are thus an extreme or limiting case of invariant generalizations, with the latter being the more generic notion. From this perspective, invariance seems a more promising notion for understanding the empirical psychology of human causal cognition, including cognition in science, than law-based notions.

Consider another example: a large rock is dropped on a glass that then shatters. Let *E* be the exact pattern of shattering (the micro-details of the shapes and positions of the shards, etc.). To derive or predict *E*, we would need an extremely precise specification of the initial state of the rock and the glass, the laws L operative in this situation, including those governing the trajectory of the rock, the force laws that characterize the interactions among the components of the glass and how these react to the impact of the rock, and much more besides. If we knew these laws, we could use them to construct an explanation of *E* that appeals to highly invariant generalizations. But in realistic circumstances, we don't know the needed laws or initial conditions. If we were to attempt to formulate an explanation in these terms, we would very likely find ourselves appealing to generalizations that are mistaken and non-invariant in some respects and to assumptions about initial conditions that are not exactly correct.

One way of avoiding this outcome is to replace the original explanandum *E* with a much more coarse-grained explanandum *E** that takes just two values: {shatters, does not shatter} and to adopt an equally coarse-grained characterization of the cause variable—for example{dropping any rock with a mass greater than *m* grams from a position directly over the glass versus not dropping any rock} along with a generalization relating these two variables. Such a generalization sacrifices a great deal of precision, but it may have a better chance of being true and it may be relatively invariant. It will also generalize better to other cases. Again the reasons why the second approach (or something like it) might be preferable are only apparent if we make realistic assumptions about our epistemic and calculational limitations.[40] This is one of many cases in which our epistemic and calculational limitations help to "shape" our causal thinking.[41]

5.7. More on Invariance and Exceptionless Generalizations

I noted earlier that claims that some relationship is invariant in some respect should not be identified with the claim that the relationship is exceptionless. In this section I want to develop this theme in more detail. First let me repeat

[40] The superior *in fact* reliability of calculations based on "upper level" generalizations in comparison with attempts to calculate purely on the basis of lower-level information is a theme that is emphasized in Wilson 2017. By "in fact" reliability I mean reliability that is realistically attainable given calculational and epistemic limitations.

[41] Some writers such as Weslake (2010) argue that recognizing invariance as a desideratum in causal analysis always leads to a preference for "low level" explanations such as those framed in terms of fundamental physics. I discuss such arguments in more detail in Woodward 2018b, but one basic point is that they neglect the role of our epistemic and calculational limitations and that invariance-based considerations should be thought of as employed in the context of such limitations.

that exceptionlessness is not sufficient for invariance. Suppose that *X* and *Y* are joint effects that are deterministically caused by a single common cause *C* (with no causal or nomic connection between *X* and *Y*, and with *X* and *Y* having no other causes besides *C*). Then the generalization linking the values of *X* and *Y* will be exceptionless, but it will not be invariant at all (in the sense of meeting our minimal threshold for invariance), since all interventions on *X* or *Y* will disrupt this relationship. Similarly, as argued earlier, that some generalization is exceptionless is not necessary for its being invariant (under some interventions and in some respects).

More interestingly, degree or range of invariance does not track exceptionlessness or degree of approximation to exceptionlessness. As noted in Section 5.2, the key element in invariance is the notion of sameness of operation or applicability of a relationship across changes. One can have a collection of generalizations G_i, each of which is exceptionless and which together cover a wide range of different phenomena with very little sameness of this sort. To return to an example used previously, imagine that some candidate gravitational force law G holds only when gravitational forces but no non-gravitational forces are present, that a candidate electromagnetic force law C holds only when electromagnetic forces and only these are present, and that when both forces are present, a new law (or laws) L holds, having no particular connection with either G or C. All of these laws G, C, L, and so on might be exceptionless in the sense that they truly describe what happens, without exception, under the very restricted circumstances in which they hold. Nonetheless, this is a situation involving generalizations with rather restricted ranges of invariance in comparison with the (actual) situation in which the gravitational force law holds invariantly whether or not an electromagnetic force is present and vice versa. Similarly, in connection with value-invariance, one might imagine a situation in which the relationships among variables that hold under interventions are very different from the relationships holding among the same variables when no interventions are present, even though, in each context, both sets of relationships are exceptionless. Again one would have a failure of these relationships to exhibit a particular sort of invariance despite their exceptionlessness. As these examples illustrate, invariance is bound up with notions of generalizability and with the existence of a kind of *unified* behavior across different circumstances (the cause or explanatory factor operates in the same way across such differences), with different varieties of invariance being associated with different sorts of particular claims or expectations about what sorts of behavior generalizes or where unification is to be found.[42]

[42] Is talk of invariance then just another way of talking about exceptionlessness plus additional stuff, like generality and unification? Although I lack the space for detailed discussion, I think that

This suggestion is likely to meet with the following response: suppose that, in accord with the earlier suggestion, some generalization G is claimed to be invariant in a range of circumstances *C* but to break down outside of *C*, in range of circumstances *C**. Then (the response goes) this information about *C* and *C** can always be built into the antecedent of G, leading to a new generalization G* that *is* exceptionless. For example, if G says that all *A*s are *B*s, then G* might say that in *C* all *A*s are *B*s. Woodward (2003) calls this *the exception-incorporating model* for laws and causal generalizations.

The exception-incorporating model contrasts with the following alternative (which I have been assuming), called the independent specification model in Woodward 2003. First, we have some target relationship or generalization G whose invariance we are assessing. Second, we have a range *C* of changes or conditions over which G is claimed to be invariant, where—this is important for reasons described subsequently—this range may be very incompletely or inexactly specified. In this model, G and *C* are (or are permitted to be) specified separately or independently: that is, in contrast to the exception-incorporating model, we do not always or automatically build information about *C* into G, although in some circumstances it may be appropriate to do this.

Woodward 2003 describes a number of reasons for preferring this independent specification model to the exception-incorporating model. I will not repeat these arguments here, but I want to briefly emphasize the way in which the independent specification model fits better with what users of invariant generalizations need to know. (Again this illustrates ways in which epistemic considerations enter into how we learn and think about invariance.) Although there are certainly cases in which a candidate invariant generalization itself can provide information about the conditions under which it breaks down,[43] this is far from always the case. Instead, in many cases the generalization by itself will not tell us about its exceptions—to the extent that these are known at all, they typically have to be discovered or specified independently. A very common situation is that although some candidate generalization is invariant over some changes and although we know some of the circumstances under which it is invariant, the information we have about the exact boundaries of its domain of

the answer is no, at least if these latter notions are understood in the way that they usually are in the philosophy of science literature. Roughly this is because on the standard characterizations, notions like generality and unification are understood in ways that are not distinctively causal or nomological, so that these notions apply to non-causal structures as well, such as classificatory schemes. One result is that these characterizations have difficulty distinguishing scientifically valuable kinds of generality and unification from varieties that are trivial or bogus or at least have no explanatory significance. Invariance considerations impose more specific requirements and structure.

[43] Quantum field theory provides examples of generalizations that signal their own conditions of breakdown.

invariance is often vague, incomplete, and perhaps in part mistaken. Because of this, such information is often inappropriate for incorporation into the generalization itself, which, for explanatory and modeling purposes, we generally want to be as clear and precise as possible. The independent specification model allows us represent this imprecise information separately in the domain description, where such imprecision is more tolerable, rather than attempting to incorporate it into our candidate generalizations. We segregate off the vagueness and imprecision by putting it into the specification of the domain and keeping it out of generalization itself. (Illustrations will be provided shortly.)

Another closely related consideration is that the independent specification model fits naturally with the idea that one can use a generalization G (otherwise appropriately related to the behavior *B* of some target system) when one knows that the behavior in question falls within the range *C* of invariance of G even if one does not know the exact boundaries of this domain (because of the vagueness and imprecision of the available information about this, as described earlier) or even if one has mistaken beliefs about those. That certain conditions are within the range of invariance of G is often something that is discovered empirically or by trial and error.[44] For example, a mid-twentieth-century physicist could justifiably use general relativity to model various astrophysical phenomena within its domain (knowing or believing correctly that these phenomena fall within the domain of GR) without knowing or believing that GR breaks down at small length scales (as is currently thought to be the case) or at least without having any very exact information about where such breakdowns occur. In such a case the physicist may have difficulty formulating a true, informative, exceptionless generalization involving the field equations in accord with the exception-incorporating model. If knowledge or true belief about such an exceptionless generalization is required for explanation, the physicist may be unable to explain even those phenomena that do fall within the domain of invariance of GR. The independent specification model says that the physicist does not need to formulate such an

[44] In other cases, information about the range of invariance of a generalization may rely on more theory-supported background knowledge that does not give us precise boundaries but nonetheless provides guidance of a more qualitative or rule-of-thumb sort. Researchers in many areas of science often have a good (although far from perfect) sense for when use of a generalization is inside or outside the domain of invariance of a generalization even if they cannot describe the boundaries of this domain precisely. Often such judgments are based on general information about the circumstances under which the influence of certain factors (or variation of those factors within a certain range) may be expected to be negligible. In disciplines like physics, this is frequently closely tied to information about the temporal, spatial, or energy scales at which those factors operate or become important. For example, a generalization G about the behavior of gases that neglects the role of intermolecular forces may accurately model some features of such behavior for a sufficiently dilute gas, although it will break down under circumstances in which intermolecular forces become important. One can know this general fact and that some target system is sufficiently dilute for G to apply without knowing either what the exact conditions are under which G breaks down or what the correct replacement is for G when intermolecular forces are important.

exceptionless generalization in order to explain. Instead all that is required is the field equations, relevant information about initial and boundary conditions characterizing the system of interest, and information that the behavior of interest is within the domain *C* of GR, all of which one can have without having information about the exact boundaries of *C* or what the correct replacement is for G when intermolecular forces are important.

One of the themes running through this portion of my discussion is the importance of having ways of representing causal information that contribute to understanding how one can learn and conduct causal analysis in circumstances in which one has *some* information but is very far from knowing everything. Information about invariance (and the accompanying independent specification model) is, so to speak, naturally built for representing and dealing with such situations. It is thus a notion that fits with the psychology of real agents with calculational and epistemic limitations. I will add that in this chapter I have stressed the role of invariance considerations in scientific contexts but, as Chapters 6 and 7 will attempt to show, many of the features on which I have focused carry over to more ordinary contexts involving causal reasoning as well. In particular, the various strategies discussed earlier for representing and learning about causal relationships in circumstances in which we do not have complete information in the form of exceptionless generalizations carry over as well to more ordinary contexts involving causal reasoning.

5.8. The Role of the Discovery of Invariant Relationships in Inquiry

I have stressed the utility of discovering invariant relationships for purposes like prediction, manipulation, and generalization to new situations but have said little about how these relate to other features or goals of inquiry such as truth. Nor (with the exception of some brief remarks in Section 5.4) have I discussed how a concern with invariance might structure inquiry or figure in discovery. Here I will draw on some very interesting and insightful ideas in Carroll and Cheng 2010—see especially in Section 5.8.2—, although the use I will make of these ideas will be my own. I stress the following, interrelated themes: (1) The discovery of invariant relationships is an important goal of inquiry, but we should not think of this as a goal that we trade off against truth.[45] Nor should we think of invariance as evidence for truth. Instead we want to discover relationships for which it is *true* or nearly true that they are invariant in various

[45] Depending on the circumstances, invariance can trade off against precision, as described earlier, but an imprecise generalization can still be true.

ways. Thus the requirement of truth (or at least near or approximate truth) is built into the goal of identifying invariant relationships. This is one reason why the discovery of invariant relationships is also not a "merely pragmatic" goal in any sense of "pragmatic" that is disconnected from what is true of nature. Put differently, invariance in relationships is a matter of the holding of certain kinds of truths—truths that we regard as particularly important to discover, rather than something that competes with truth or is evidence for truth. (2) The identification of invariant relationships is important in causal analysis and explanation, but this goal (and assumptions about which relationships are invariant) is also important more generally in structuring and organizing inquiry—it has what Wilson 2017 calls an "architectural" role.

5.8.1. Invariance as a Goal

To motivate these claims, I digress briefly to a related set of issues concerning the goal of *explanation* in inquiry. A number of writers advocate some form of what they call inference to the best explanation (IBE). This comes in many different varieties, but a straightforward, unvarnished version connects explanation and truth (or evidence for truth) in the following way: the fact that some claim *E* would if true best explain some known explanandum *M* is evidence for the truth of *E*—*E*'s potential explanatoriness in the sense that it would explain if true is a sign of its truth.[46]

A thorough exploration of this claim would require engagement with lots of distinctions and subtleties. I think, however, that at least in many cases, this is the wrong way to think about the role that the goal of finding explanations plays (and ought to play) in inquiry.[47] Instead, the following picture is in many cases more nearly correct: first, unless additional special conditions happen to be satisfied, the fact that (i) *E* would if true best explain *M* is by itself no reason to think that (ii) E *is* true. Assume for purposes of present discussion that we do want our explanations to appeal to premises that are true (or nearly so). According to the view that I favor, to show that they *are* true we need to appeal to independent evidence in support of such truth claims—and (i) by itself is not such evidence. Note that this view is consistent with explanation (or finding better or best explanations) being a goal of inquiry—it is just that its role is not that

[46] Note that, as I have characterized IBE, it is very different from inferring to the only possible explanation that is consistent with the empirical evidence—all other possible explanations having been ruled out empirically. I have no objection to this form of inference—indeed I think plays a central role in science and everyday life.

[47] Or at least we don't *have* to think about the goal of finding explanations in this way and in many cases we should not do so. For a philosophical discussion of IBE that reaches a similar conclusion see Nyrup 2015.

of providing evidence for the truth of what the explanation asserts. Instead its role is to direct us to particular sorts of truths—those truths that can figure in explanations. Among the truths we care about discovering are explanatory truths—and these will include truths about invariant relationships, since many of these figure centrally in successful explanations.

This idea can be motivated in a slightly different way. It has often been observed that even assuming that truth is an aim of science, not all truths are scientifically interesting or valuable. A detailed and precise record of the exact dimensions of a very large number of grains of sand obtained from a particular beach might contain lots of truths, but (aside perhaps from some very special contexts) these are not the sorts of truths science generally aims at discovering. So the idea that science aims at truth, even if otherwise unexceptionable, is incomplete: it needs to be accompanied by an account of *which* truths (of the many possible truths we might discover) matter for scientific purposes. The idea that among the truths we should aim at[48] are those that can figure in explanations provides one answer to this question.

I see the discovery of invariant relationships (which on my view play a central role in explanation) as having a similar status. Science aims at the discovery of such relationships and, when this is possible, at the discovery of relationships that are invariant under some large or relevant range of circumstances. But this is *not* to say that invariant relationships are more likely to be true than less invariant ones or anything similar. This way of putting it gets things the wrong way around—invariant relationships are just relationships regarding which certain truth claims (about the respects in which they are invariant) hold. Again, invariance isn't evidence for truth; its role is rather that of characterizing a particular sort of truth (truths about which relationships are invariant), the discovery of which is a goal of science.

Another analogy may help to make this claim clearer. Consider the discovery of predictive relationships as one goal of science, where we mean by this any relationship that can be used for successful prediction, whether or not the relationship in question is causal. It would be wrongheaded to argue as follows: if hypothesis h relating the value of *X* to the value of *Y* were true, it would allow for successful prediction of *Y* from *X* (or, alternatively, h if true would permit more accurate predictions of *Y* from *X* or perhaps predictions of more other quantities in addition to *Y* than various alternative hypotheses). Therefore (the argument goes) this fact—that h if true would be a good predictor—is evidence that h is true. (Call this inference to the best predictor.)

[48] Of course I don't claim that these are the only truths we should aim at. Non-explanatory description can be a legitimate goal.

This argument seems obviously defective. A postulated deterministic relationship between the detailed shape of the leaves in my morning tea and the behavior of the US stock market for the rest of the day would, if that relationship were true, be highly predictive, but of course it does not follow that such a relationship *is* true or actually predictive. Again this is consistent with successful prediction being among the goals of science. However, this goal characterizes a certain class of truths that we seek—those that can be used for predictive purposes. Its role is not that of providing evidence for truth. Similarly for invariance.

Let me now turn to another role for invariance—a role that is consistent with the one just described, but which is distinctive enough to deserve separate discussion. This has to do with the role of invariance considerations in the organization and direction of inquiry and the discovery of new hypotheses. As philosophers are fond of observing, given any finite body of evidence, there are a large number (perhaps an infinitude) of hypotheses that are consistent with this evidence. After observing two green emeralds, one might conjecture that the next one will be green as well, but it is also consistent with observation to conjecture that the next one will be red or grue. For learning and successful inquiry to proceed at all, some strategy for efficiently navigating or searching through this space of hypotheses, all of which are consistent with the evidence so far observed, is required. Such a strategy should tell us, among other things, which hypotheses to consider first, given the goals of inquiry and the constraints we face and how we should revise our hypotheses in the face of conflicting evidence.

Of course, specific subject matter considerations will often play an important role in suggesting particular strategies, but there are often more general things to be said. As an illustration (cf. Schulte 2017), suppose that in the preceding example there are only two possibilities—either (i) all emeralds are green or, alternatively, (ii) at least one emerald is non-green. Then a simple learning theoretic analysis shows that in this case, if one follows a strategy of conjecturing that all emeralds are green if the first emerald examined is green and only discards that hypothesis if one observes a non-green emerald, and then concludes (ii), this will be a strategy that eventually converges to the truth about which of (i) or (ii) is correct. Alternative strategies will either not converge to the true hypothesis in all circumstances or will require more retractions (changes of mind in whether (i) or (ii) is conjectured) in doing so. If our goal is adopt a strategy that converges to which of (i) or (ii) is true for all possible sequences of colored emeralds and that minimizes the largest number of retractions one is forced to make, then the common-sense strategy of conjecturing that all emeralds are green and only changing that conjecture when forced to do so by the evidence achieves this goal. Notice that this argument does not appeal to the claim that it is *probable* that all emeralds are green or some uniform color and does not involve an inference that such a claim is true or likely, given the observed evidence.[49] Beginning with the hypothesis that all emeralds are green and treating this as a default (to be rejected only when there is evidence forcing one to do so) is instead justified on the basis of considerations having to do with efficiency of search and hypothesis revision.

As Carroll and Cheng 2010 in effect observe,[50] assumptions about invariance can also be thought of as playing a broadly similar role in the organization of inquiry. Suppose that it is observed that C causes E in background circumstances B_1 and that it is assumed on this basis that C will also cause E in different background circumstances B_2. This is an assumption about the invariance of the $C \rightarrow E$ relationship across changes from B_1 to B_2. In the general case this assumption may well be mistaken. There are plenty of cases in nature in which C interacts with B_1 to produce E but does not produce E in the presence of B_2. But if we adopt the assumption of invariance (of the $C \rightarrow E$ relationship) in the new circumstances B_2 as a default, one of two things will happen. The first possibility is that the invariance assumption turns out to be correct, as far as B_2 goes, so that we have successfully identified a relationship with a larger range of invariance than we originally supposed—we've learned a new invariance-related truth or the truth about a "more invariant" relationship. Suppose, alternatively, that $C \rightarrow E$ fails to hold in B_2. If we value invariance as a goal, this suggests that we should consider revising our original $C \rightarrow E$ hypothesis so that it holds invariantly across both background contexts. In other words we consider modifying our hypothesis in such a way as to restore invariance. For example, we conjecture initially, after observing the motion of the planets, that all objects move in elliptical trajectories and that this is caused by some force specific to the sun. However we then observe other objects such as cannon balls near the earth moving along parabolic trajectories, so that our original conjecture is falsified by these cases (or alternatively does not apply to these cases if we restrict the conjecture to planets). We replace it with a set of more invariant generalizations (those of Newtonian mechanics) that covers both sets of cases. Of course nothing guarantees that it will be possible to find such a more invariant generalization—the point is that there are ceteris paribus considerations in favor of looking for it.

This is a very simple example, but it illustrates several more general points about the way in which assumptions about invariance can serve as defaults in a way that helps to efficiently organize inquiry (efficiently, that is, as long as finding invariant relationships is one of our goals). First, assumptions about invariance create or correspond to default expectations about how nature will behave.

[49] I don't mean that it is wrong to claim this hypothesis is probable but merely that the argument just described does not require this.

[50] I should note that they don't make an explicit connection with learning-theoretic arguments in the way that I have. What is similar is the idea that default assumptions can play a useful role in the organization of inquiry.

Second, because we have these expectations, we notice when they are violated and view such violations as indicating the need for some sort of revision—also guided by invariance-based considerations—in our original hypothesis. It is important to this process that such invariance-based considerations can often generate relatively specific suggestions about possible revision.[51] Moreover, as Cheng and colleagues show, in work discussed in Chapter 7, ordinary subjects (and not just scientists) often behave in the way just described. For example, they tend to assume, as a defeasible default, that different causes of the same outcome operate independently or do not interact (this corresponds to an invariance assumption, as noted earlier) and revise this assumption only if forced by the evidence to do so. This assumption of independence or non-interaction in turn leads to a default preference for certain functional forms for describing causal relationships over others—additive functional forms for continuous variables and alternative functional form, described earlier, for binary variables. Moreover, the presumption that causes operate invariantly can lead one to conjecture that new causes, not previously recognized, are operative.[52] Again I emphasize that to play such roles the invariance assumptions we make need not always be correct—even when incorrect, they can still play a useful role in organizing inquiry.

There is a strong tendency in the philosophical literature to either attempt to connect various possible goals of inquiry to truth so that explanatory power, simplicity, and other presumed virtues are thought of as valuable because (or to the extent that) they are evidence for truth or else to understand these goals as potentially or actually in conflict with the value of truth—for example, we adopt hypotheses that are not entirely true because they are "simple" and we value simplicity along with truth. My earlier remarks are meant to suggest that neither of these ideas does full justice to the various roles that different values and goals can play in inquiry. First, such goals can play the role of directing our attention to particular kinds of truths at which we should aim, in which case achievement of the goals is neither evidence for truth nor in conflict with it. Second, the way in which inquiry is organized and how we go about searching through spaces of possible hypotheses—which hypotheses we consider first or as defaults, what we do next when we discover contrary evidence and so on—is also crucial. Goals (like the discovery of invariant relationships) can help to effectively structure inquiry without our needing to assume that hypotheses possessing features corresponding to these goals are more likely to be true than alternatives.

[51] An obvious difficulty with the suggestion that hypothesis change in science is (or ought to be) guided by some generic goal like "simplicity" is that it usually yields no precise or specific guidance about what form hypothesis revision should take or which of a number of possible revisions should be tried first. By way of contrast, invariance-based considerations can offer much more specific guidance about what to do next.

[52] See Carroll and Cheng 2010 and Chapter 7.

6
Invariance Applied

6.1. Introduction

In this chapter, I apply some of the ideas about invariance from Chapter 5 to some examples involving common-sense causal thinking. I will also engage with some philosophical ideas about causation deriving from David Lewis and some of his students. For most of my discussion (unless indicated otherwise) I will assume that we are dealing with cases involving a very simple causal structure, described in what follows, that may take either a token or type form. The causal relata are events, which may either occur or fail to occur (and are thus representable by binary variables). In the token case the assumed structure is

$$\text{Token event } c \text{ deterministically causes event } e. \tag{6.1}$$

I also assume that no complicating larger structure is present—there is no backup cause that will cause *e* if *c* does not, *e* is not causally overdetermined by some second cause *c*′, and so on. The notion of causation in play will be a minimal notion that satisfies a token cause version of the interventionist condition (M) or something similar such as a theory like Lewis's in which non-backtracking counterfactual dependence is sufficient for causation. In this case it is reasonable to assume that the following counterfactuals are true:

$$\text{O}(c) \,\Box\!\!\rightarrow \text{O}(e) \tag{6.2}$$

$$\text{not O}(c) \,\Box\!\!\rightarrow \text{not O}(e) \tag{6.3}$$

where O(*c*) etc. are the propositions that *c* etc. occur, and $\Box\!\!\rightarrow$ is the counterfactual conditional. For the purposes of discussion in this part of this chapter I'm going to follow Lewis in assuming that counterfactuals of form (6.2) are true

Causation with a Human Face. James Woodward, Oxford University Press. © Oxford University Press 2021.
DOI: 10.1093/oso/9780197585412.003.0007

when c and e both occur—that is, that strong centering holds.[1] Later I will consider what will happen if this assumption is relaxed.

The counterfactuals (6.2)–(6.3) may be understood as either non-backtracking counterfactuals in the standard Lewis-Stalnaker framework or as interventionist counterfactuals—for much of this chapter the difference between these two possibilities will not matter, although in other contexts it is important. As a working example, imagine that Suzy's throw of a rock causes a glass bottle to shatter by striking it, no other causes of shattering are present, and the relation between the throw and shattering is deterministic, so that, in accord with (6.2)–(6.3), if she were to throw, the bottle would shatter, and if she were not to throw, the bottle would not shatter.

In the type-causal analogue to this case, a similar structure is present at the type level: Cs deterministically cause Es, and no other causes or potential backup causes of E are present. (When a bottle is shattered it is replaced by a new, unshattered bottle, so that the causal relationship is repeatable.) If $C = 1$ or 0 corresponds, respectively, to the occurrence or nonoccurrence of the throw and $E = 1$ or 0 corresponds to the occurrence of a shattering or not, then we may write $E = C$, where this is understood as meaning

$$\text{If it were the case that } C = 1\text{, then it would be the case that } E = 1 \qquad (6.4)$$

and

$$\text{If it were the case that } C = 0\text{, then it would be the case that } E = 0. \qquad (6.5)$$

Following Lewis (1986), I will say that the occurrence of e is counterfactually dependent on the occurrence of c if and only if the two counterfactuals (6.2) and (6.3) hold. And although Lewis does not apply his framework to type-causal claims, I will similarly say that E is counterfactually dependent on C if and only if the counterfactuals (6.4)–(6.5) hold.[2] Lewis holds that counterfactual dependence between token events is sufficient for causation and for cases having the structure described here; the interventionist account has a similar implication for both type and token claims.

[1] For our purposes we may think of strong centering as the claim that the possible world most similar to the actual world is the actual world itself. It follows from this that if both p and q are true, $p \square\!\!\rightarrow q$.

[2] In this case the counterfactuals will have a straightforward interventionist interpretation. I will not speculate about how they might be understood within a Lewis-style possible world framework.

More recently, Lewis (2000) has drawn attention to other patterns of counterfactual dependence having to do with what he calls "influence." These describe not just the counterfactual dependence of the occurrence of the effect on the occurrence of the cause, but also the counterfactual dependence of the time and manner of occurrence of the effect on the time and manner of the occurrence of the cause. Thus, as Lewis notes, in the example just described, it ordinarily will be true that if Suzy were to throw the rock in a sufficiently similar way, but slightly earlier or later, the shattering would occur slightly earlier or later; if Suzy were to vary the direction or momentum of the throw, but in such a way that the rock still struck the bottle sufficiently hard, corresponding variations in the effect (in the manner of shattering, the dispersal of the glass, and so on) would result. The counterfactual dependence of the occurrence of effects on their causes is such an obvious feature of many examples of causation that it is easy to miss the fact that there are other features, including those just described, having to do with counterfactual structure that play an important role in causal judgment.

The feature on which I will focus in this chapter has to do with the invariance of the causal relationship (and, more specifically, since we are operating with a counterfactual account of causation, the invariance of certain counterfactuals associated with it) to changes in various other factors—particularly invariance under changes in background conditions, as this is characterized in Chapter 5. I will call this background invariance for short.[3] (In other words, we will be considering relationships that are minimally causal in the sense of satisfying (M) or (6.2)–(6.3) or (6.4)–(6.5) and asking about the extent to which those relationships are invariant under changes in background conditions.) Interestingly, the role of this consideration in causal judgment is noted by Lewis himself. He describes causal claims that (in our language) hold in the actual circumstances but are relatively non-invariant under changes in background conditions as "sensitive" and those causal relations that would continue to hold under various sorts of changes in the actual circumstances as "insensitive." Here is how Lewis describes sensitivity:

> When an effect depends counterfactually on a cause, in general it will depend on much else as well. If the cause had occurred but other circumstances had been different, the effect would not have occurred. To the extent this is so, the dependence is sensitive. (1986, 186)

[3] Elsewhere I have called this feature "stability," and it is also called this in some of the psychological literature discussed subsequently. In this book I use "stability" as an alternative general term for invariance and restrict "sensitivity" in the way indicated.

Clearly sensitivity/insensitivity captures one aspect of invariance—at least roughly the same aspect as background invariance (or so I will assume). Lewis uses his notion of sensitivity to advance a proposal about the difference between "killing" and "causing to die"—he claims that killing is causing to die by relatively insensitive causation. He offers the following example of sensitive causation: he writes a strong letter of recommendation that causes someone—call him X—to get a job he would not otherwise have gotten (again in this context this just means that X's getting the job is counterfactually dependent on whether Lewis writes the letter), which in turn causes someone else, Y, who would have gotten the job in the absence of Lewis's letter, to take another job instead. In consequence Y meets and marries someone she would not otherwise have married. This couple has children who would not have existed in the absence of the letter and who eventually grow old and die. So their deaths are counterfactually dependent and hence (since on Lewis's view counterfactual dependence is sufficient for causation) caused by the recommendation. Lewis (1986) describes this as "comparatively sensitive causation" because, as he puts it, "there are many differences that would have deflected the chain of events. But if you shoot your victim point blank, only some very remarkable circumstances would prevent his death." This second action—the shooting—involves "insensitive" causation.

I will return to Lewis's suggestion about the connection between killing and insensitive causation of death at the end of this chapter, but for the present I want to develop some themes concerning the connection between sensitivity = non-invariance under background changes (background non-invariance) and causal assessment in more detail. Lewis's examples exploit the fact that we tend to think that there are important differences of some kind between relatively sensitive and relatively insensitive causal claims, but he does not (to my knowledge) further explore the significance of those differences.

I begin by suggesting that (as an empirical matter) many people will agree that there is some important difference between Lewis's pair of examples, even if they disagree about how best to characterize that difference. My own (entirely unsystematic) observation suggests that people (including philosophers) have a range of more specific responses to examples of the sort Lewis discusses. Some may think that the relatively non-invariant relationship present in the letter of recommendation example (and also in some of the examples described subsequently) shows that these relations are not causal at all, so that, for example, it is just false that Lewis's writing the letter caused deaths. They thus take the example to show that counterfactual dependence is not after all sufficient for causation. Others may think instead that although causal claims involving highly sensitive relationships are literally true, such claims are misleading or inappropriate or non-perspicuous in some way, at least if offered without additional qualification or elucidation, so that they involve some violation of conversational "pragmatics." Still others may think that the relationship in the letter of recommendation example (and similar cases) is genuinely causal but will then add that such highly sensitive causal relationships are defective in a way that goes beyond mere pragmatic failure—for example, they are relatively uninformative, shallow, or unilluminating in a way that reflects on their status qua causal claims.

My own view, as explained previously, is that from a normative perspective, the best systematic account judges that sensitive causal claims like that associated with the letter of recommendation example are true, assuming that they are intervention-supporting in a way that conforms to (M) or some token analogue to (M). Thus differences in sensitivity among intervention-supporting causal claims are differences among true causal claims rather than differences distinguishing true from false causal claims. Although I will employ this analysis as a way of organizing my discussion, my interest in this chapter is *not* in arguing for it, and much of what I say will not depend on it. For the most part all that I will assume in what follows is that, other things being equal, we tend to regard relatively background-invariant/insensitive relationships as preferable in some way, from the point of view of causal analysis, to less background-invariant relationships. "Preferable" here means (at least) that we value the discovery of more background-invariant relationships and, other things being equal, are less than fully satisfied with less background-invariant relationships. This is intended to cover a number of the possibilities described earlier: that highly sensitive (relatively non-background invariant) causal claims are false or that, alternatively, they are deficient in some other way. (I insist, however, that their deficiency does not consist just in their violating some domain-general rule of conversational pragmatics—see subsequent discussion.) For ease of reference, I will sometimes cover these various possibilities by saying that highly sensitive or relatively non-invariant causal claims are "non-paradigmatic."

To the best of my knowledge, there have been no empirical psychological studies of Lewis's particular examples. However, as we shall see in more detail in Chapter 7, there are empirical studies of ordinary subjects' causal strength judgments regarding causal claims in which sensitivity/background invariance is manipulated, with sensitive / relatively background-non-invariant claims regularly receiving lower causal strength ratings. In fact, as we shall also see, causal strength judgments seem to track a number of distinguishable features of causal claims (including different aspects of invariance) but *part* of what they track is degree of sensitivity / background invariance. One natural way of thinking of strength judgments is that they measure (at least in part) the extent to which subjects take causal claims to be paradigmatic or non-paradigmatic in the sense described previously. Moreover, although this interpretation is not strictly required, it is also natural to think of the fact that subjects assign variations in strength judgments to causal claims as capturing or reflecting the idea that subjects think of causal claims that satisfy a minimal condition (such as (M)) as genuinely causal but also as capable of varying along other dimensions, reflected in their strength judgments, having to do with their "goodness" qua causal claims.[4] If (in accordance with Chapter 3) we think of the role of these examples in Lewis's discussion as relying at least in part on his expectation that others will judge similarly, then his expectation is, as an empirical matter, correct. That is, his own responses to the examples correctly track a general pattern in causal judgment.

6.2. Sensitivity and Background Invariance

With this as background, let me turn to a more detailed discussion, focusing first on the sensitivity / background invariance of the counterfactual (6.2) associated with the token-causal claim (6.1). (My reason for focusing on the counterfactual (6.2) as opposed to (6.3) will become clearer later. Roughly, the reason is that in assessing (6.1), the invariance of (6.2) is more important than the invariance of (6.3).) As I have already explained, I take the background invariance of (6.2) (or, alternatively, an analogue to it, relating *c*-like events to *e*-like events) to have to do with whether it would continue to hold under various sorts of changes or departures from the actual state of affairs which are such that *c* or a *c*-like event continues to occur.[5] In the case of a token-causal claim and the associated counterfactuals, there will of course be a single actually obtaining set of background circumstances B_a. In assessing the invariance under background changes of (6.2), we thus ask whether counterfactuals of the following form are true:

> If *c* or a *c*-like event had occurred in circumstances B_i
> (where B_i is different from the actual background
> circumstances B_a), *e* or an *e*-like event would have occurred. (6.2*)

[4] As noted in Chapter 7, recent empirical work has attempted to distinguish judgments of causal strength from what are called judgments of "support": the latter has to do with judgments of whether a causal relationship is present at all, while strength judgments have to do with how "strong" the causal relationship is, given that one is present. This fits with the idea, adumbrated earlier, that we can use (M) and similar criteria to determine whether a causal relationship is present, regarding it as a further question how invariant that relationship is. To the extent that subjects are willing to judge that a causal relationship is present and regard it as a separate question how "strong" it is, they are following the normative analysis I recommend.

[5] This reference to whether "analogues" to (6.3) would continue to hold (and to *c*-like events) is intended for the metaphysically inclined and is meant to allow for the possibility that some departures from actuality—for example, those that alter the time or place of occurrence of *c* and *e*—may be such that they would affect the identities of *c* and *e*. To simplify the exposition, I will sometimes suppress reference to *c*-like and *e*-like events, speaking instead simply of what happens with respect to *e* when *c* occurs in circumstances B_i, but readers who care about such things can supply the "analogue" interpretation as they think necessary.

There are at least two different dimensions to this assessment. First, when such comparisons are possible, we can ask whether (6.2*) holds under "many" different background circumstances B_i or a "large" range of these. (Recall Chapter 5 and also see my subsequent discussion of what this might mean.) Second, when such comparisons are possible, we can also consider invariance under background circumstances that represent a relatively "small" departure from actuality as opposed to those that represent what Lewis calls a "large and substantial" departure. (The latter will include what I later call "far-fetched" or "improbable" departures.) Recall also from Chapter 5 that subject matter-specific considerations can be relevant to whether a change or departure is regarded as "substantial." A causal claim like (6.1) will be sensitive or relatively background non-invariant to the extent that many changes in background circumstances and/or changes that are not far-fetched or substantial departures from the actual circumstances will result in (6.2*) being false or non-applicable (in the sense described in Chapter 5, Section 5.2). Such a claim will be insensitive / non-background invariant to the extent that (6.2*) continues to hold under such changes.

In Lewis's framework for assessing counterfactuals (assuming an interpretation that adopts the requirement of strong centering), the counterfactual (6.2) relating the occurrence of *c* to the occurrence of *e* will be automatically true as long as *c* and *e* occur.[6] Thus, within this framework, there is an obvious sense in which, if we confine ourselves to contexts in which *c* and *e* occur, the counterfactual (6.2) does no further work over and above what follows just from the occurrence of *c* and *e*. Instead it is the truth of (6.3) that is crucial to whether *e* is counterfactually dependent on *c*. But even if we accept that (6.2) is always true if *c* and *e* occur, it is a further and nontrivial question whether and to what extent (6.2) would continue to hold under various departures from the actual circumstances—that is, whether to what extent counterfactuals of form (6.2*) are true. To the extent that we find it plausible that counterfactuals relating the occurrence of the cause on the occurrence of the effect (and not just counterfactuals relating the nonoccurrence of the cause to the nonoccurrence of the effect) are relevant to the assessment of causal claims, it is natural to focus on considerations having to do with that aspect of the sensitivity of (6.2) that is captured by (6.2*). In other words, considering the sensitivity of (6.2) (or its analogues) allows counterfactuals relating the occurrence of the cause to the occurrence of the effect to do real, nontrivial work in capturing aspects of causation.

6 The general claim that the occurrence of *c* and *e* is sufficient for (6.2) to be true does not hold within an interventionist framework for evaluating counterfactuals that rejects strong centering (Woodward 2016c; Briggs 2012). Nonetheless, we are assuming that in the example described, (6.2) is true when evaluated according to interventionist standards.

In view of these considerations it is not surprising that, although when given a causal claim of form (6.1), *c* caused *e*, we can assess the sensitivity (background invariance) of both of the associated counterfactuals (6.2) and (6.3), the sensitivity or insensitivity of the counterfactual (6.2) typically carries more weight than the sensitivity or insensitivity of the counterfactual (6.3) in our assessment of the overall sensitivity of (6.1). (This was our reason for focusing first on (6.2).) In consequence, a causal claim of form (6.1) for which the counterfactual (6.2) is highly sensitive will, ceteris paribus, strike us as sensitive / background non-invariant and hence non-paradigmatic, even if the counterfactual (6.3) is relatively insensitive. By contrast, if (6.2) is insensitive and (6.3) is sensitive, (6.1) will (at least often) strike us as unproblematic. In other words, other things being equal, causal relationships for which counterfactuals relating the *occurrence* of the cause to the *occurrence* of the effect are relatively insensitive tend to strike us as paradigmatic (unproblematic, particularly valuable to discover, likely to receive higher strength ratings, etc.), but we tend to care less about the sensitivity of the counterfactual relating the nonoccurrence of the cause to the nonoccurrence of the effect—this has less impact on our judgments about the associated causal claim. This is not to say, however, that we do not care at all about the sensitivity of the latter counterfactual. (Of course, we do care about the *truth* of 6.3 in the context under discussion.) In particular, as we shall see, if (6.2) is relatively sensitive and (6.3) is sensitive too, this may make us even more inclined to judge that the associated causal claim (6.1) is sensitive than if (6.2) is sensitive and (6.3) is insensitive. But it is the sensitivity of (6.2) that plays the primary role; the sensitivity of (6.3) is secondary.[7]

To illustrate these ideas, return to Lewis's examples, beginning with the letter of recommendation. Assume that Lewis writes the letter, and let N be some person who will exist in the future, eventually dying, who would not have existed if Lewis had not written the letter. Both of the counterfactuals

$$\text{If Lewis were to write the letter, N would die} \tag{6.6}$$

and

$$\text{If Lewis were to not write the letter, N would not die} \tag{6.7}$$

are true. Thus N's death is counterfactually dependent on and, on Lewis's (1986) view, caused by the letter writing. (Again, the interventionist account concurs

[7] I intend these as (in the first instance) empirical claims. But, as my accompanying discussion suggests, I also think that they make normative sense—another illustration of the intertwining of the empirical and normative that results when there is some normative justification for the behavior people in fact exhibit.

with this assessment, although it views this relationship as only minimally causal, in virtue of satisfying (M).) What about the sensitivity / background non-invariance of this causal relationship? Consider first the counterfactual (6.6). This counterfactual is highly sensitive: if Lewis had written the letter, but the actual circumstances had been different in any one of a number of ways that are not at all far-fetched, N would not have existed and hence would not have died. For example, if contrary to actual fact, N's father Z had not decided to move to city A where the job was located that N's mother Y got as a consequence of Lewis's letter, or if Z had lingered a little less long in the bar in A where he met Y, or if the department at University B had not decided to offer W a position with the result that the position at A became available to Y, and so on, then N would not have existed. Similarly for many other entirely realistic and not at all far-fetched changes. By contrast, the counterfactual (6.7) is far less sensitive. In most scenarios in which Lewis does not write the letter of recommendation, N does not exist and hence does not die. This illustrates my earlier claim that it is the sensitivity of the counterfactual linking the occurrence of the cause to the occurrence of the effect, rather than the sensitivity of the counterfactual linking the nonoccurrence of the cause to the nonoccurrence of the effect, that matters most for the overall sensitivity of the causal claim.

6.3. Additional Illustrations

To further illustrate the notion of sensitivity, let us return to the example of the bottle shattering. The sensitivity / background invariance of the counterfactual

> If the event of the rock thrown by Suzy striking
> the vase were to occur, the shattering of the vase would occur (6.8)

has to do with whether (6.8) or its analogues would continue to hold under circumstances that differ in various ways from the actual circumstances. Put slightly differently and considering the counterfactuals that we are taking to be associated with (6.8) as we vary background conditions, what we are interested in is whether (and which) counterfactuals of the form

> If the rock thrown by Suzy were to strike the bottle
> in circumstances Bi different from the actual
> circumstances, the bottle would (still) shatter (6.9)

are true for various B_i.

Some of the circumstances B_i for which claims of form (6.9) are true are so obvious that they will seem trivial. If Suzy's rock strikes the glass in Boston at the moment at which someone sneezes in Chicago, then presumably if that person had not sneezed but the world had remained relevantly similar in other respects, the bottle still would have shattered. Similarly, if we vary the color of Suzy's blouse or the price of tea in China at the time of the impact. Other claims about the insensitivity of counterfactual (6.9) are more interesting. Readers who found plausible my (and Lewis's) claim that in the original example the timing of the shattering is counterfactually dependent on the timing of the impact and in turn on the timing of Suzy's throw (recall Lewis's remarks about influence) presumably did so because they thought that if Suzy had thrown the rock (and the impact had occurred) slightly earlier or later, counterfactuals of form (6.9) involving small changes in temporal background but nothing more would still be true. That is, they assumed that if Suzy had thrown the rock a bit earlier or later but in other respects the background circumstances remained the same, this wouldn't affect whether (6.8)/(6.9) was true but only the time at which the shattering occurred. Similarly, if both the bottle and the throw had been displaced exactly ten feet to the right, then barring the presence of obstacles or other complications in the new situation, (6.8)/(6.9) would still be true. Similarly also for some range of possible variation in temperature, wind conditions, and so on. We thus see that Lewis's ideas about influence are (at least in many typical situations) closely bound up with assumptions about the relative invariance or insensitivity of certain relationships.[8]

Note also that the token judgments just described are closely related to—indeed, they are at least in part motivated or supported by—similar patterns of insensitivity that hold at the type level. Of course, when thrown rocks strike bottles, the bottles do not always shatter. Nonetheless, impacts of thrown rocks on bottles will be followed by shattering in many different circumstances—at different times, different spatial locations, and for a wide variety of other variations in background conditions. Indeed, it is at least in part because of this that my original description seemed so natural—my claim that the impact of the rock caused the shattering seemed unsurprising in part because the introduction of such impacts will be followed by shatterings in many different circumstances, a fact that we sum up in a claims like "Impacts of rocks cause bottle shatterings." If I had instead said that the shattering was caused by Suzy's scratching her nose, this would have been more puzzling. There are certainly possible circumstances in which bottle shatterings are counterfactually dependent on nose scratchings, but these circumstances are rare and rather specialized. Typically, they

[8] I do *not* claim, though, that influence is nothing more than sensitivity / background invariance—influence is also bound up with one aspect of causal specificity. See Woodward 2010 for additional discussion.

involve considerable stage setting: for example, Billy promises to throw a rock at the bottle if and only if Suzy scratches her nose; Suzy scratches, Billy keeps his promise, throws, hits the bottle, and the impact causes shattering. This dependence (and, more specifically, the counterfactual relating Suzy's scratching to the shattering) would be disrupted if the stage setting were altered—for example, if Billy were not present or if no promise had been made. In contrast to the connection between the impact of Suzy's rock and the shattering, counterfactuals connecting nose scratchings to bottle shattering are likely to be rather sensitive, and this seems connected to our puzzlement when the shattering is attributed (without further elucidation) to Suzy's nose scratching. This last causal claim strikes us as non-paradigmatic and explanatorily unsatisfying. We can make the sensitive dependence between Suzy's nose scratching and the shattering more intelligible by breaking it down, as I just did, into intermediate links of dependence that are individually less sensitive—the existence of the promise between Billy and Suzy, the link between Billy's throwing and the shattering, and so on.[9] This illustrates one way in which more insensitive or invariant relationships tend to be more satisfying from the point of view of explanation.

As discussed in more detail subsequently, the counterfactuals (6.2)–(6.3) (recall that 6.2 relates the occurrence of c to the occurrence of e and 6.3 relates the non-occurrence of c to the non-occurrence of e) and assessments of their sensitivity correspond to different strands or elements in the way we think about causation. One of these (following Hall 2004) we might label *dependence* and the other (somewhat more tendentiously)[10] *production*. Counterfactual (6.3) captures the idea that in simple cases in which other actual or potential causes

[9] As noted in Chapter 5, this is part of what is done when mechanisms and mediating variables are identified.

[10] This label will strike some as tendentious because they will hold that genuine cases of "production" require something more than relative invariance/insensitivity—perhaps something like a connecting physical process between cause and effect. This raises the question of whether all and only such those causal relationships in which there is a connecting process are relatively invariant or insensitive. I discuss this issue in more detail later, suggesting that the answer is no. I'm thus willing to label insensitive relationships in which there is no connecting process as cases of production—absence of oxygen produces death.

Although I lack space for detailed discussion, let me add that in my view it is very plausible that many cases of causal relationships in which there is a connecting process between cause and effect will also be cases in which that relationship is relatively invariant. (This empirical claim is defended in more detail in Woodward 2018b.) I speculate that we may tend to judge cases in which there is a connecting process between cause and effect as, other things being equal, more paradigmatically causal because these are typically relatively invariant rather than regarding cases in which there is a connecting process as more paradigmatically causal independently of whether the relationship is relatively invariant. Considered as an empirical thesis, this might be explored experimentally by presenting subjects with scenarios in which invariance is manipulated independently of whether a physical connection is present. Double prevention and other sorts of scenarios give us cases in which no connecting process is present and in which invariance can be manipulated independently and here the experimental evidence described in Chapter 7 as well as some more anecdotal armchair evidence described later in this chapter suggest that invariance has an independent effect on causal judgment. For example, as an empirical matter double-prevention relations in which no connecting

of *e* are absent, the occurrence of *e* depends on the occurrence of *c* in the sense that, assuming both *c* and *e* occur, *e* would not have occurred without *c*. In this straightforward sense *c* is a difference-maker for *e*.[11] Although it is well known that this simple version of the idea that causes are difference-makers breaks down in complex cases involving preemption and overdetermination, it is nonetheless true that in many paradigmatic cases, the intuition is correct—witness the rock-throwing example with which this chapter began.

Turning to cases in which a simple version of the idea that causes are difference-makers breaks down (because of preemption, overdetermination etc.), as noted in Chapter 2, a number of researchers (e.g., Halpern and Pearl 2005; Hitchcock 2001; Woodward 2003; Halpern 2016) have defended the claim that a more subtle version of this idea is defensible in such contexts: causes are difference-makers for their effects in such contexts when other causes or potential causes are appropriately controlled for. As an illustration, consider a structure in which Suzy's throw (c_1) shatters the bottle (*e*) but Billy, who does not throw, would have thrown (c_2) if Suzy had not and Billy's throw would have shattered the bottle in that case. Fixing the fact that Billy does not throw (the actual situation), if Suzy had not thrown, the bottle would not have shattered. Moreover, given that Suzy does throw, the bottle shatters, so in this sense Suzy's throw is a difference-maker for the shattering. Similarly, in a case of symmetric overdetermination in which Suzy's and Billy's rocks strike the bottle simultaneously (both being sufficient for shattering), if Billy had not thrown, the shattering would not have occurred if Suzy had not thrown.[12] The fact that the presence of other causes of *e* can make it the case that simple counterfactuals of form (6.3) linking the nonoccurrence of *c* to the nonoccurrence of *e* do not hold is one respect in which such counterfactuals are relatively sensitive / background non-invariant and one reason why such sensitivity (since it is common and unavoidable) often carries less weight in causal assessment than the sensitivity of counterfactuals of form (6.2).

While the counterfactual (6.3) is linked to the difference-making aspect of causation, I will suggest later that facts about (in)sensitivity/invariance of

processes are present receive higher causal strength ratings to the extent such relations are relatively invariant. Also of interest are cases in which a physical connection is present but the causal relationship is highly sensitive to variation in background conditions. Unpublished experimental data from Rose, Danks, and Machery (in preparation) suggest that adult subjects assign lower causal strength ratings to such relationships, despite the presence of a physical connection, again suggesting an independent effect of invariance. (For additional discussion see Woodward 2018b.)

[11] Similarly at the type level, the fact that there is some value of *C* such that if it had been different, the value of *E* would have been different, captures a notion of dependence.

[12] The accounts of Hitchcock, Halpern, and Woodward differ in detail but all provide more precise rules for what it is allowable to hold fixed or change in making such difference-making apparent. These accounts do not capture the full range of actual cause judgment but they do cover the examples described above.

counterfactuals of form (6.2) linking the occurrence of the cause to the occurrence of the effect instead correspond to one way of understanding the idea that the introduction of a cause should "suffice" for its effect (which I take to be another way of describing the feature that we think is present in "production"). Here "suffice" does not mean that the cause c as described is all by itself sufficient for the effect but rather something like the following: although conditions B_i in addition to c must be present for the effect to follow, there are many such additional conditions, one or more of which is likely to be present (their presence would not be far-fetched or a substantial departure from actuality) and which are such that if such a B_i is present, the effect follows if c is present.[13] I suggest that this feature influences whether we find it natural to describe a cause as "producing" (or as having the power to produce) an effect. Shooting someone through the heart "produces" death not only because, in the absence of other causes, the victim would not have died (dependence is present) but also because, in most non-far-fetched circumstances, such shooting is followed by (suffices for) death. By contrast, because of the sensitivity/non-invariance of the counterfactual (6.2), we find it far less natural to describe writing a letter of recommendation as a producer of death, on either the token or type level.

It is worth noting that the strategy for revealing dependence between c and e when other causes of e are present, which involves controlling in some way for the effects of these other causes (as described previously), itself assumes some degree of invariance or insensitivity in the operation of both c and these other causes. In a case of symmetric overdetermination with both c_1 and c_2 causing e, with no interaction between c_1 and c_2, it is because (or to the extent that) we think c_1 operates in the same way with respect to e, regardless of whether c_2 is also operative, that we think that we can understand the relation between c_1 and e when c_2 is present by looking at situations in which c_2 is absent. In other words, in following this strategy we assume invariance of the $c_1 \rightarrow e$ relation across changes in whether or not c_2 is also present, so that the appearance of a difference-making relation between c_1 and e when c_2 is absent is taken to tell us something about the relation between c_1 and e when c_2 is present. A similar analysis applies to cases of linked overdetermination in which Billy will throw if Suzy does not: we assume that the causal relation between Suzy's throw and the shattering is the same whether or not Billy is present and ready to throw. When Billy will throw if Suzy does not, the dependence relation between Suzy's throw and the shattering is more subtle (we need to consider a situation in which Billy will not throw to

13 Halpern 2016 describes such cases in terms of "strong causation" or "robust sufficiency." We might also drop the assumption of determinism and require only that the occurrence c raise the probability of e substantially across a range of background conditions, yielding a probabilistic version of production.

bring this relation out), but it still tracks the presence of the same causal relationship that would be present if Billy were absent.[14]

In many paradigmatic cases of causation, such as Suzy shattering the bottle, both elements—difference-making or dependence in the sense captured by (6.3) and insensitivity / background invariance of the counterfactual (6.2)—are present to a high degree. Nonetheless, there are also many examples in which these two elements can come apart or vary somewhat independently of each other. In the recent literature on causation a great deal of attention has been focused on cases involving various forms of overdetermination in which simple difference-making relations of a sort captured by (6.3) fail to be present. Cases where there is straightforward counterfactual dependence (of form (6.3)) but variation along the dimension of sensitivity/invariance of the counterfactual (6.2), have received considerably less attention, but they are also interesting and important, as this chapter attempts to illustrate.[15]

[14] The idea that in the cases under discussion the causal relationship between *c* and *e* should not depend on whether other causes or potential of *e* are present (*c* acts in the same way with respect to *e*, whether or not these other factors are present) is related to and has a somewhat similar motivation to what Paul and Hall call the instrinsicness thesis (cf. Lewis 1986) and their blueprint strategy (2013, 124ff.), which relies on the notion of an intrinsic duplicate. However, Lewis and (if I understand them correctly, Paul and Hall) hope to provide a reductive characterization of "intrinsic duplicate" (i.e., a characterization that does not itself appeal to causal notions). For reasons that should be familiar to the reader, I am skeptical that this is possible. But if interpreted non-reductively, I think the basic idea behind the blueprint strategy is correct. It is another way of trying to capture the idea that causes should to some extent behave in the same way when other causes are present as when they are absent.

[15] Given the role of both strands in causal judgment it is natural to wonder whether one of them is dispensable or at least less important than the other. The framework I have adopted, according to which a relationship must satisfy (M) in order to qualify as minimally causal, arguably gives a kind of priority to the dependence strand. I think this element is not dispensable: we need reference to something like the difference-making or dependence aspect of causation to exclude candidate causes that contain superfluous elements. These can be present even in invariant relationships. For example, the relationship between ingestion of birth control pills by males and failure to become pregnant is highly invariant but the relation between ingestion and non-pregnancy is not a causal relationship because whether or not a male becomes pregnant does not depend on whether he ingests birth control pills. As noted earlier, we think that the dependence aspect of a causal relation can be hidden by the presence of other causes, but (I claim) we also think that if the relationship is genuinely causal, the dependence aspect can be made to reappear under the right sort of control for these other causes. On the other hand and going in the other direction, I agree that it is also hard to make sense of a relationship as causal when dependence is present but the relationship (and particularly the counterfactual relationship between the occurrence of the cause and the occurrence of the effect) is, so to speak, sensitive to virtually all possible background circumstances—at the very least this seems to be a limiting and highly non-paradigmatic case of causation. As an illustration, consider the counterfactual "If the collapse (*r*) of the Roman Empire had not occurred in the fifth century, then the French Revolution (*f*) would not have occurred." (Here *r* and *f* refer respectively to the occurrence of the collapse and the occurrence of the revolution.) The consequent of this counterfactual is vague, but it is arguable that the counterfactual itself is true (perhaps because if the antecedent were true, there would have been no France) or at least not obviously false. Moreover both *r* and *f* occurred. Although we have (or may have) dependence, the "If *r*, then *f*" counterfactual is (at best) extraordinarily sensitive, since under many possible variations in background conditions, it is not true that *r* would have been followed by *f*. This is reflected in the observation that few people will accept "*r* caused *f*" despite the apparent truth of the counterfactual.

6.4. Sensitivity / Background Invariance and Departures from Actuality

I remarked in Chapter 5 that judgments of background invariance/sensitivity (perhaps particularly but by no means only in the case of common-sense causal judgment involving token-causal claims) are influenced by (among other considerations) whether the variations under which a relationship continues to hold seem to reflect a large or small departure from actuality. I want now to say a bit more about this idea since I will rely on it in portions of what follows.

One nicely behaved and straightforward possibility is that we have a theory describing a state space representing a range of alternatives to the actual state and a natural associated metric specifying the distance between the actual state and these alternatives. Consider a driver who comes upon an icy patch in the road, begins to skid, loses control of the car, and then regains it, ending up on the shoulder of the road and not in the ditch beyond it. Suppose the causal claim of interest has to do with the relationship between encountering the icy patch and the subsequent trajectory of the car. Among the alternatives to the actual state of affairs are those in which the driver hits the icy patch at a different speed or from a different angle. If his actual speed is 60 mph, an alternative in which he is traveling 55 mph is naturally viewed as closer to actuality than alternatives in which he is traveling 30 mph or 90 mph. Similarly, if the driver has had three glasses of wine, an alternative in which he has two is closer to actuality than an alternative in which he has none. If the outcome of the encounter between the driver and the ice would have been very different under small variations in the driver's speed or alcohol consumption—if the driver would have gone into the ditch if he had been traveling at 65 mph or had 3.5 glasses of wine, then the causal relationship between hitting the ice patch and the outcome is (in these respects) relatively sensitive or background non-invariant.[16]

In other cases, while there may be considerable agreement about whether some departure from actuality is substantial, this judgment may rest on considerations that are multifaceted, not reliant on some small set of metrics, and perhaps more difficult to make precise. If the actual situation is one in which Suzy throws a rock and it travels via an unimpeded path toward the bottle, striking and breaking it, most people would agree that variations in which there is a thick,

In addition, as observed earlier, some degree of invariance/insensitivity seems to be required if we are to make sense of dependence relations when overdetermination and similar phenomena are present. So to this extent, both dependence and some degree of invariance appear to be non-dispensable elements in causal claims.

[16] As this example suggests, spatial and temporal distance and functions of them like velocity are fruitful sources of metrics for comparisons of how "close" alternative variations in background conditions are.

solid steel barrier between the rock and the bottle or in which a second person is present who throws a rock that deflects Suzy's rock in flight represent relatively large departures from actuality, at least in comparison with alternatives in which the rock Suzy throws still strikes the bottle but with a slightly different momentum because a slightly stronger breeze is present or the bottle is moved a tiny distance to the right. If, were Suzy to throw, the bottle would shatter in alternative scenarios in which the wind speed, the orientation of the bottle, and so on do not vary too much from actuality, we are likely to think of the connection between the throw and the shattering as relatively insensitive/invariant, even though it is true that under other more far-fetched variations (the presence of the steel barrier and so forth), this counterfactual would no longer hold.[17]

Other considerations that seem to influence assessments of far-fetchedness include the following: how improbable or uncommon the changes in question seem, either in general or given the particular circumstances of the example; whether the changes require alterations in background structures or institutions that are ordinarily stable and/or difficult to change; and so on. In addition, it is also worth noting—this is a point to which I will return—that, as this example illustrates, there seems to be an important asymmetry between, on the one hand, adding new background circumstances to the actual situation and, on the other, removing circumstances that are present in the actual situation. The former tends to generally strike us as more of a departure from actuality and hence as mattering less for the assessment of sensitivity/invariance. In the nose-scratching case, at least part of the reason why we regard the causation as sensitive is that the bottle shattering would not have occurred in the absence of Billy's promise—that is, removing Billy's promise seems like an appropriate change for the purposes of assessing sensitivity. By contrast, we are considerably less likely to judge that the causal relation between Suzy's throw and the shattering is sensitive on the grounds that had the steel barrier been present, the throw would not have led to shattering. We view the former change as a matter of removing something (Billy's promise) from the original situation, rather than adding something and hence, ceteris paribus, as a smaller or more allowable change. By contrast,

[17] This is a claim that, like many others in this chapter, might be tested systematically. When a causal claim would continue to hold under small or nearby departures from actuality (where these are understood along the lines described previously), are subjects more likely to assign it a higher causal strength rating (or a higher rating on some other measure of the goodness of the causal claim) than they do to a similar claim that would fail to hold under even small departures from actuality? If we just ask subjects to verbally evaluate how "close" or similar various possibilities are to one another, are they guided by the considerations described here? Do their closeness judgments influence their judgments of causal strength in the way suggested? Is there some other way of accessing subject evaluations of closeness besides verbal behavior?

the insertion of the barrier is viewed as adding to the original situation and hence as a larger change or at least as less appropriate for the assessment of sensitivity.[18]

So far I have focused mainly on cases involving token causation. Here the actual circumstances in which the cause and effect occur are the obvious baseline for assessments of the extent of departures from "actuality" in background conditions. In the case of type-level causal claims, this built-in baseline is not available, but (as noted in Chapter 5) we nonetheless often operate with notions of what counts as a more or less substantial departure from a "normal" range of possibilities. Given, say, a candidate psychological generalization G linking mental state M_1 to mental state M_2, we might ask whether G continues to hold under various changes in subjects' external environment that are normal or occur with some frequency for human beings—such background conditions will not seem far-fetched or a major departure from actuality. On the other hand, asking whether G would continue to hold under a change in subjects' brains that involves replacing half of their neurons with silicon alternatives does seem like a far-fetched departure. As with token-causal claims, at least in many contexts we care more about the degree of sensitivity of type-level claims under normal, non-far-fetched variations in background conditions than under variations that are highly abnormal.

6.5. Sensitivity/Invariance Distinguished from Other Features

Next a remark about the relationship between sensitivity and other features a relationship may possess. Bearing in mind examples like Lewis's, it may be tempting to think that the relative invariance of the relationship between *C* and *E* is simply a function of the "length" of (or the "number" of intermediate steps in) the causal chain from *C* to *E*. When Lewis writes his letter of recommendation, there are many intermediate steps along the causal chain leading to deaths—the causal relation between writing the letter and the deaths is highly indirect and also relatively non-invariant. In the case of shooting someone, the connection between shooting and death seems much more direct and relatively invariant. This may prompt the thought that perhaps our different reactions to the two

18 A related observation is that while removing or holding fixed certain other causes *c** of an effect *e* so as to make a difference-making relation between *c* and *e* apparent (where that relation would not otherwise seem apparent) is a natural strategy, it is not a plausible strategy to try to make a difference-making relation apparent by adding new causes to *e*. That is, if there is no apparent difference-making relation between *c* and *e*, we don't think that adding other causes of *e* is a way of making such a relation visible, for the obvious reason that any difference-making that is present may be due to the new causes. Of course the considerations that lead a change to be regarded as an addition rather than a removal merit further exploration.

examples are really driven by our tendency to regard relatively direct causal relations (with few intervening steps) as more paradigmatically causal than very indirect causal relationships.[19] Although there is a defensible idea lurking in the neighborhood of this response that I will come to shortly, the response as described is incorrect in several respects. First, length and number of intermediate steps (like the notion of the "directness" of a causal relationship) are clearly relative to how we "grain" things or to the choice of vocabulary for describing causal relationships. With respect to a list confined to variables like "shooting" and "death," the causal link between shooting and death looks short and direct; with respect to an expanded list of variables that might be used to describe the chain of physiological changes in the victim that eventuate in death (damage to the heart, loss of its ability to pump blood, failure of oxygen to reach the brain), the causal chain from shooting to death looks longer and less direct. However, including these intermediate variables will not by itself make the overall causal relationship between shooting and death any more (or less) sensitive. Roughly speaking, the overall background invariance of a relationship between *C* and *E* that involves intermediate links $C_1 \ldots C_n$ depends on how background invariant the intermediate links are and how the ranges of invariance of those links relate to one another. At one extreme one might have a set of intermediate links $C \rightarrow C_1$, $C_2 \rightarrow C_3, \ldots C_n \rightarrow E$ and so on, each of which is relatively background invariant but where the conditions under which each link is invariant are quite different from one another so that the overall relationship between *C* and *E* is relatively non-invariant. (The overall invariance of the $C \rightarrow E$ link will reflect the *intersection* of the circumstances under which each of the individual links is invariant, and this intersection may be quite small.) Alternatively, there may be considerable overlap among the circumstances under which the individual links are invariant, in which case the overall $C \rightarrow E$ link may be nearly as invariant as each of the individual links. Something like this may well be the case in the example of a shooting causing death. By contrast, there is likely to be little or limited overlap between the range of invariance of the individual links following the letter of recommendation, so that the overall relation is judged far more non-invariant than many of the individual links, which may themselves be relatively more invariant.

What is true (and illustrated by the letter of recommendation) is this: when an overall candidate causal or explanatory relationship is relatively non-invariant, but the individual links involving intermediate steps are more invariant, we can

[19] I am not aware of any empirical research that bears directly on how the subjects' strength judgments regarding individual links in a causal chain relate to their judgments about the overall strength of the chain. Are judgments of overall strength some function of judgments of the strength of individual links? A function of the weakest link only? These are questions that may be worth investigating. Philips and Shaw (2015) explore a somewhat related question concerning responsibility ascription. See also Murray and Lombrozo 2017.

gain intelligibility (or improve matters from the point of view of causal analysis or perhaps explanation) by replacing the overall relationship with one exhibiting the intermediate links—exhibiting the intermediate links is better because we are now appealing to relationships that are individually more invariant. Thus in Lewis's example, while the overall relation between writing the letter and the deaths seems completely opaque and non-explanatory (and to this extent non-paradigmatic), this relationship becomes understandable when we are presented with an account detailing the intermediate links. By contrast, if (as I have suggested) the intermediate links in the shooting case are unlikely to be substantially more invariant than the overall shooting-death relationship, this implies that there is less of an explanatory gain to be obtained by exhibiting those links, which is why in many circumstances there will seem to be no particular need to provide them. Indeed, much of the explanatory appeal of exhibiting "mechanisms" or mediating variables consists in the fact that in doing so we replace an overall input-output relation that may be relatively non-invariant with an account exhibiting mediating links that are individually more invariant—see Woodward 2013.[20]

6.6. Absences

I turn now to applying some of the ideas just described to some puzzle cases about causation discussed in the philosophical literature. One can think of what follows as a suggestion about why, as an empirical matter, we make some of the causal judgments we do but also as an account of why, at least in some cases, those judgments are normatively justified. There is an ongoing debate about whether omissions or absences can be causes. Focusing first on cases in which the candidate causal relations involve token causation, there are many cases in which *e* occurs, *c* does not occur, and the occurrence of *e* is counterfactually dependent on the nonoccurrence of *c* (in the sense captured by the two counterfactuals (6.2) and (6.3)) but in which we are reluctant to judge that the nonoccurrence of *c* caused *e*. This suggests that in at least some cases involving candidate causes that are characterized as nonoccurrences, counterfactual dependence is not sufficient

[20] What is the relationship between the background invariance of a generalization and whether it is deterministic? These are very different notions. A probabilistic generalization G, which specifies that if certain conditions C are satisfied, some outcome O follows with probability p, with $0 < p < 1$ might be highly invariant: G might hold for a range of interventions and across a wide range of background circumstances in the sense that O follows with stable probability p in all of those circumstances. Note also that a probabilistic generalization might be highly invariant even if the probability p is very low. Alternatively, G might be very sensitive / relatively non-invariant even if p is high. For more on how background invariance might be distinguished empirically from other features a generalization might possess, see Chapter 7.

for causation. To use a well-worn example, it might be the case that the death of certain flowers in the White House garden depends counterfactually on Trump's failure to water them—if he had watered them they would have survived and if, as was in fact the case, he hadn't watered them, no one else would have and they would have died. Nonetheless most people would resist judging that Trump's failure to water caused the death of the flowers.[21] Some theorists go further, claiming that, for purposes of normative theory, absences or omissions are never causes, at least for what they take to be the literal or primary notion of "cause." There are at least two broad motivations for this position, which need not be mutually exclusive. One appeals to the idea that there is a fundamental metaphysical distinction between presences and absences and that only presences can figure in causal relationships. For example, Beebee 2004 holds that causation is a relation between events and that absences, being literally "nothing" or mere privations, are not events and hence cannot stand in causal relationships.[22] A second and

[21] There are some obvious complications and confounds here. Perhaps—this is just one possibility—when people make this judgment they interpret the question as about whether Trump is morally responsible for the death of the flowers, although in other contexts they distinguish moral responsibility from causation. So perhaps their responses reflect nothing about their judgments concerning whether Trump's omission caused the death of the flowers but only their judgment that Trump is not morally responsible for this outcome. (This would be an example of ambiguity in the verbal probe, similar to examples discussed in Chapter 7.) This suggests the need for verbal probes and other strategies that disentangle causal and responsibility judgments. However, I will ignore this issue in what follows and take the causal judgment at face value.

[22] Presumably, if this argument or the following Salmon/Dowe argument is cogent, it also follows that absences cannot be effects. There are obvious issues here, which I will largely ignore, about the basis on which some possibility gets coded as an absence or omission rather than a presence. For example, is death an absence (of life)? If this is the case and if causal relata must be events and absences are not events, it would follow that death cannot be either a cause or an effect and that inquiries into causes of deaths are conceptually incoherent. This would come as a surprise to coroners and epidemiologists. Or, alternatively, is death a positive presence or the occurrence of an event rather than a mere privation, and if so, is continued life an absence (of death)? How would one go about how about answering these questions? How about starvation? Is that an absence? Can it nonetheless cause death (as it certainly seems to)? I will not try to address such issues here and will confine my focus to cases in which the application of the presence/absence distinction is generally regarded as unproblematic. Needless to say, to the extent that the distinction is unclear or its basis problematic, strong claims about absences never being causes or effects will seem less appealing.

It is also worth adding that, as nearly as I can see, the argument that absences are not events and hence correspond to nonexistent relata and cannot stand in causal relationships cannot be coherently formulated within an interventionist framework. Within that framework the relata of causal relationships are what variables or values of variables correspond to in nature. For example, the claim that the gardener's failure to water caused the plants to die is interpreted as a claim relating a value taken by a watering variable W, with values {water, not water} to a variable representing whether the plants die. When the variable W takes the value = not water, this corresponds to something "real" in the world—the gardener's not watering—and not to something nonexistent or unreal. In general within an interventionist framework one thinks of causal relationships as dependency relations that always relate two or more values of the cause variable to different values of the effect variable. There is nothing problematic about one or both of the cause and effect variables having a value corresponding to "absence," and the dependency relation incorporates what happens when this is the case just as it does when the variables take the value "present." In other words, the claim about what happens under "absent" and what happens under "present" are both part of the dependency relationship—it is not as though only the latter counts as causal.

related argument, alluded to at several points in previous chapters, is that if *c* is to cause *e*, there must be a causal process connecting *c* to *e*. Assuming that absences are not connected to other events via such processes, it follows that they can be neither causes nor effects.

As an empirical matter, though, it seems uncontroversial that people often do judge that absences and omissions are causes (and effects).[23] For example, people readily judge that the absence of oxygen caused a particular death and, also, at the type level, that the absence of oxygen causes death. Similarly for deaths caused by starvation, malnutrition, and dehydration.[24] Most people think it uncontroversial that lack of vitamin C causes scurvy. Moreover, as these and earlier examples illustrate, again as an empirical matter, people discriminate *among* relationships of counterfactual dependence involving absences, judging that some are causal (or perhaps are more paradigmatically causal) and others are not (or perhaps are less paradigmatically causal). It is natural to wonder what underlies (again as an empirical matter) the distinctions we make concerning the causal role of absences and whether there is any normative basis for such distinctions. Neither of the two views I have described, according to which absences are never causes, provides an answer to these questions.

In what follows, I propose to explore some possible answers to this question, asking, first, about what as an empirical matter underlies the distinctions people make concerning the causal role of absences[25] and then exploring whether there

[23] For empirical evidence that in the case of non-waterings people do sometimes judge that omissions are causes, see, for example, Henne et al. 2019 and Clarke et al. 2013. Both papers report evidence supporting the claim (which seems uncontroversial just on the basis of common knowledge of practices of judgment) that subjects are more likely to agree with the statement that an actor's omission is a cause of some outcome if the omission involves violation of a norm. For example, the gardener's failure to water the plants is more likely to be judged a cause of their deaths (or a "stronger" cause) if the gardener is required by some norm to water the plants. While I fully agree that norms influence causal judgment in cases of causation by omission and in many other cases, I would also emphasize that it is very plausible that they do not do so independently of invariance/sensitivity considerations. For example, when there is a norm that the gardener waters the plants, others not subject to the norm are presumably less likely to water and their doing so is more likely to be somewhat abnormal. In such cases the counterfactual "If the gardener were to omit, the plant would die" is more insensitive than in a case in which there is no norm requiring the gardener to water. As Thomas Blanchard has pointed out to me, this also helps to explain why the claim that Trump's failure to water the plants caused their death strikes us as less acceptable than the claim the gardener's failure caused their death. In the usual case in which the norm is that the gardener waters, the counterfactual linking Trump's failure to water to the death of the plants will be relatively sensitive since in that case the plants are likely to be watered by the gardener (and hence survive). For relevant additional discussion of the role of norms in causal judgment and attribution see Hitchcock and Knobe 2009; Icard et al. 2017.

[24] Indeed these causes of death are sufficiently paradigmatic that they appear on death certificates. Malnutrition, presumably involving an absence of nourishment, is cited by the World Health Organization as a leading cause of childhood mortality in poorer parts of the world.

[25] Large portions of this chapter including the proposals that follow are based on Woodward 2006. Although intended as empirical (as well as normative) claims, the proposals themselves were definitely originally formulated from the armchair and not as a result of any systematic empirical investigation. For some experimental results bearing on the proposals see Chapter 7.

may be some normative rationale for these distinctions. I should acknowledge at the outset that depending on the example under discussion, a number of different kinds of considerations may be at work, including (as a number of writers have observed) ideas having to do with moral or other sorts of responsibility especially when human actors are involved.[26] (See also footnote 21.) I want to focus, however, on the following possibility: that in a number of cases in which we are reluctant to say that some absence is a cause (even though the right sort of counterfactual dependence, represented by (6.2)–(6.3) is present), an important reason for our reluctance is that the relationship between the absence and its putative effect is relatively background non-invariant/sensitive. By contrast, the absences that we more readily accept as causes often involve relationships that relatively invariant.

I begin with a contrast between two examples. First, consider

My writing of this very chapter was caused by my not being
hit by a large meteor prior to beginning it (6.10)

and the associated counterfactuals

If I were not struck by a large meteor before beginning it,
I would have written this very chapter (6.11)

and

If I were struck by a large meteor before beginning it,
I would not have written this very chapter. (6.12)

Assume that in fact I have written this chapter and that I have so far not been struck by a large meteor. That (6.12) is true seems uncontroversial. It is arguably less obvious that (6.11) is true—I will return to this shortly—but let's assume for the moment that it *is* true so that my writing this very chapter is counterfactually dependent on my not being struck by a large meteor before beginning it.[27] This

[26] As noted previously, yet another consideration that (as an empirical matter) plausibly influences causal assessment, both in connection with absences and elsewhere, has to do with which possibilities regard as "serious." (See also Woodward 2003.) When the occurrence of *c* is not regarded as a serious possibility, many will reject the claim that the absence of *c* caused *e*, even when counterfactual dependence is present. Thus one might reject the claim that Trump's failure to water the flowers caused their deaths on the grounds that his watering was not a serious possibility. The remarks that follow about the role of invariance in assessing causation by absence claims are intended as complimentary to this observation about the role of serious possibility rather than in competition to it. That is, my claim is that invariance considerations influence assessment of causation by absence claims in addition to whatever role is played by serious possibility considerations.

[27] The Lewis/Stalnaker framework with a commitment to strong centering judges that (6.11) is true. As noted later, if one rejects strong centering, one might conclude that (6.11) is false.

pattern of counterfactual dependence tempts some to say that (6.10) is literally true. On the other hand, this claim strikes others as quite counterintuitive—so much so that they regard examples like (6.10) as clear counterexamples to the contention that counterfactual dependence is sufficient for causation. Moreover, many who think that (6.10) is literally true will also think that there are differences of some significant sort between (6.10) and other more paradigmatic cases of causation.

Focusing first on (6.11) (in this case the counterfactual relating the occurrence of the putative cause—my failure to be struck—to the effect), we see that even if it is true, it is highly sensitive/non-invariant. Even if I had not been struck by a meteor, there are many relatively small changes that might have occurred under which I would not have written this very chapter. I might not have been invited to give various talks that prompted me to begin writing the essay on which the content of this chapter is based, I might not have had certain conversations that have influenced the content of this chapter, and so on.

This relative non-invariance of (6.11), I suggest, plays a role in people's reluctance to accept (6.10) or in their inclination to think that there is something about (6.10) that separates it from more paradigmatic cases of causation. In contrast, (6.12) is presumably quite insensitive. In virtually all close-by scenarios in which I am struck by a large meteor, I die.

The asymmetry between the sensitivity of (6.11) and the insensitivity of (6.12) provides further illustration of my claim that it is the sensitivity or insensitivity of counterfactuals relating the occurrence of the putative cause (in the case of (6.10), the nonoccurrence of the meteor strike is the occurrence of the putative cause) to the occurrence of its putative effect rather than the sensitivity or nonsensitivity of the counterfactual relating the nonoccurrence of the cause to the nonoccurrence of the effect that exerts the primary influence on our judgments regarding the causal claim. Even if true, (6.10) strikes us as nonparadigmatic because (6.11) is sensitive even though (6.12) is insensitive.

We can provide further support for this analysis by reversing the facts of the case. Suppose that the counterfactual claims (6.11) and (6.12) remain true but that now I am struck by a meteor and die before writing this chapter. Consider the claim that

Being struck by a meteor caused me not to write this chapter. (6.13)

Now our focus shifts to the counterfactual (6.12) since this is now the counterfactual relating the occurrence of the putative cause in (6.13) to its effect. Since (6.12) is insensitive / relatively invariant, we judge that there is nothing odd or nonparadigmatic about (6.13) qua causal claim, despite the fact that (6.11), relating the nonoccurrence of the cause to the nonoccurrence of the effect in

(6.13), is sensitive. Again this assessment matches most people's intuitive judgment about (6.13).

At this point let me confront an objection.[28] To the extent that ordinary judgment regards (6.10) as a deficient or non-paradigmatic causal claim, the account I have provided would explain it. But (the objection goes) ordinary intuition/judgment regards (6.10) not just as non-paradigmatic but as false. By contrast (the objector continues) my account, insofar as it relies on (M) or something similar, judges (6.10) as true. So insofar as the goal is to capture ordinary judgment/intuition about omissions, my account does not fully achieve this. There are a number of possible responses to this objection. Perhaps it is not so clear that ordinary judgment regards (6.10) as false. Perhaps, as suggested earlier, when *c* (in this case, being struck by a meteor) is highly unlikely, this adds (beyond invariance-based considerations) to our willingness to regard claims in which the absence of *c* is treated as a cause as false. In any case, the methodology I advocate does not regard ordinary judgment as sacrosanct or incapable of being mistaken. Perhaps to the extent that people regard (6.10) as false, they are mistaken—but their mistake is easily explained in a non-ad hoc way since it is easy to confuse "false" with "true but highly non-paradigmatic." Next recall that so far we have been assuming strong centering for the evaluation of counterfactuals. If we reject strong centering (which is certainly an option on an interventionist account of counterfactuals), it is arguable that (6.11) is false. If (6.11) is false, we don't have the counterfactual dependence that is required for (6.10) to be true. So we can recover the falsity of (6.10) in this way if we wish.

Note also that the rejection of strong centering seems very close to requiring that for a counterfactual to be true it must exhibit some degree of insensitivity/invariance. In requiring such insensitivity we are in effect proceeding as though there are other worlds besides the actual world that are as relevant to the assessment of the counterfactual as the actual world—the actual world no longer has a privileged position with respect to the similarity relation since there are other worlds that may be equally close to the actual world. Requiring that the consequent of a counterfactual must hold in all such closest worlds in which antecedent holds plausibly leads to the conclusion that (6.11) is false. Thus the difference between, on the one hand, rejecting strong centering and taking (6.11) to be false and, on the other hand, taking it to be true but then judging it to exhibit failures of sensitivity/invariance, which renders it non-paradigmatic, is perhaps not so very large. In both cases we are picking up on the failure of (6.11) to be true in

[28] Thanks to Zina Ward for raising this objection and for suggesting that (in this particular respect) my account amounts to an error theory of a number of ordinary judgments about absences.

worlds other than the actual one, but expressing this somewhat differently, either in terms of the falsity of (6.11) or its sensitivity.[29]

Let us now contrast these variants on the meteor example with two other causal claims involving absences:

Absence (of access) to oxygen caused N's death (6.14)

Many German civilians were caused to die from starvation (that is, from absence of food) by the British naval blockade of 1919. (6.15)

I take it to be uncontroversial that we are much more likely to regard (6.14) and (6.15) as acceptable than (6.10). Indeed many people will regard (6.14) and (6.15) as unproblematic, despite the fact that they seem to attribute causal efficacy to absences and despite the fact that there is no connecting causal process running from the putative cause to its effect. Certainly when an agent acts so as to bring about the absence described in (6.14)—I place a plastic bag over your head or remove all of the oxygen from an airtight room into which you have wandered, and death results—we seem to have little hesitancy in thinking of these as acts that cause death.[30] Similarly for the blockade described in (6.15) or if I were to lock you in a cell and refuse to feed you. I trace the greater acceptability of (6.14)–(6.15) in comparison with (6.10) to the relative insensitivity of counterfactuals like

If N were to lack access to oxygen, N would die (6.16)

and

If German civilians were denied access to food in 1919, they would die. (6.17)

[29] In other words the worldly facts on which the assessment of (6.10) rests seem to be largely the same under both analyses. Let me also acknowledge that this puts some pressure on my earlier organizing assumption that there is a relatively sharp distinction between "binary" judgments about whether *c* causes *e* and judgments about how "good" or paradigmatic this causal claim is, if true, with the latter judgments being more graded. What can I say? Things are complicated and it is not useful to pretend that they are not.

[30] Suppose that we hold that absences can't be causes or effects. How then might we describe the intervening links when the act of placing the plastic bag over the victim's head (the act itself being something "positive") leads to death? Obviously we can't say that the act causes the absence of oxygen, which causes death. The question is whether there is any way of describing the situation in a more fine-grained way that avoids attributing a causal role to absences at some point.

Under most non-far-fetched cases in which background conditions vary from the actual circumstances, people who are deprived of oxygen or food die. Moreover, the type-counterfactual claims corresponding to (6.16)–(6.17) and the corresponding type-causal claims are also true.

Next let us contrast all of the preceding cases with yet another case, also involving an absence or omission: Y is a critically ill patient whose care is the exclusive responsibility of Doctor Z. The nature of Y's disease is such that he will die if he is not given various medications, and no one else is permitted to administer the medications if Z doesn't. Z fails to administer the medications and Y dies. Here the relevant counterfactuals are

If Z omits to administer the medication, Y dies (6.18)

and

If Z does administer the medication, Y does not die (6.19)

and the causal claim of interest is

Z's omission to administer the medication caused Y's death. (6.20)

How sensitive is the counterfactual (6.18)? On the one hand, it looks considerably less sensitive than the counterfactual (6.11), relating the failure of the meteor to strike me to my writing this chapter (claim (6.10)). Given Z's omission, Y's death will occur under a significant range of non-far-fetched scenarios, as long as we continue to suppose that the disease is present in the same form and no one else besides Z is permitted to administer the medicine. Moreover, (6.19) is also (we may assume) relatively insensitive, and this may lend some additional, but secondary, support to the judgment that (6.20) is relatively insensitive. My suggestion is that these facts help to explain why we are considerably less reluctant to accept (6.20) than to accept (6.10). However, while (6.20) is less sensitive than (6.10), there are nonetheless variations in background conditions that, although they depart in some significant ways from the actual situation, do not seem widely farfetched or implausible and under which (6.20) (and (6.18)) would no longer hold. Most obviously, it is not at all a far-fetched possibility that Y's disease might have been absent or present only in a milder, nonlethal form, and if so, Y would not have died, even given Z's omission. (This contrasts with, for example, (6.14), where it is not a serious possibility that the physiological conditions that in conjunction with the absence of oxygen are sufficient for death might be absent in a normal human being.) Similarly, Y would have survived even given Z's omission, if someone else had administered the medication. In many medical contexts, this is not just possible but likely. Often hospitals are set up in such a way that others will become aware of Y's condition and will administer the medicine if Z does not. In both these respects, (6.20) differs from (6.14) and (6.15). The upshot is that while (6.20) is less sensitive than (6.10), it is more sensitive than (6.14) and (6.15). This is reflected, I believe, in our reaction to these claims—(6.14) and (6.15) strike us as more paradigmatically causal (more readily acceptable, and so on) than (6.20), which is in turn less problematic than (6.10).

For purposes of additional comparison, consider a final example. Suppose that you omit to send $10 to a famine relief organization O, X who lives in country A dies of starvation, and X would have lived if you had sent the money. Letting *c* be your omission to send the money and *e* be X's death, the relevant counterfactuals are

If *c* were to occur, then *e* would occur (6.21)

and

If *c* had not occurred (that is, you send the money), then *e* would not have occurred. (6.22)

Assuming that these counterfactuals are true, *e* counterfactually depends on *c*. As before, I take it that while some people may hold that

Your omission to send money caused X's death (6.23)

others will be reluctant to accept this claim, at least without some further qualification. The causal claim (6.23) will strike most people as at least different from paradigmatic, true causal claims. Again it is natural to trace these reactions, at least in part, to the fact that even if true, (6.21) is rather sensitive. Even supposing that you omit to send the money, if any one of a number of things had happened differently—if the corrupt dictator who runs country A had stolen a little less foreign aid from other sources, if the food transportation network in A had not been disrupted by war, if X had not been weakened by previous malnutrition, or if someone else had donated money, then X would not have died.

Moreover, in this case, unlike the other examples in this section, it is also plausible that the counterfactual (6.22) linking the nonoccurrence of the cause and effect is also quite sensitive. Even if (6.22) is true (as we are assuming), it is very plausible that if background circumstances had been different in various ways, then even if you had sent the money, X would still have died. This would happen, for example, under small variations in the behavior of organization O itself (they decide to spend a bit more money in country B and less in country A) or in the food distribution network within A. This sensitivity of (6.22) may contribute in a secondary way to the overall judgment that (6.23) is sensitive.[31]

The sensitivity of (6.22) also has the consequence that in the variant case in which you do send the money to O and X survives, the causal link between your sending the money and X's survival is sensitive: it is more like the link between Lewis's letter of recommendation and the deaths it "causes" than the link between shooting someone point-blank and the death this causes. Again, this seems to reflect our intuitive judgment: the dual sensitivity of both of the counterfactuals (6.21) and (6.22) is associated with the judgment that the causal claims connecting the omission to X's death and sending the money to X's survival are both sensitive.

Let me now try to relate these remarks to some observations from the empirical literature in psychology due to Patricia Cheng (1997). Cheng notes that it is a feature of common-sense causal thinking that we distinguish between what she calls generative and preventive causes. There is an asymmetry between these two kinds of causes. Roughly speaking, the operation of a preventing cause requires the presence of a generative cause but not vice versa. A generative cause can produce an effect even if no preventive cause is present. By contrast, preventive causes prevent by interfering with or blocking generative causes; when no relevant generative cause is present, a candidate preventing cause, even if present, is not viewed as operative or efficacious. Thus, if someone has ingested poison (a generative cause of death), the ingestion of an antidote may prevent death, but if no poison has been ingested, then although someone who takes the antidote will (let us assume) not die, we do not think of this as a case in which the antidote acts as a preventer of death, or as a cause of survival, presumably because the outcome of survival would be the same regardless of whether the antidote is ingested. As Cheng shows, there is considerable empirical evidence that ordinary subjects make causal judgments in accord with the generative/preventive distinction. What is true of prevention also seems true of causation by absence

[31] Let me acknowledge an obvious issue: to the extent that (6.21) and (6.22) are sensitive, in real-world circumstances we may be uncertain about whether these counterfactuals are true or at least think that we are not in a position to know that these counterfactuals are true. Even though we have stipulated that the counterfactuals are true, this uncertainty may influence our reaction to them. This is a general problem with the use of hypothetical scenarios: merely stipulating that p is true in some scenario may not be enough to ensure that subjects judge about the scenario on the basis of the assumption that p is true. Their knowledge that in the real world it will be uncertain whether p is true may intrude and influence their responses (another case in which subjects whose intuitions are solicited may be unaware of or not able to fully control factors that influence their judgments). I will not pursue this problem here, but in some cases more careful verbal probes may help ameliorate it.

(or omission): we usually think of it as parasitic on the presence of some additional generative cause, presumably (or so I conjecture) because causation by absence usually or always involves the nonoccurrence of some cause that would have been a preventer if it had occurred. For example, in the example of Doctor Z, the role of the doctor's omission in causing the patient's death clearly depends on the presence of some additional generative cause (the presence of a disease) that will produce death if not interfered with. As noted earlier, this has the consequence that the counterfactual linking the omission to the patient's death will at least be sensitive to changes that remove (or sufficiently modify) this generative cause. In the absence of the lethal form of the disease (we assume) the patient's death will not occur even if the doctor omits to administer the medication. More generally, this fact—that the causal efficacy, such as it is, of omissions depends on the presence of an additional generative cause—means that there is an important respect in which causation by omission will always be somewhat sensitive: omissions will no longer cause their effects under variations in which the relevant generative process is no longer present (or sufficiently modified), and this will show up in the fact that the counterfactual relating the omission to its effect is sensitive to such changes.

At this point, the reader may well wonder about how this analysis applies to (6.14) and (6.15). What are the generative causes present in these examples? In fact, such causes are present—they are just hard to see. In both cases, the generative causes are those physiological processes P that in the presence of food and oxygen sustain normal life and in their absence produce toxic effects—cell death and so on. But while the presence of the lethal disease strikes us as an adventitious and readily modifiable feature of the situation in (6.20), the processes P are "normal" features, the removal or relevant modification of which would be very far-fetched. We have little idea, for example, of what sorts of changes in P would be required for a human being to lack access to oxygen for an extended period of time and yet survive and no conception of how to bring about such changes—at the very least they would have to be "extraordinary" (to use Lewis's description).

Next, let me elaborate on an observation made previously: the connection between sensitivity and Lewis's notion of influence. Suppose that the causal relationship between *X* and *Y* is relatively insensitive in the sense that in the presence of *X*, *Y* would continue to occur at a time, place, and manner determined by *X* (that is, in accord with whatever relation between *X* and *Y* governs the time, place, and manner of *Y*'s occurrence) and that this is also true under some substantial range of changes in other causes of (including background conditions that are causally relevant to) *Y*. Then these other causes or background conditions will have relatively little influence in Lewis's sense on *Y*, at least in the presence of *X*. This is reflected in many of the examples already discussed. Suppose that my decisions concerning the content of a certain paragraph of this chapter at a time *t* shortly before its composition at *t* + *d* have a great deal of influence on the content of that paragraph in the sense that different decisions will lead to different contents, with this *decision → content* relation being relatively insensitive. (That is, the relation itself is insensitive even though the content depends on my decision.) Then for the most part, other causes of the content of the paragraph are likely to have relatively little influence.

6.7. Summary of Empirical Predictions regarding Omissions and Other Conclusions

I've framed the preceding discussion in a way that is common in philosophy—as a set of claims about the role of sensitivity / background invariance in judgments "we" make about the causal role of omissions, with these claims illustrated by application to a set of examples, regarding which, I am assuming, at least many others will judge as I do. But we can also express these claims as a set of empirical predictions about how people (where, as explained in Chapter 3, this means many or a substantial number of people, not all) will judge regarding claims in which absences/omissions are taken to be causes.

Here then are some predictions.[32] As we will see in Chapter 7, there is evidence that at least some of them are correct, although more systematic testing would be desirable.

(6.7a) Most people will not regard all claims in which an outcome counterfactually depends on an absence as a paradigmatic causal claim (where, as before, a non-paradigmatic claim covers such possibilities as that the claim is false, true but defective in some way, and so on). But at the same time most people will regard some claims in which absences figure as causes as paradigmatic (or at least relatively unproblematic). So neither of the two extreme positions—that when the right sort of relation of counterfactual dependence is present, absences are never regarded as causes, or that they always are (or are always treated as just as paradigmatic as other sorts of causes)—describes most people's patterns of causal judgments.

(6.7b) Whether a causal claim in which an absence is said to be a cause is regarded as true, acceptable, non-misleading, paradigmatic, etc. is, ceteris paribus, influenced, in the manner previously described, by the extent to which

[32] You might think that if I have evidence that some of these claims are correct, they are not really "predictions." I'm using "prediction" here the way it is often used in science, in which explanation of a known effect counts as a prediction. In addition, many of the claims in this chapter were formulated in my 2006 paper, well before there was any experimental evidence in support of them.

that causal claim is invariant or sensitive. (I don't claim that this is the only influence on such judgments—there are others.) In other words, we should expect to find that people more readily endorse (as expressed, for example, in higher strength ratings) causal claims involving absences for which the cause-effect relationship is relatively invariant and that they are less ready to endorse claims involving absences that are more sensitive.

(6.7c) The same pattern holds for causal claims not involving absences, as illustrated by Lewis's letter of recommendation example.

(6.7d) In simple cases in which no alternative causes of the effect are present, so that the relation between cause and effect conforms to the two counterfactuals (6.2) and (6.3), the invariance (or not) of the counterfactual (6.2) linking the occurrence of the cause to the occurrence of the effect has a greater impact on causal judgment than the invariance of the counterfactual linking the nonoccurrence of the cause to the nonoccurrence of the effect.

I explained my use of the phrase "paradigmatically causal" earlier in this chapter. As we will see in more detail in Chapter 7, it is not entirely clear what verbal probes are best suited to get at the kinds of distinctions that are associated with predictions (6.7a)–(6.7d) or that best track the extent to which a claim is regarded as paradigmatically causal. As noted previously, one common kind of probe that has been used for this and related purposes asks about causal strength. For example, one such probe asks, "On a seven-point scale, how appropriate would it to be describe this relationship as causal?" However, there are many other causal strength probes with alternative wording. In general my prediction would be that causal claims that are relatively invariant (including claims involving omissions as causes that are relatively invariant) would score higher on appropriately designed causal strength probes than less invariant claims. In fact, this is what is found in experiments discussed in Chapter 7. I emphasize, however, that it is an empirical question what sorts of verbal probes are best suited to detect and explore the extent to which subjects make distinctions of the sort described in this chapter.

Before leaving this section, let me remind the reader about the overall structure of my argument regarding omissions. I am *not* claiming that if the empirical claims made here about the role of invariance/sensitivity in causal judgments involving omissions are true, it follows that such judgments are normatively correct or appropriate. I do think that these claims are normatively justifiable, but their normative justification comes from the general considerations described in previous chapters (including, especially, Chapter 5) about the normative significance of invariance in causal judgment. We value invariance in causal relationships for all of the reasons described previously, and this provides a normative rationale for distinguishing between more or less invariant causal claims, including claims having to do with the causal role of absences. To the extent that, as an empirical matter, people, judge in accord with this rationale (judging more invariant claims regarding absences as more acceptable, etc.), this shows that they are behaving in normatively appropriate ways, which is what we would expect if they are behaving "rationally" in connection with these judgments. Put the other way around, to the extent that people value the discovery of invariant causal relationships and behave rationally with respect to this goal, it is plausible to expect they will judge in accord with (6.7a)–(6.7d). Thus rather than the fact that people judge in accord with (6.7a)–(6.7d) providing a normative justification for the claims about the role of invariance made previously (which is effectively what many appeals to "intuition" or ordinary judgment as a basis for normative justification assume), the normative claims (independently justified) show how such judgments can be rational or make sense. This in turn provides a reason to expect that, as an empirical matter, people's judgments will exhibit such features, to the extent that they are rational.

Two final points. First recall that when I claim that certain judgments (about the causal role of absences or anything else) are normatively correct or appropriate, all that I mean is they have a means/ends justification. That is, they make normative sense given that our immediate goal is the identification of invariant relationships, with this in turn being linked to goals having to do with manipulation, control, and generalization to a range of different circumstances. As suggested earlier, I do not think that, over and above this, there is any sense to be given to the question of whether absences are "really" causes or not. We should replace this question with the question of whether (and under what circumstances) it makes normative sense to treat absences as causes (and under what account of "normative sense" this is so). Second, recall from Chapter 3 that, on this picture of how the normative and empirical are related, as far as normative justification goes, it does not matter so much whether everyone or virtually everyone judges in the same way with regard to the causal significance of absences. Although my guess is that many people judge in accord with (6.7a)–(6.7d), I also assume that some people will judge that absences are never causes (at least literally speaking).[33] (It is an interesting question, which I will not pursue here, whether there are defensible normative justifications for these patterns of judgment.) On the view sketched earlier, we might reasonably hope to explain why judgments in accord with (6.7a)–(6.7d) are widespread, but we should not expect this to be a universal pattern. We also should not expect that everyone who has mastered "the concept of causation" to judge in accord with

[33] After all, some philosophers claim this.

(6.7a)–(6.7d).[34] Since, on my view, normative justification does not derive from a demonstration that so and so is how nearly everyone judges, I do not regard this lack of universality as a problem.

6.8. Double Prevention

Cases of double prevention are cases in which the occurrence of some event or process, c_1, would prevent some outcome, *e* (which would occur in the absence of c_1), but in which some second event, c_2, prevents the occurrence of the potential preventer, c_1, thereby allowing *e* to occur. In such cases, the occurrence of *e* depends counterfactually (in Lewis's sense) on the occurrence of c_2 (since c_2 and *e* occur, and if c_2 had not occurred, c_1 would have occurred and would have prevented *e*). Both Lewis's original theory and the version of interventionism represented by (M) (at least when accompanied by strong centering) will conclude that c_2 causes *e*. Nonetheless, many people will feel at least some resistance to this conclusion, at least in some cases.

Recall, from Chapter 1, Hall's (2004) example, which illustrates the basic idea. Suzy's plane will bomb a target (*e*) if not prevented from doing so. An enemy pilot will shoot down Suzy's plane (c_1) unless prevented from doing so. Billy, piloting another plane, shoots down the enemy pilot (c_2), and Suzy bombs the target. Suzy's bombing counterfactually depends on whether Billy shoots down the enemy pilot. Nonetheless, some—perhaps many—find problematic the unqualified claim that

Billy 's shooting down of the enemy pilot caused Suzy 's bombing. (6.24)

In this connection, Hall notes that the relationship between Billy's action and the bombing lacks several important features that (he thinks) we ordinarily associate with causation. One of these is what Hall calls "locality": Billy's shooting may occur at a great spatiotemporal distance from the bombing—hundreds of miles away and hours earlier, with no intervening events connecting the firing to the bombing in a spatiotemporally continuous way. In addition, the relationship of counterfactual dependence between Billy's shooting and the bombing itself depends on features that seem "extrinsic" to that relation. For example, in a variant on the original example in which the enemy's superiors were about to

[34] Since apparently competent users of causal language seem to disagree about whether absences can be causes and many do not assign a uniform causal role to absences, it seems unlikely that any appeal to what is built into our concept of causation can be used to settle questions about the causal significance of absences. The methodology I recommend is an obvious alternative.

order him not to attack Suzy when Billy shot him down, Suzy's bombing would no longer depend on Billy's action, even though the variant seems "intrinsically" just like the original example, differing only in the far-off intentions of the enemy's superiors. Even if we don't share Hall's assessment, it seems hard to deny that as far as ordinary judgment goes we recognize some important difference between the connection of Billy's action to the destruction of the target and, say, the connection between shooting and death. Similarly, the connection between Suzy's pressing the button that releases the bombs and the destruction of her target seems to differ from the connection between Billy's action and the destruction of the target, even though the former, like the latter, involves the removal of a preventer (whatever holds the bombs in place) and no transfer of energy/momentum from cause to effect.

Hall's own diagnosis is that double-prevention examples show that we operate with two distinct concepts of causation. One concept (which Hall calls "dependence") involves counterfactual dependence but is not transitive and does not require a spatiotemporally continuous process connecting cause and effect or the determination of causal structure by intrinsic features—in these respects, Hall's notion has many of the features I associated with the "dependence" counterfactual (6.3). The other concept ("production") satisfies suitably formulated conditions of locality, intrinsicness, and transitivity but need not involve counterfactual dependence.[35] Billy's firing is a cause of the bombing in the dependence sense but not in the production sense. By way of contrast, shooting someone at point-blank range and breaking a bottle with a thrown rock are paradigmatic cases of production.

Where Hall sees two distinct "concepts" of causation, I see instead a single, more unitary notion with two different but interrelated strands or elements, both of which can be understood in terms of the (different) counterfactual commitments they carry. As explained previously, one of these strands has to do with the counterfactual dependence of the effect on the cause (given the actual circumstances), particularly as expressed by counterfactuals of form (6.3) or by more sophisticated variants of this when other preempting or overdetermining causes are present. The other strand has to do with the sensitivity/invariance of the counterfactual linking the occurrence of the cause to the occurrence of the

[35] I will not try to discuss Hall's very interesting positive characterization of production in any detail. For what it is worth, however, I am inclined to think that transitivity is not a reasonable condition to impose on any concept of causation, whether production or dependence, for reasons I describe in Woodward 2003, 57–59. Note in particular that the intermediate causal links in McDermott's well-known dog bite example (1995) are intuitively cases of production, but transitivity appears to fail in this case. On the other hand, I think that Hall is right that something in the neighborhood of intrinsicness matters to causal judgment, although, as argued earlier in connection with the "blueprint strategy," I would understand relative absence of intrinsicness in terms of relative absence of invariance.

effect, as captured by the sensitivity of the counterfactual (6.2) under variations in background circumstances.[36] As suggested earlier, we tend to judge that a cause involves production when it is true both that the counterfactual linking the occurrence of the cause to the occurrence of the effect is relatively insensitive and that the dependence counterfactual (6.3) holds for the relation between cause and effect when other ("off route") causes of the effect have the value "absent."

In Hall's example both counterfactuals connecting Billy's firing and the bombing are relatively sensitive/non-invariant. First, although the bombing depends (in the sense captured by the counterfactual of form (6.3)) on the firing, given the actual circumstances, as noted earlier, that dependence would be disrupted by many changes in background conditions, including different orders to the enemy pilot. Second, the counterfactual of form (6.2) linking the occurrence of the firing to the bombing is also sensitive since even given the firing there are many possible variations in background conditions under which Suzy's bombing would not occur. I suggest that it is these sensitivity-based features that, at least in part, lead us to judge that despite the dependence present in the example, it lacks aspects that characterize paradigmatic cases of causation Or to put the matter another way, it is the combination of dependence and relative sensitivity/non-invariance of both of the counterfactuals of form (6.2) and (6.3) that leads many of us to have the ambivalent reactions that we do to Hall's example: the presence of counterfactual dependence leads us to judge that the case involves genuine causation; the relative sensitivity of associated counterfactuals inclines us to deny this judgment or at least to have misgivings about it.

Thus, on my view, Hall's two distinct concepts instead represent variations along two different dimensions of a single way of thinking about causation: paradigm cases of production (that is, what I have proposed to call production) involve the presence of a relatively insensitive (or more generally, invariant) causal relationship which is also such that dependence of the sort captured by the counterfactuals (6.2) and (6.3) will be present when other causes of the effect are absent, as is the case for Lewis's shooting example. However, one can have such dependence with sensitivity / relative non-invariance of the counterfactual (6.2) (causation that does not look very productive), as illustrated by many of the examples discussed in this chapter, including those involving double prevention and many cases of omission. Moreover, one can have insensitivity of the counterfactual (6.2)—and thus the presence of a productive cause—without dependence

[36] As suggested subsequently, the insensitivity of counterfactual (6.2) tracks something like the "sufficiency" of the cause for the effect, under one understanding of "sufficiency," and the dependence counterfactual, or some more complex version of it, tracks something like the "necessity" of the cause for the effect. Many accounts of causation from Mackie (1974) onward make use of some version of the idea that causal relations are to be understood as some combination of "sufficiency" and "necessity" but with different ways of understanding what these notions involve. See also Icard et al. 2017 for a version of this idea applied to causal selection.

of the sort expressed by the counterfactual (6.3), as illustrated by cases of preemption and overdetermination, although, as noted earlier, dependence of the effect on the cause can in such cases (or at least many of them) be made to reappear by controlling appropriately for other causes of the effect. In general, both strands or dimensions seem relevant to (and, as argued in footnote 15, some form of dependence seems required for) judgments of causation. The presence of a relatively insensitive relationship between *c* and *e* (or between *C* and *E*, in the type-causal case) is not sufficient for *c* to cause *e*, if *e* is or would not be dependent on *c* in the absence of other causes of *e* or when these are appropriately controlled for. In addition, as argued previously, the device of making dependence between *e* and *c* "reappear" in complex structures involving preemption or overdetermination requires that there be some degree of insensitivity present in causal relations. In particular, this requires insensitivity of the $c \rightarrow e$ relationship to certain variations in the presence of other causes. So neither strand or dimension is dispensable in causal reasoning or judgment—both work together.

Returning to the Billy/Suzy example, let's explore in a little more detail why the connection between Billy's firing and Suzy's bombing is relatively sensitive (for both of the associated counterfactuals). First, there are many possible changes—changes that do not seem at all unlikely or far-fetched—under which the counterfactual

If Billy were to shoot down the enemy pilot,
Suzy would drop the bombs (6.25)

would no longer be true. This would happen if, for example, Suzy had changed her mind about carrying out the bombing or if a second fighter who would shoot down Suzy unless interfered with in some way had been present. More generally, as is true of all cases involving prevention, the causal efficacy of Billy's preventive activities are parasitic on the presence of another, potentially generative cause (the action of the enemy fighter in shooting down Suzy). In addition, the role of this cause, qua preventer of Suzy's bombing, were it to be effective, would be parasitic on the potentially generative cause represented by Suzy's activities. Moreover, in this case, we also have sensitivity of the counterfactual relating the nonoccurrence of the cause to the nonoccurrence of the effect,

If Billy were not to shoot down the enemy fighter,
Suzy would not drop the bombs (6.26)

and this may reinforce our judgment about the sensitivity of the overall causal claim (6.24). Counterfactual (6.26) is sensitive because, for example, if the enemy pilot had eluded Billy and then changed his mind about pursuing Suzy, or Suzy had been able to elude the enemy pilot and so on for many other possibilities, Suzy's bombing would presumably still have occurred and hence (6.26) would be false.

In addition to the sheer number of contingencies that would disrupt (6.25) and (6.26), their character matters too. Of particular relevance in this context is the fact that we are inclined to attach a special significance to the sensitivity of causal and counterfactual claims under changes in human decisions and actions. Other things being equal, to the extent that a relationship would be disrupted if human actors were to act or choose differently in any one of a large number of different ways that seem not at all unusual or far-fetched, we will be particularly inclined to regard that relationship as sensitive. In the Billy/Suzy bombing example (as well as in the aid-to-charity example), this sort of sensitivity is present in spades. If Suzy had changed her mind about the desirability or morality of the bombing, the counterfactual (6.25) would not hold. If the enemy had not decided (or had not been ordered) to fly his plane that morning, or if he had eluded Billy but then decided not to attack Suzy, the counterfactual (6.26) would be false.

In a discussion of this example, Halpern and Pearl (2000) contend that part of our reluctance to regard Billy's action as straightforwardly a cause of the bombing stems from our sense that this is to treat Suzy (and the enemy pilot) as automatons rather than as agents for whom there is a genuine possibility of choosing differently. They suggest that if we were to substitute mechanical preprogrammed drones for the enemy and Suzy in the example (so that the enemy drone will automatically and inevitably shoot down Suzy if not intercepted by Billy and the Suzy drone will automatically carry out the bombing unless shot down by the enemy drone), resistance to treating Billy as a cause of the bombing would become less pronounced. This is an empirical claim that strikes me as plausible, but it needs to be tested. If people do exhibit the pattern of judgment described by Halpern and Pearl, an account along the lines I have suggested would help to explain this pattern—drone behavior is more invariant than human behavior.[37]

Let us now consider, by way of contrast, another system involving double prevention: the synthesis of the enzymes that metabolize lactose in *Escherichia coli*. This occurs when and only when lactose is present in the bacteria's environment. Simplifying greatly, the mechanism involved in this synthesis works in the following way: when lactose is absent, a repressor protein is synthesized

[37] This seems true as an empirical matter, but we also seem to suppose (and I take this to be what Halpern and Pearl are suggesting) that we think that when an outcome is causally influenced by a voluntary human act, there is always the possibility that the agent could have chosen otherwise. In fact, there is considerable empirical evidence for significant stochasticity in human choice. This may lead us to judge that causal chains in which human actions figure as intermediate variables are less invariant than corresponding chains involving mechanical devices. This in turn may help to explain our well-known willingness to trace causal chains back to human actions but (in a number of cases) not through them.

that prevents the genes that would otherwise synthesize the enzymes involved in lactose synthesis from being transcribed; when lactose is present, the repressor is inactivated and the relevant genes contribute to the production of the enzymes. The presence of the enzymes counterfactually depends on the presence of lactose in the environment, but this is because the presence of lactose prevents the presence of something that if present would prevent the synthesis of the enzymes. In this case, needless to say, human agency plays no role in the operation of the system. It is also well buffered against a range of environmental contingencies, at least of the sort that are likely to occur or involve modest departures from the actual environment. In addition, the combination of the role of lactose in the synthesis of the repressor protein and the role of the latter in turning off and on the genes involved in the synthesis of the lactose enzymes is not an accidental or ad hoc feature of this system but is, rather, a normal, "designed" feature that is due to natural selection: the presence of just one component of the system (e.g., the synthesis of the repressor protein in the absence of lactose) without the other would confer no selective advantage at all.[38] For all of these reasons, the counterfactual dependence between lactose and the synthesis of the enzymes strikes us as relatively insensitive or invariant, despite the fact that it involves double prevention and despite the fact that there is no connecting process from cause to effect.

Consistently with this, we seem to have little hesitation in concluding that this is a straightforward case of causation—indeed genetics texts describe this as a case in which the presence of lactose "controls" (but via "negative" rather than "positive" control) the presence of the enzyme. I take examples like this to suggest that it is not the presence of double prevention per se that makes us

[38] I owe to Chris Hitchcock the suggestion that it may be true in general that we are more willing to regard relationships of double prevention as causal when this relationship seems present by "design"—for example, because there has been natural selection for the relationships or because the relationship is the intended result of human design. Thus, to take an example from Schaffer (2000), we have no problem with the suggestion that pulling the trigger of a gun causes it to fire even if the gun works by double prevention. Similarly, it seems unproblematic to claim that Suzy's pushing the bomb bay button in her plane causes the bombs to fall, even if this also involves the removal of a preventer: the doors to the bomb compartment. In both cases, the overall relationship of counterfactual dependence is rather stable and designed to be so. Changes that would disrupt the dependence would need to be rather "abnormal." Experiments by Lombrozo described in Chapter 7 provide empirical support for this suggestion. She and colleagues (e.g., Lombozo 2010) find that relationships that are relatively invariant receive stronger causal strength ratings, that people are more willing to consider relationships causal when an association involves a direct physical connection (perhaps because these relationships tend to be more invariant) rather than double prevention, and—when double prevention is involved—are more inclined to regard such relations as causal or assign them higher causal strength ratings when they are more invariant. When agents are involved, the relation between intentional actions and outcomes receives higher strength ratings than accidental outcome/action relation and similarly for relations between designed features and outcomes they are designed to produce (whether by human agents or natural selection) as opposed to "accidental' feature outcome relations. As argued previously, typically the intentional and designed relations will be more stable or invariant.

reluctant to regard a relationship of counterfactual dependence as causal; instead, it is the relative sensitivity/non-invariance of the associated counterfactuals that prompts this reaction.[39] In addition, this example reinforces the conclusion that we reached in previous sections of this chapter: that relatively insensitive causation may be present even when there is no spatiotemporal connecting process or transfer of energy and momentum from cause to effect. It may well be true, as an empirical matter, that such connecting processes are often realizers of insensitive causation—see footnote 10 and Woodward 2018c for more on this possibility—but they are not the only such realizers. This is one reason why the project of trying to find specific physical realizers (or underlying physical natures) to identify with (all) causal relationships seems misguided. If what characterizes causal relationships is the presence of certain dependence relationships that possess invariance-related features, many different underlying physical processes, not all of which involve direct spatiotemporal connections between cause and effect, can realize these.

These remarks may prompt the following reaction in some readers: what you, Woodward, say about the role of sensitivity/invariance in people's *judgments* about the causal status of double-prevention relations may or may not be correct (presumably that is an empirical question), but what I (the reader) would like to know is whether the relations in questions *really are* causal or not and not just what influences people's judgments about such cases.

My response parallels what I say about the causal role of absences in Section 6.7 and follows from my methodological remarks in Chapters 1 and 3. Relations involving double prevention have a feature possessed by more paradigmatic causal relations—dependence—and possess another relevant feature (invariance) to differing degrees, in some cases having far less of it than more paradigmatic causal relationships. If I am correct, these features influence our judgments about such cases and, moreover, at least within an interventionist framework, this influence seems normatively appropriate. Thus someone who is interested in manipulation and control and associates these with causal relationships will find it reasonable to classify some relationships involving absences and some double-prevention relations as causal (or even paradigmatically causal when these are relatively invariant). Other double-prevention relations that are not relatively invariant will not be classified as paradigmatically causal. Once we recognize this, there is no obvious motivation for supposing that there must be some further fact, over and above the features just described, that determines which of these relationships are "really causal" and which are not—instead there are just the facts about the features we have described. For this reason, it is hard to see

[39] Or at least sensitivity considerations have a significant independent effect on causal judgment in such cases.

how a debate about which of these relations is truly causal could be resolved in a non-arbitrary way. Or rather, it seems that the only fruitful way of pursuing such a question is to reconstrue it so that it becomes an issue about whether various distinctions we might make among different cases of double prevention (as well as absences) make sense and have some recognizable normative rationale, thus replacing the metaphysical question about which of the relations is "really causal" with methodological questions. This in effect is what I have tried to do. Sometimes philosophical progress is made by replacing a question with a somewhat different, better question.

6.9. More on Insensitive Causation and (Some Aspects of) Control

As argued previously, the reasons for caring about whether a causal relationship is insensitive are connected to the general reasons we have for valuing the identification of invariant relationships that have been described in Chapter 5. In this section, I want to say a bit more about the connection between invariance and control. This will also connect up with Lewis's remarks about the difference between killing and causing death.

Consider an agent who wishes to bring about some outcome *E* by manipulating the state of a cause *C* of that outcome. Typically, whether *E* occurs will depend not just on the state of *C* but on many other circumstances besides. Think of these circumstances as capable of assuming any one of a large number of values $B_1, \ldots, B_n$. For some of these values B_i, *E* will follow if *C* is introduced, while for other values, *E* will not follow when *C* is introduced. Even if *C* is under the control of the agent, it will often not be under the agent's control which of the values B_i is realized. Furthermore—this is important—the agent may not know the value of B_i that obtains in circumstances in which she acts or whether that value is such that it will enable the occurrence of *C* if *E* is introduced or will instead frustrate it. Moreover, even if the situation is one in which the various possible values of B_i are described by a probability distribution $Pr(B_i)$, the agent may not know this probability distribution—from the agent's perspective, she may face a situation of uncertainty or "ambiguity" rather than risk. Alternatively, there may not be any well-behaved probability distribution for these background conditions. In situations of this sort, it often will be a good strategy for the agent to look for a cause *C* of *E* that is such that the relationship between *C* and *E* is relatively insensitive to different values of B_i (or at least insensitive to values of B_i that are believed "close" to the actual circumstances or are likely to occur in the environment in which the agent acts). In the ideal case, the agent will be able to find a cause *C* that makes *E* unavoidable or inevitable—that is, a cause

C that is followed by E for all possible values of B_i.[40] Then it will not matter at all which value of B_i happens to be realized. Even if there is no such cause at hand from which E follows in all circumstances, the agent can still look for a cause that makes E *relatively* inevitable or unavoidable: that is, a cause C that is followed by E for a large range of different values of B_i (or for many such values that have a significant probability of occurring, if that is known), hence a causal relationship between C and E that is relatively insensitive.[41]

Consider the choice between a robust poison that is capable of causing death in a wide variety of circumstances—regardless of diet, physical condition of the victim, and so on—and a fragile poison that will cause death only in very special circumstances. Imagine a would-be murderer who does not know whether these special circumstances obtain and cannot control whether they obtain. The murderer might be well advised to employ the robust poison, even if, were the special circumstances present, the probability of death would be higher with the fragile poison and even if the murderer thinks it plausible (but is not certain) that these special circumstances are present. As this example illustrates, employing a C that is a relatively insensitive cause of E can give one a kind of control over whether E obtains and a kind of insurance against the possibility that nature or other agents might thwart one's plans that a more sensitive cause does not provide. We see this feature—or its relative absence—in many of the examples discussed earlier. In Lewis's example, shooting someone through the heart is a good way of making his death nearly inevitable—inevitable not just in the sense that it makes death highly likely given the actual circumstances but also in the sense that death is very probable for most non-far-fetched variations on those circumstances. Thus shooting through the heart gives one a high degree of control over whether the death occurs—given the shooting, there is not much that nature or any other agent can do to prevent death. By way of contrast, writing a letter of recommendation or withholding money from a charitable organization is not a way of making someone's death nearly inevitable in this proof-against-changing-circumstances sense, even if, as it happens, in the actual circumstances the death is counterfactually dependent on withholding the money or writing the letter. There is just too much else that easily might have happened and in happening would have disrupted the counterfactual link between the withholding or

40 We noted earlier that Halpern (2016) calls this "robust sufficiency."

41 If the agent knows the probability distribution of the background circumstances B_i and also has information about the probability of the desired outcome E for each choice of a possible cause C_k that might bring about E in each of the B_i, the agent can choose the C_k that maximizes the expectation that E will occur. Needless to say, the agent will often not have all of this information. My suggestion is that if the agent does not know the probability distribution of the B_i, it may be a good strategy to choose a C_k that has a relatively invariant substantial probability of leading to E across a range of different B_i. This is one way in which the strategy of choosing an invariant cause for E may differ from the strategy of choosing a cause that maximizes the probability of E. For additional discussion see Chapter 7.

the writing and death. In this sense, the strategy of choosing withholding money does not give one control over whether any particular death occurs. Similarly for writing the letter of recommendation.[42]

I suggest that these considerations underlie and provide a rational motivation for Lewis's inclination to describe the shooting but not writing the letter as a "killing"—"killing" seems like the right description when a causing of death in the counterfactual-dependence sense, along with a high degree of control (that is, relative insensitivity), is present.[43] Parallel considerations provide at least part of the explanation for why biologists find it natural to say that the presence of lactose in the environment of *E. coli* "controls" whether it produces enzymes that synthesize lactose, despite the fact that the mechanism of control involves double prevention, while, by way of contrast, it does not seem natural to describe Billy's activities as "controlling" whether Suzy drops her bombs. As I have argued elsewhere (Woodward 2010), a similar elucidation suggests itself for claims that some trait is under "genetic control." To the extent that we value causal relationships that provide a basis for a relatively high degree of control, we should also value relatively invariant causal relationships.

[42] For experimental results that support this general picture of the relationship between whether an agent chooses a relatively insensitive action leading to an outcome and responsibility ascription, see Grinfeld et al. 2020.

[43] Of course whether an action is a case of killing or a mere causing of death in the sense described also has important implications for moral assessment. Cases of killing in the sense described involving employment of an insensitive cause of death are much more likely to involve outcomes that are foreseen and intended (and perhaps planned) and thus agents who are more culpable. Agents who aim at killing by choosing insensitive causes are typically more dangerous in virtue of choosing means that make death as inevitable as possible. Note also, however, that if killing is insensitive causing to die, then, in line with my preceding arguments, some omissions can count as killings—for example, failing to feed a small child for whom one has responsibility—and should be evaluated accordingly. Similarly for some actions that cause deaths but not via a connecting process, as with the British naval blockade described earlier.

7

Experimental Results Concerning Invariance: Cheng, Lombrozo, and Others

7.1. Introduction

The previous two chapters focused on developing an analytical framework for understanding the role of invariance-based considerations in causal judgment, largely drawing on anecdotal illustrations drawn from ordinary life and some bits of science. In this chapter I turn to an examination of some experimental work that explores the impact of invariance considerations on ordinary causal judgment in a more systematic way.

7.2. Causal Strength: Some Preliminary Considerations

I begin, however, with some methodological remarks on the notion of causal strength and its measurement, topics that figure centrally in the discussion that follows. As I noted in previous chapters, a very common procedure in empirical studies of human causal cognition is to ask subjects to make judgments of (what is called in the literature) "causal strength" where this is expressed on a graded scale and to then attempt to model how such judgments (or perhaps some underlying psychological state for which such judgments provide evidence) vary as subjects are exposed to different data or scenarios. These might provide, for example, information about the frequency with which the effect occurs in the presence and absence of some putative cause.

The particular verbal probe used in connection with the elicitation of strength judgments varies with different researchers, often considerably. For example, in different experiments subjects have been asked such questions as "How appropriate would it be (on a seven-point scale) to describe this relationship as causal?" "On a seven-point scale, how much do you agree with the following statement about what caused *e*?" as well as "On a scale from 0 to 100, how much does cause *c* raise the probability of outcome *e* when *c* is introduced into a situation in which no other causes of *e* are present?" (used in Buehner et al. 2003). As discussed in more detail later, as an empirical matter, different verbal probes directed at subjects making judgments about the same scenarios can lead to

Causation with a Human Face. James Woodward, Oxford University Press. © Oxford University Press 2021.
DOI: 10.1093/oso/9780197585412.003.0008

different results in the sense of different strength ratings—an outcome that is not very surprising, given the different semantic content of the probes. There is some tendency in the literature to treat the various probes (or at least many of them) as all directed at measuring the same quantity (some unitary notion of causal strength) with the issue then being which probe is best suited for this purpose—that is, which probe is the best way of "measuring" causal strength judgments. But an obvious alternative possibility is that some of the different probes measure or correspond to non-trivially different things that might be meant by "causal strength" and/or that they measure different features that people judge are present in causal relationships. A closely related concern is that there may not be any single unitary notion of causal strength that subjects operate with and that asking subjects via a single verbal probe to provide strength ratings may conflate different notions that ought to be separated.

One distinction among different aspects of causal strength that is now generally accepted in the empirical literature has to do with the contrast between (i) the subject's uncertainty about whether a causal relationship exists at all (sometimes called "confidence" or "support") and (ii) the subject's judgment of how "strong" that causal relationship is (on some understanding of strength, which may be specified in a number of different ways), given that the relationship exists. Consider an experiment in which a subject is asked to provide a judgment of causal strength between *C* and *E* on the basis of information provided in a contingency table but where the sample size is very small. The small sample size may lead the subject to not be very confident about whether any causal relationship exists at all, even if the evidence, as far as it goes, suggests a very strong relationship between *C* and *E*—that is, that if a relationship exists, it is strong. The subject's degree of uncertainty about whether any causal relationship exists between *C* and *E* is (it seems natural to suppose) conceptually distinct from questions about the characteristics of that relationship (whether *C* is a weak or strong cause of *E*) assuming that some relationship does exist. One would expect at least some subjects to be sensitive to this distinction. For example, if, in the case of binary events, subject strength judgments track at least in part something like the degree to which the cause increases the probability of the effect in comparison with some alternative (as claimed by a number of researchers), then it seems entirely possible that subjects might be relatively confident (perhaps because they see lots of data) that a weak causal relationship exists in the sense that the cause raises the probability of the effect only by a slight amount. Thus in this case as well as others, confidence/support and strength come apart.

As I remarked earlier, this point is now generally recognized. Thus in experiments in which the dependent variable is a strength rating, it makes good methodological sense to separate the impact of degree of confidence judgments from judgments bearing on other aspects of strength, rather than using some probe that conflates confidence and strength. One way of doing this is to first ask subjects for judgments of how confident they are that a causal relationship exists and then ask only those who exceed some threshold about whether such a relationship is present to make a second causal strength judgment about this relationship in response to a differently worded probe. The case for distinguishing these two sorts of judgment is developed in detail in Griffiths and Tenenbaum 2005. These authors also provide formal models of subjects' degree of confidence judgments (which they call "support"). They suggest one way in which support might be measured in a Bayesian framework is by means of the posterior probability that some causal link exists in a graphical model, conditional on some data set generated by the model. Alternatively one might compare the log likelihoods of two models, one of which asserts the existence of a causal link and the other saying there is no such link. In either case, the support statistic computed is quite different from the usual measures of causal strength. As an empirical matter, this statistic better fits confidence judgments than standard strength measures.

However, even separating out the impact of support judgments, different experiments have used the same or similar causal strength probes to track what seem, analytically, to be quite different features of causal scenarios. As noted earlier, Cheng has used a causal strength probe to try to capture something like judgments of the degree to which a cause boosts the probability of its effect in a situation in which no other causes of the effect are present. She claims that such judgments are well explained by her causal power theory. (See subsequent discussion.) By contrast, Lombrozo and colleagues (Lombrozo 2010 and Vasilyeva et al. 2018), in experiments also described later, use a different strength-rating probe to present evidence that subjects give higher ratings to causal relationships that are relatively insensitive or background invariant in roughly the sense described in Chapter 6. Although, as we shall see, Cheng's causal power measure incorporates some invariance-based considerations, it does not (just as a matter of how it is defined) fully capture variations in the extent to which a relationship is background invariant. Intuitively, one reason for this is that a cause might provide a big boost in the probability of the effect in highly specialized background circumstances but not behave similarly in other background circumstances, so that the relation is not very background invariant, even though it may have, averaging across different backgrounds, relatively high causal power (and may receive relatively high strength ratings using Cheng's verbal probe).[1] Given that causal strength judgments (or, more precisely, ratings elicited by the verbal probes used by Lombrozo et al.) are influenced by such background invariance

[1] As noted earlier, it is also possible for a cause to raise the probability of its effect by only a small amount, thus having low causal power in Cheng's sense, even though this probability raising is highly invariant across different background contexts. So this is another respect in which power and background invariance come apart.

considerations, it is thus not surprising, as Lombrozo and colleagues find, that such judgments are not fully captured by Cheng's causal power model. Thus in these examples causal strength probes are used to elicit judgments about somewhat different aspects of causal relationships.

These observations raise the following general question about what we should want in a verbal probe of causal relationships and what it makes sense to try to model. Suppose, as seems to be the case, that responses to a generic undifferentiated causal strength probe like "To what extent is it appropriate to describe this relationships as causal?" are influenced by a variety of different factors, including degree of confidence that a causal relationship exists, degree to which the cause boosts the probability of the effect, degree of background invariance, and, very likely, other factors as well,[2] with the relative influence of these possibly varying across different subjects. One possible response is to stick with the generic strength judgments and try to model all of the factors that influence them. Another possible strategy is more dissective. Given that a variety of different factors seem to influence causal strength judgments, researchers should look for or create different verbal probes that are differentially sensitive to those factors, as has already been done in the case of support. Thus one might devise a verbal probe that tracks probability boosting, a probe that tracks degree of background invariance, and so on. One might then try to construct computational accounts that model separately the influence of variations in each factor on the differentiated verbal probe that is associated with measuring it.[3] In other words, one attempts to do for the other factors that influence generic strength judgments what has already been done in connection with support.

Something like this dissective strategy has been followed by Cheng and colleagues in papers I discuss subsequently. These authors replace previous probes for causal strength judgments with a more differentiated verbal probe that they claim more directly measures those aspects of causal strength that their power PC model is intended to capture. They then focus on modeling the judgments produced by this probe. I suggest later that there are arguments for taking this procedure further in connection with other aspects of causal strength (or with regard to other factors influencing how good or paradigmatic a causal claim is judged to be). In other words, at least to some extent it seems desirable to

[2] See also Kominsky et al. 2015 and Icard et al. 2017 for recent studies showing that judgments of causal strength vary with the typicality of the cause and the background conditions; also see the discussion of proportionality in Chapter 8. In the former papers, strength judgments are used to probe practices of causal selection—yet another use that differs from the ways in which Cheng and Lombrozo use the notion.

[3] This is a generally recognized methodological strategy in science: when confronted with evidence that some outcome measure O reflects the operation of a number of different causal factors, it often makes sense to look for measures and experimental procedures that separately track those factors—or at least those that are of interest—and then model these, rather than trying to directly model O.

have additional verbal probes that further disentangle the various influences on strength judgments. As we shall see, failure to do this has made it harder to compare and assess experiments. On the other hand, as we shall also see, too much dissecting has its own limitations.

7.3. Cheng's Power PC Model

With this as background I turn to a discussion of Patricia Cheng's power PC model and some accompanying experimental results, as well as some dissenting claims by Shanks and coauthors. Cheng's model provides a very striking illustration of the power of invariance assumptions. I should also say at the outset that what follows is only a very partial discussion of a very large and sophisticated body of work, with much of importance being omitted. Cheng's model[4] represents causes and effects as binary events, which can either be present or absent. Causes can be either "generative"—they can promote their effects—or they can be "preventive," interfering with the operation of generative causes. The "power" of a cause i to produce effect e is represented by p_i, the probability with which i produces or causes e if i is present. Note that this *not* the same as Pr(e/i)—the conditional probability of e, given i. Among other considerations, the latter quantity reflects the influence of other causes of e (not on the route from i to e but independent of it) that are present when i is. Let a represent such other causes of e, and assume that when present they produce e with probability p_a and that these are all generative rather than preventive causes of e. Assume also that i and a are all the causes of e and that e does not occur when it is not caused.[5] Assume too that subjects have access to data about the frequencies of occurrence of i and e (and hence also to not-i, not e, and conditional frequencies involving these), but they do not direct observe p_i and p_a—these have to be inferred, to the extent that they can be. (Cheng describes p_i and p_a as theoretical entities that contrast with frequencies of i, e etc. which are observable.) Cheng makes the following two additional assumptions about i and a:

(7.1) i and a influence the occurrence of e independently[6]

[4] As Glymour (1998) and others have noted, Cheng's model can be regarded as a particular parameterization of a causal Bayes net involving a so-called "noisy or gate."

[5] Thus we are assuming that we dealing with a structure for which it is known that i causes e (rather than their merely being correlated), that a represents alternative causes of e that influence e along a different route from i, and that i does not cause a or conversely. What we are trying to figure out is the "strength" of the causal relation between i and e, on the assumption that there is such a relation and the preceding conditions hold.

[6] (7.1) does not mean that i and a are statistically independent but rather that the probability p_i with which i causes e if present is independent of the probability p_a with which a if present causes e. In other words, the causal influence of each cause on e remains the same regardless of whether e is influenced by the other cause.

(7.2) The causal powers with which *i* and *a* influence the occurrence of *e* are independent of the probability with which *i* and *a* occur so that, for example, the probability that *i* occurs and causes *e* is just $Pr(i).p_i$.

Note that both (7.1) and (7.2) are invariance assumptions that we encountered in Chapter 5. (7.1) says that p_i—the power of *i* to cause *e* when *i* is present—is invariant under changes in p_a and similarly that p_a is invariant under changes in p_i. (7.2) says that p_i and p_a are invariant across changes in the probability with which *i* and *a* occur. Cheng thinks of (7.1) and (7.2) as "default" assumptions that people bring to causal learning situations—as explained in Chapter 5, nothing guarantees that such assumptions will be true in the situation of interest, but they are nonetheless useful points of departure for reasoning that can be relaxed as the empirical evidence warrants.

Given these assumptions, Cheng derives an expression for the representation of causal power as well as an account of the circumstances in which it can be estimated. Of course she does not assume that subjects consciously reason in accord with the algebra she describes. What she presents is a computational-level rational reconstruction, with subjects judging as if they computed and represented causal power in the manner described. As she puts it at one point, her analysis describes "what reasoners compute"; I take it that this means that her model describes the input data subjects use, and a function that subjects compute in going from this input to their output judgments. Cheng's derivation then proceeds as follows:

First, $Pr(e)$ is given by the union of the probability that *i* occurs and causes *e* and that *a* occurs and causes *e*:

$$Pr(e) = \Pr(i).\, p_i + Pr(a).\, p_a - Pr(i).\, p_i.\, \Pr(a).\, p_a$$

Conditionalizing on the presence of *i*, we obtain (recalling that p_i and p_a are independent of $Pr(i)$)

$$Pr(e/i) = p_i + Pr(a/i).\, p_a - p_i.\, Pr(a/i).\, p_a.$$

Conditionalizing on the absence of *i*, we obtain

$$Pr(e/not\ i) = Pr(a/not\ i).\, p_a$$

Defining (7.3) $\Delta p(i)$ in the usual way = $Pr(e/i) - P(e/not\ i)$, it follows that

$$\begin{aligned}\Delta p(i) &= p_i + Pr(a / i).\, p_a - p_i.Pr(a / i).\, p_a - Pr(a \mid not\ i).\, p_a \\ &= [1 - Pr(a / i).\, p_a].\, p_i + [(Pr(a / i) - Pr(a / not\ i)].\, p_a.\end{aligned}$$

Thus

$$p_i = \frac{\Delta p(i) - [Pr(a / i) - Pr(a / not\ i)]\, p_a}{1 - Pr(a)\, p_a}.$$

Now consider the special case in which *a* and *i* occur (statistically) independently, so that

$$(Pr(a / i) = Pr(a / not\ i).$$

Then

$$p_i = \Delta P(i) /\ 1 - Pr(a)\, p_a$$

p_a cannot be estimated from the frequency data available, but since *e* is caused either by *i* or *a*, we can replace $Pr(a).p_a$ with $Pr(e/not\ i)$, yielding

$$p_i = \Delta p(i)\ /\ [1 - Pr(e / not\ i)] \tag{7.4}$$

Thus if the assumptions described previously are satisfied, p_i, the causal power of *i* to produce *e*, can be estimated from frequency data concerning the co-occurrence of *i* and *e*. Suppose that experimental subjects are presented with such data and are asked to estimate the causal strength with which *i* causes *e*. Cheng claims that as an empirical matter such causal strength judgments (when elicited by an appropriate probe) will track p_i at least qualitatively.

This leads to a number of specific predictions, including the following: first, let us compare causal power with Δp, which as explained earlier is one of the most common measures of causal strength used by psychologists. As is apparent from (7.4), p_i and Δp diverge unless $P(e/not\ i) = 0$. Assuming that one of the main goals of each model is to predict causal strength judgments, then, if Cheng's model is correct, p_i should track judgments of causal strength more closely than Δp.

Consider, for example, situations in which $Pr(e/i) = Pr(e/not\ i) = 1$. In such situations $\Delta p = 0$ if *i* is a generative cause. Thus subjects guided by Δp in their causal judgments should report that in this situation *i* does not cause *e* (assuming that a strength rating of 0 corresponds to absence of causation). By contrast the denominator of (7.4) is zero when $P(e/not\ i) = 1$, so that (7.4) is undefined in this circumstance. Thus it is plausible that subjects guided by p_i in their strength judgments should report that they are unable to reach any conclusion about the causal strength of *i* in this situation. Note it is arguable that the latter judgment rather than the judgment based on Δp is the normatively correct one: When $P(e/not\ i) = 1$, a "ceiling effect" is present: since *e* is always present (produced by the alternative cause *a*) the power (if any) of *i* to cause *e* cannot reveal itself in any differential probability of occurrence of *e* in the presence versus the absence of *i*. An appropriate experiment or set of observations that would provide evidence relevant to whether *i* causally influences *e* is one in which *e* sometimes fails to occur in the absence of *i*. In the absence of such evidence, there is an obvious argument that the normatively correct judgment is that one has no basis for judging, one way or the other, whether *i* causes *e*. In fact, when ordinary subjects are given this option, this is the alternative they tend to choose.

A second prediction, if subjects are guided by Cheng-type causal power in their strength judgments, is this: as is apparent from (7.4) for a fixed value of Δp, p_i (and so judgments of causal strength) will increase as $Pr(e/not\ i)$ increases. As an empirical matter, many subject judgments do exhibit this feature, but this has often been treated as an "irrational bias" of some kind. It is certainly normatively inappropriate if the correct normative theory for judgments of causal strength is given by Δp. By contrast, this feature is both predicted and shown to be normatively reasonable by the power PC theory.[7]

It is worth reflecting a bit more on the normative differences between the power PC model and Δp. Intuitively speaking, Δp is normatively deficient as a measure of the causal strength of *i* in producing *e* because it does not correct for confounding: both $Pr(e/i)$ and $Pr(e/not\ i)$ reflect not just the relationship between *i* and *e*, but also the extent to which other causes of *e*, captured by *a*, are operative. If, for example, Δp is relatively low, this might reflect the fact that *i* is not very effective in causing *e* (when *i* occurs, its probability of causing *e* is low), but it might also reflect the fact that many instances of *e* are caused by the alternative cause *a*, so that even if *i* is a strong cause of *e*, it has limited opportunities to

[7] On the other hand, when $Pr(e/not\ i)$ is close to zero, p_i will be close to Δp. This is also holds empirically: $-\Delta p$ is a better predictor of judgments of causal strength under such circumstances. I'll add that the fact that judgments of causal strength of *i* with respect to *e* increase as $Pr(e/not\ i)$ increases is a striking example of a judgment that may initially seem normatively unreasonable to many but which has a somewhat non-obvious normative rationale. We may view this as another example in which observations of actual behavior may suggest a previously unconsidered normative theory.

boost the probability of *e* (which of course can never exceed 1) since the presence of *a* by itself raises the probability of *e* to a high level. One consequence is that the Δp measure will not generalize appropriately to new situations in which the distribution of other causes of *e* is different from in the original situation. That is, the value of Δp for the relation between *i* and *e* will vary depending on whether additional causes are present and on their efficacy in causing *e*, and moreover this will happen even if there is no interaction between *i* and these other causes with respect to *e*. The invariance assumptions built into (7.4) correct for this (in effect by normalizing Δp to correct for the influence of other causes of *e* besides *i*), assuming of course the applicability of the assumptions that go into its derivation. In fact, in other experiments, Cheng and coauthors have shown that, as an empirical matter, causal power does a much better job of predicting which causal judgments (and associated measures of strength) subjects will generalize to new situations with different distributions of new causes than alternative measures like Δp. Again, to the extent that we value such generalizability, judgments based on causal power will be normatively reasonable.

To further illustrate these ideas about causal power and their connection to generalizability, I turn to some experiments conducted by Liljeholm and Cheng (2007). In the first experiment, one group of participants (the causal power group) saw the results of hypothetical studies in which randomly selected patients were (i) either not exposed to any medicine or exposed to medicine A (left-hand panel in Figure 7.1) or alternatively (ii) either not exposed to any medicine or exposed to both medicines A and B (right-hand panel in Figure 7.1). The results are summarized in Figure 7.1, with smiling faces representing headache-free patients and frowns patients with headaches. Note that the causal power of medicine A as computed by (7.4) is .75, which is also the causal power of the treatment consisting of A and B in combination, so that causal power is the same across the left-hand and right-hand panels. On the other hand, Δp is different across the two panels—it is .25 for the left hand panel and .75 for the right-hand panel.

A second group (the constant Δp group) was presented with scenarios with the opposite profile, in which causal power varied across the two panels but Δp was constant, with results shown in Figure 7.2.

Subjects in both groups were then asked to judge whether medicine B was a cause of headache.[8] Two-thirds of the subjects who saw Figure 7.1 (the constant causal power group) judged that B was non-causal, while only a fifth of those in the Δp constant group who saw Figure 7.2 did. Given some very natural background assumptions, both these majority responses are normatively correct.

[8] Thus in this case the verbal probe did not ask explicitly about causal strength, although causal power is explicitly calculated and compared across the different scenarios and used to interpret subject behavior. Thanks to Thomas Blanchard for bringing this to my attention.

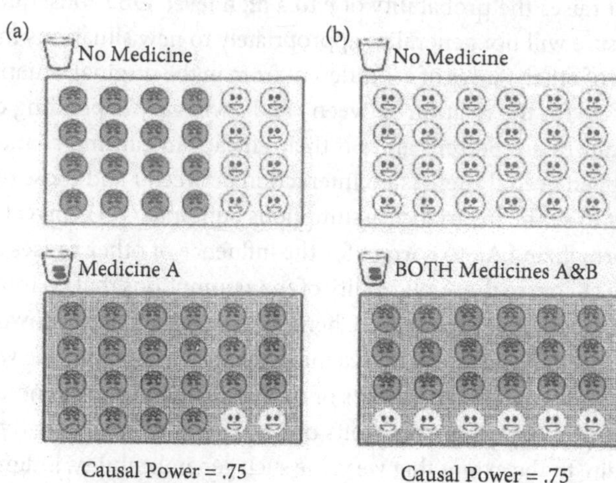

Figure 7.1

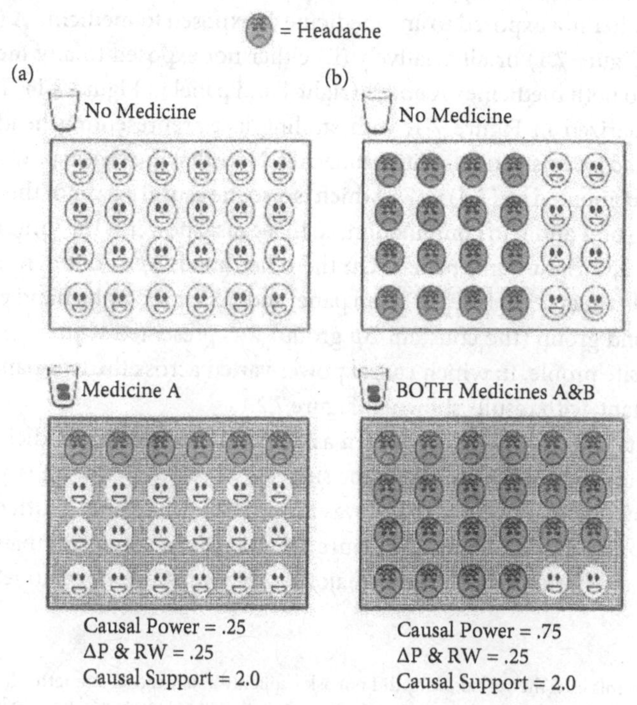

Figure 7.2

These include the causal power assumptions that medicine A has a fixed propensity, power, or probability of causing headache for each patient receiving it, which is the same across the two panels and which is not affected by how many patients receive the medicine. (Note that this is related to the SUTVA invariance assumption discussed in Chapter 5.) As noted previously, the left-hand panel in Figure 7.1 shows that among those who do not have headaches, 75 percent develop headaches when they take A, which suggests that the causal power of A causing headache for an arbitrary patient receiving A is .75, which is also the result from (7.4). In the right-hand panel (b) in figure 7.1, 75 per cent of those who do not have headaches develop headaches when they take medicine B (in conjunction with A). This is just what we would expect if all of the headaches in the lower part of right panel were due to medicine A, assuming that A has the same power to produce headaches across the two situations described by the left- and right-hand panels—an invariance assumption, obviously. Thus there is no basis for supposing that any of these headaches were due to medicine B. Indeed supposing that some of the headaches were caused by B would require further gratuitous assumptions for which there is no evidence—for example, that B suppresses the ability of A to cause headache while at the same time causing headaches itself but in just such a way that the total proportion of headaches among those who did not previously have headaches remains constant or, alternatively, that some portion of the headaches caused by A are also caused by B but B causes no headaches in subjects who are not also caused to have headaches by A.

As Liljeholm and Cheng (2007) put it, what this first experiment shows is that participants' expectations about what will generalize across contexts is captured by the causal power notion rather than Δp. It is the fact that the incidence of headache in the lower right panel is just what we would expect on the basis of the causal power attributed to A, that leads subjects to judge that B is non-causal. Note by contrast that in Figure 7.2 it *is* reasonable to judge that B has an effect on headache on the basis of causal power considerations—from the right-hand side of Figure 7.2, the incidence of headaches among those who did not have them and who received both A and B is much higher (. 75) than the incidence of headaches among those who did not have them and received just medicine A (.25), indicating that B has an effect on headache over and above that produced by A.

One way of thinking about these experiments is as follows: what does it mean for a cause to "act in the same way" across different contexts[9]? In Figure 7.2 it might be argued that there is sameness of action across the left- and right-hand

[9] Recall the remarks in Chapter 5 linking invariance to the notion of a factor acting in the same way across different contexts.

sides in the following sense, which corresponds to what is captured by Δp: in both cases ingestion of medicine A adds six headaches to the total. However, if medicine exhibits sameness of action at all across different contexts, this is not the kind of sameness we expect it to exhibit. For A to exhibit the sameness of behavior just described, it would somehow have to keep track of the total number of headaches it caused among groups of different size, all members of which were exposed to A, and then limit this total to exactly six. Needless to say, we don't expect medicines to behave in this way.[10]

In yet another set of experiments by these authors the same method was employed except that subjects were presented with three hypothetical studies in which they were asked to judge whether a single medicine interacted with varying background factors across the studies. Recall that in the derivation of the formula for causal power (7.4) it was assumed that the cause *i* did not interact with the background causes represented by *a*. Clearly this is a strong assumption that will not be satisfied in many realistic contexts. However, Cheng's framework also provides a natural account of what it is for an interaction to be present—this will be reflected in the fact that the causal power of the cause of interest *i* to produce *e* varies depending on what other causes of *e* are present.[11] This second set of experiments probed whether subjects judged in accord with this conception of interaction: one group of subjects saw a set of studies in which causal power remained constant, while another group saw studies in which causal power varied, as described in table 7.1.

Δp varied across the studies for both groups. Consistent with the result from the first experiment, subjects in the power constant group tended to judge that no interaction was present, while those in the power varying group tended to respond that an interaction was present.[12]

One of the lessons we may extract from Cheng's experiments is that a model according to which people represent causal relations in terms of causal power

[10] One might find this behavior in other sorts of situations: imagine an agent who hands out money to members of groups of varying size under the constraint that exactly six people from each group receive money regardless of the size of the group. It is interesting to reflect on the status of our assumptions about how the behavior of medicines is likely to generalize across contexts. On the one hand, these are far more specific than the mere expectation that the behavior will generalize in some way or other (which is close to vacuous). On the other hand, these assumptions are still relatively generic in comparison with still more specific claims that might be formulated. The assumptions we expect to generalize are, so to speak, at an intermediate level of generality. As argued in Chapter 5, this is characteristic of invariance assumptions. They are general but specific enough to provide guidance.

[11] Recall that causal power reflects the power of *i* alone to produce *e*. Of course the incidence of *e* will change depending on what other causes of *e* are present, even if no interaction is present. Interaction involves the presence of other causes changing the causal power of *i* to produce e.

[12] The verbal probe employed to test whether subjects judged that an interaction was present was "Do you think that Medicine A interacts with some factor that varies across experiments or do you think that the medicine influences the patients in different experiments in the same way?" (2007, 1019).

Table 7.1 Relative Frequencies of Headache for the Three Hypothetical Studies in Experiment 2

Subject group	Study 1		Study 2		Study 3	
	e|no A	e|A	e|no A	e|A	e|no A	e|A
Power-constant	16/24	22/24	8/24	20/24	0/24	18/24
Power-varying	0/24	6/24	0/24	12/24	0/24	18/24

based on some relatively simple invariance assumptions seems to provide a good fit with at least some aspects of lay causal judgment. This suggests that some substantial number of people operate with assumptions about or representations of causation in which some invariance-based notions play an important role. We also see that to the extent that an invariance-based model of causation is a normatively appropriate one, people's causal judgments are, at least in some respects, normatively defensible. Moreover, this is a case in which experimental results (like the tendency to judge that *i* is a stronger cause of *e* when *e* occurs more frequently) can suggest normative possibilities that may otherwise not have occurred to the theorist—again (as emphasized in previous chapters) not because we should regard people's causal judgments or "intuitions" as automatically normatively correct, but because it turns out that sometimes such judgments can be provided with a validation in terms of normative theory and because there may be selective forces of various kinds at work in shaping such judgments.

Also worth noting is the following point, which echoes observations made in Chapter 3: suppose that we think of people's judgments of causal strength as not very different in status from the intuitions or judgments about causes that the armchair philosopher makes. (Of course armchair philosophers don't usually make judgments about causal strength[13] but rather make judgments about whether various relationships qualify as causal. What I'm supposing is that both sorts of judgments can be treated as not different in principle.) Presumably judges of causal strength have access to their own strength judgments/intuitions, which, after all, they report. In some cases, the armchair philosopher may be well attuned to ordinary people's strength judgments or be sufficiently similar in response to ordinary people that her own judgments/intuitions about strength, should she make them, will reflect judgments shared by others. So it is not terribly far-fetched to suppose that the philosopher may have (empirically fallible)

[13] You might wonder why not. After all, ordinary subjects are quite willing to make causal strength judgments about cases. My guess is that philosophers don't think of such judgments as revealing anything about the nature or concept of causation and don't consider them for this reason. If so, this is another respect in which a focus on nature or concept questions is unduly restrictive.

access to other people's causal judgments, including their strength judgments, both in the experimental scenarios employed by Cheng and elsewhere. By contrast, it is considerably less clear that the typical philosopher will have any particular insight into *why* people's causal strength judgments exhibit the patterns they do—what the factors are that causally influence such judgments and what computations lead to the judgments and whether these are best described by Cheng's model or by some different model. This reflects the general point, made in Chapter 3, that even if people (including philosophers) are good judges of how others will judge, they often don't have reliable information about the causes of those judgments.

By this I don't mean that it is *impossible* that the correct causal/computational model might be discovered by some combination of armchair reflection and intuition; after all, a priori reflection on the consequences of various invariance assumptions presumably played a role in Cheng's invention of her model, so that there is an important "armchair" (in this case mathematical) component to this discovery. However, discovery/invention is different from empirical confirmation, and it seems clear, particularly in light of the controversies described later concerning the empirical status of Cheng's model, that confirmation (or not) requires careful experiment. That is, the empirical status of the model as an account of aspects of human causal judgment is not something that can be decided from the armchair. Moreover, although there are ideas in the philosophical literature that resemble Cheng's in various ways,[14] it is certainly not the case that philosophers theorizing about causation (or notions of causal power) have converged on anything like Cheng's notion or even on notions in which invariance-linked ideas play an important role. Cheng's model may be incorrect, but if it is, armchair-based methods don't seem sufficient to discover this.

There is, in addition, a more general philosophical point that emerges from Cheng's proposals and experiments that is worth emphasizing. As Cheng notes, measures like Δp have the advantage that they are defined entirely over "observable" (or at least "measurable" events)—in the case of Δp, these are observed frequencies. However, this is accompanied by a serious disadvantage—these "observable" measures and relationships defined in terms of them are not invariant or generalizable in the way we would like them to be, and this detracts greatly from their usefulness. As noted earlier, when the strength or power of i to cause e is measured by Δp, this will vary depending on the extent to which other causes of e are operative, even if those other causes operate completely independently of e, in the sense that there are no interaction effects. When strength is measured by Δp, it is no longer a feature of the $i \rightarrow e$ relationship taken in

[14] Including, importantly, Cartwright's (1989) notion of a causal capacity—a connection acknowledged by Cheng.

itself (no longer an "intrinsic" feature) and does not export to new situations in which the distribution of other causes of *e* is different from in the original situation.[15] Conceptualizing the strength of the $i \rightarrow e$ relationship in terms of causal power (or some other invariance-based measure) avoids these limitations, but this comes with a cost—it requires thinking of causal relationships in such a way that they are not definable in terms of or fully reducible to relations among observable (or for that matter, non-modal) quantities. So we have a trade-off: what is observable (or most readily observable) as a measure of causal strength does not generalize or satisfy natural invariance requirements, and what satisfies such requirements leads to a conceptualization in terms of relationships that are not (or may not be) directly observable. As an empirical matter, people (ordinary folk and scientists) accept this cost regarding observability, and it seems normatively appropriate for them to do so, given that they value certain kinds of invariance.[16]

In saying that we are prepared to sacrifice observability for gains in invariance in our thinking about causal relationships, I do not of course mean to suggest that we think about causal claims (or claims about associated invariance-based features) in such a way that they are in every case untestable. As we noted previously, there are circumstances in which causal power, as characterized by Cheng, can be estimated from observable data, provided appropriate additional assumptions hold (e.g., the assumption that *i* and *a* statistically independent), besides those that go into Cheng's derivation. A similar point holds for other invariance-based claims and for causal claims conceptualized in terms of invariance. As noted in Chapter 2, this reflects a general pattern present in other standard treatments of causal relationships: reliable inferences about these generally require both observable, statistical data *and* background assumptions that go beyond these statistical data in various ways. (In Cheng's case these background assumptions include but are not limited to various invariance claims.) When these background assumptions are not satisfied, causal relationships, and

[15] Recall my remarks about "intrinsicness" in Chapter 6. This is a hard notion to characterize, but one way of thinking about it is in terms of the idea that there should be features of causal relationships that they "carry around" (Cartwright's phrase) across contexts—this amounts to an invariance or insensitivity condition of some kind, where the features in question are features that continue to hold across variations in which other causes are operative, so that in this sense the features don't depend on the presence of these other causes. Cheng-type causal power is a better candidate for a feature of causal relationships that is at least somewhat "intrinsic" in the sense just specified than Δp. Note though that the intrinsicness of causal power in Cheng's model is not (in contrast to other philosophical attempts to characterize the notion of intrinsicness) specified reductively, since "causal" assumptions go into this characterization.

[16] To put the point slightly differently, philosophers with a empiricist or Humean bent may wonder why we should make use of a notion of causation that departs at least in some respects from what is directly observable or that is not fully characterizable in terms of non-modal observables. My answer is that such a notion has a number of advantages described here and in Chapter 5 and that the world, often enough, supports the use of such a notion.

measures of causal strength associated with them, may not be "identifiable" from the available statistical data. That is, we may not be able to use the data to determine which causal relationships hold or with what strength. To the extent that one models causal relationships in terms of invariance- or interventionist-based ideas, however, we do not conclude that causal relationships do not exist when they are not identifiable but rather hold that the available evidence does not suffice to discover whether or not they obtain. To return to an example of Cheng's, in circumstances in which *e* always occurs even when *i* does not occur (because *e* is always produced by some background cause *a*), it will be impossible to determine from the available frequency data whether *i* causes *e*. To determine this, we need to intervene (or do something similar, such as finding additional appropriately discriminating observations) to vary whether *i* occurs in circumstances in which *a* does not cause *e* to always occur and observe whether there is a change (increase) in the incidence of *e*. If we find that this is the case, then (assuming that *a* and *i* act independently), we can use invariance-based considerations to reason in the following way about the case in which *e* always occurs and is caused by *a*: we assume as a default that *i* acts in the same way in the situations in which *e* occurs and is not caused by *a* and the situation in which *e* always occurs and is caused by *a*. We thus infer that when *i* is present, it causally influences *e* in the latter situation.

In this way, conceptualizing causal relations as invariant leads us to countenance the possibility that sometimes they may be "hidden" by the operation of other causes, in the sense that they remain operative when other causes are present even if unambiguous evidence for this does not show up in available frequency data. As noted in Chapter 5, we see something like the same thing at work in more sophisticated scientific contexts—for example, when a total cause is resolved into components, as when the total force experienced by some object is decomposed into gravitational and electromagnetic forces. Total forces are presumably closer to being observable than component forces and hence perhaps more acceptable to empiricists (on some interpretations of that view), but generalizations relating sources to total forces are less generalizable and less invariant than generalizations concerning component forces, which is why scientific laws concern the latter. This provides another illustration of how a concern with invariance can create a drive toward abstraction, leading to the formulation of relationships that are relatively removed from direct observation.[17]

[17] Given the skepticism expressed in passing in my introduction about using notions like "power" to provide a metaphysics of causal claims, some readers may wonder how this fits with my sympathetic treatment of Cheng's causal power theory. I see Cheng as engaged in a quite different enterprise than the metaphysicians. To begin with, Cheng does not use the notion of power to explain or "ground" causal claims—instead the notion of cause is invoked in her characterization of causal power—p_i is the probability with which *i causes e* when *i* is present. In addition, the notion of causal power is characterized within a mathematical framework in which, as we have seen, invariance assumptions play a

7.4. Criticisms of the Power PC Model: Shanks and Others

So far I have discussed only evidence that seems to support the power PC model. However, the empirical situation is more complicated and contested, in ways that are philosophically and methodologically interesting. In a critical discussion of the power PC theory, Lober and Shanks (2000) agree that the theory is the correct normative theory of causal judgment in the situations to which it is intended to apply—that is, situations that satisfy the background assumptions of Cheng's theory. They claim, however, that the model is empirically inadequate as a descriptive account of human causal strength judgments. (Related criticisms are advanced in Griffiths and Tenenbaum 2005.) Both sets of authors draw attention to two patterns present in human causal strength judgments that are prima facie inconsistent with the causal power model. The first consists in the fact that a number of subjects provide positive causal strength ratings in the presence of "non-contingency"—that is, when $\Delta p = 0$, with the magnitude of these ratings being influenced by $Pr(E/not\ C)$ (an effect called "outcome density bias"). This is inconsistent with both the power PC and models based on Δp alone, which predict strength ratings of zero in such cases.

Second, recall the experiments of Liljeholm and Cheng discussed earlier in which Δp is held constant, causal power varied, and causal judgments are shown to track causal power. Lober and Shanks were able to replicate this result, but they also did the "opposite" experiment in which causal power was held constant across different experimental conditions and Δp varied. Of course the causal power theory predicts no difference in judgment across these conditions, but such judgments *are* found to vary, appearing (in this respect) to support Δp over causal power. In an extremely interesting analysis of their data, Lober and Shanks show that their subjects can be separated into two groups. One of these (the power group) seems to be guided by the normative considerations that led to the construction of the power PC theory (e.g., this group is aware of ceiling effects and tries to take them into account in their causal strength judgments and also corrects for the presence of other causes *a* in estimating the causal strength of *i*). The other group (the contingency participants) seems not to take these considerations into account. As one can see from Figure 7.3, which depicts how causal ratings vary across different values of Δp, when causal power is held constant, the power participants behave pretty much as the power PC theory predicts—their

crucial role—power is characterized in part by its connection to these assumptions. To the best of my knowledge, contemporary metaphysical accounts of powers don't have these features—the accounts are not mathematical and there is no explicit connection to invariance-based ideas. One way of putting the contrast is in terms of the difference between a common sense claim about powers—e.g., (i) aspirin ingestion has the power to relieve headaches—and (ii) a metaphysics of powers. My view is that there is nothing problematic about (i) when it is understood along the lines described by Cheng; (ii) is in a different category.

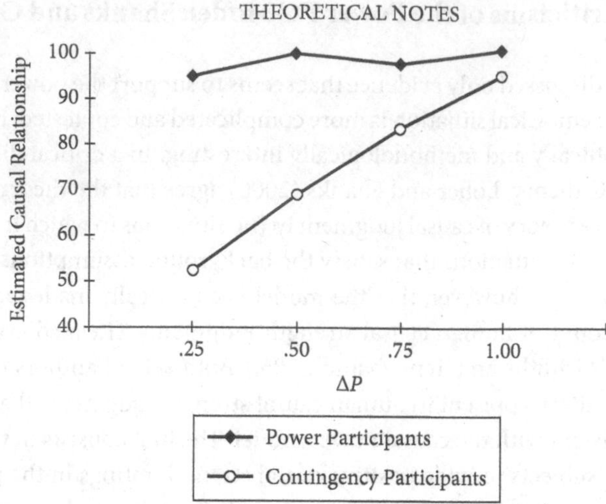

Figure 7.3
Source: Loeber and Shanks 2006, 206, Figure 6.

ratings are fairly constant across different values of Δp. By contrast, the ratings of the contingency group clearly increase with increasing values of Δp, which is what the Δp but not the power PC model predicts. Thus, as Lober and Shanks put it, different participants appear to be following "different strategies" in generating causal strength judgments.

These results provide a concrete illustration of some of the difficulties that subject heterogeneity poses for the idea that virtually all competent speakers share a single concept of causation that is reflected in their judgments. Do the power participants and the contingency participants have "different concepts" of causation or causal strength? Or do they instead share the same concept but apply this differently or make different inferences on the basis of it? Are they perhaps just answering different causal questions because of differences in how they interpret the verbal probe? (See subsequent discussion for evidence supporting this possibility.) Is it part of "our concept" of causation that participants in the contingency group are making a mistake—failing to take into account a consideration (correction for the presence of other causes in their strength judgments) that is implied by their concept, so that that they may share the same concept as the causal power group but apply this concept wrongly?[18] I don't claim that such questions

[18] We might also ask whether, if this is the case, we should disregard the judgments of the contingency group and just focus on the alleged shared concept and what it implies when applied or interpreted correctly.

are unanswerable in principle, but it is certainly not obvious how to go about answering them. If we can avoid getting entangled in them, that seems desirable.[19]

Lober and Shank's own view is that a third model—what they call the unrestricted Rescorla-Wagner model—provides a better fit to human causal judgment than either the Δp or power PC model. This is a version of the Rescorla-Wagner model that does not impose the restriction that the beta parameters in the model have the same value for both the presence and absence of the effect.[20] Unlike the "restricted" version of the RW model, in which these parameters are assumed to be equal, the unrestricted model does not converge on Δp. Lober and Shanks claim that the unrestricted model is able to fit the empirical data concerning human causal strength judgment better that either power PC or Δp. (Of course this may in part reflect the fact the unrestricted model has more free parameters.) Thus, in effect, what Lober and Shanks are claiming is that human causal judgment (or at least portions of it) is best understood within an associative learning framework, after all.[21]

In a response to these results from Lober and Shanks, Buehner, Cheng, and Clifford (2003) appeal to several considerations, as well as additional experiments. One consideration has to do with what they call "ambiguity of the causal question." First, they endorse Griffiths and Tenenbaum's (2005) concern, described previously, that standard verbal measures of causal strength may conflate subjects' degree of confidence that a causal relationship exists (Buehner et al. describe this as a matter of "reliability") with strength more narrowly conceived. However they also point to a second possible source of ambiguity in the verbal probes used to measure causal strength. When asked about the causal strength of *c* with respect to *e*, subjects might (i) interpret this as a question about "the current learning context"—that is, as asking "what difference to the effect *e* does the candidate cause make in the current learning context, in which alternative causes already produce *e* in a certain proportion of the entities" (2003, 1126). Here the question is understood as asking what additional difference *c* makes to *e*, given

[19] To be fair, by no means all of the problems posed by subject heterogeneity disappear if one avoids formulating cognitive theories in terms of claims about shared concepts—serious issues remain. My point is rather that these problems are exacerbated by the shared-concepts picture.

[20] See the discussion of the Rescorla-Wagner model in Chapter 4.

[21] If we wanted to interpret Lober and Shanks's claims in terms of concept talk, we might perhaps try to interpret them as claiming that *one* concept that underlies some human causal judgments) is the concept of an associative relationship of the sort captured by the unrestricted RW model. On the other hand, they also explicitly acknowledge that human beings are also able to reason about causal relationships in other more sophisticated ways that conform to the requirements of more normative, non-associative models (2000). So, pursuing a "concepts" interpretation, one possibility is that their view is that humans have several different concepts of causation, which they apply in different contexts. Again, however, my opinion is that a better assessment is that concept talk is not very helpful in capturing their views.

that alternative causes are already causing some instances of *e*. As should be obvious, Δp is the normatively correct answer to this question.

A second way of interpreting the causal strength question is as (ii) "What difference does the candidate cause make in a context in which alternative causes never produce *e*?" where the normatively correct answer is given by the power PC model. Buehner et al. suggest that the results for experiments with positive contingencies that appear to be inconsistent with power PC might be explained by the fact that approximately half of the subjects are interpreting the causal strength question along the lines of (i) and the other half along the lines of (ii). They support this claim with a K-means cluster analysis that does indeed show that subjects segregate into two groups along the lines just described. They also performed a second experiment in which the causal question was altered along the lines of (ii) since (as they see it) this corresponds to the notion of causal strength that the power PC model is intended to capture. In particular, they asked subjects to estimate "how many entities out of a group of 100 which did not show an outcome would now have the outcome in the counterfactual situation in which the candidate cause was introduced" (2003, 1128). As they note, this is in effect an "intervention" question, which asks what the effect of the cause *i* would be on the frequency or probability of the effect *e* if *i* were introduced by an intervention into a situation in which it was previously absent. They further argue that asking for an estimate in terms of proportion of entities also makes it clearer that the question is not about reliability or degree of confidence that a causal relation exists and thus disambiguates the strength question from questions about support. In this second experiment, subjects behaved more in accord with the causal power model.

A second set of considerations to which Buehner et al. appeal in order to explain deviations from the power PC model has to do with what they call "misrepresentations due to memory limitations." As they note, ambiguities in the causal question cannot explain outcome density bias in which non-contingent ($\Delta p = 0$) conditions receive positive ratings for causal strength. However, some of the experiments of Lober and Shanks 2000 that showed outcome density bias involved data which were presented sequentially to subjects, rather than all at once, as in a contingency table, which is the format that Buehner et al. use. (Recall that the Rescorla-Wagner model, a version of which is Lober and Shank's preferred explanation of some causal judgments, applies only to information presented in this format, which is why they employ it.) Buehner et al. argue that when data are presented in this format, subjects may not have remembered them sufficiently well to accurately compute the contingency information on which estimates of causal power (and Δp) are based. If, in non-contingent conditions, a number of subjects mistakenly believe that a small contingency is present, this might lead them to provide positive causal strength ratings. Buehner et al. collected subjects' subjective probability estimates for *Pr(E/C)* and *Pr(E/not C)* and found that the subjects did indeed misperceive non-contingent conditions as involving positive contingencies.

With these considerations in mind, Buehner et al. 2003 conducted further experiments in which efforts were undertaken to minimize possible effects of memory limitations on subjects' causal judgments and in which the question designed to elicit causal strength judgments was changed along the lines described earlier. They claim that these (what they regard as) improvements in experimental design yield results that more strongly support the power PC theory than alternatives.

In response, Perales and Shanks (2008) returned to the relationship between the power PC model and strength judgments based on the counterfactual questions employed by Buehner et al. In one set of experiments, subjects' judgments were elicited regarding a scenario in which butterflies are exposed to radiation and develop mutations. One group of subjects was asked a "causal judgment" question ("To what extent does radiation cause mutation, evaluated on a scale from 0 to 100?") and another group a "counterfactual" question similar to the question employed by Buehner et al. 2003 ("How many out of 100 butterflies, none of which would show a mutation if unradiated, do you estimate would show a mutation if radiated?"). Both groups were also asked a "confidence" question ("How confident are you in the judgment you have just made?"). The format in which data relevant to these questions were presented was again different from the contingency table format employed by Buehner et al.: Perales and Shanks presented their data to subjects sequentially (trial by trial) in the form of "records" reporting whether individual butterflies were irradiated and whether they developed mutations. Perales and Shanks report that judgments from the causal judgment group "mirror" Δp. However, in two conditions in which power is equal ($p = .5$), subjects in the counterfactual group produced different mean judgments (52.8, SD = 17.0, 75.0, SD = 19.5), which they argue is not what one would expect if the counterfactual question closely tracked causal power.[22]

In fact, Perales and Shanks report that judgments in the counterfactual group were "modulated" by *Pr(E/C)*, by which they apparently mean that these subjects were to some significant extent, but not exactly, "estimating" *Pr(E/C)* in reporting their judgments (2008, 1486). (Note that on many accounts of counterfactuals

[22] However, given their reported standard deviations (see main text), one might wonder how to interpret this difference in means. As shown in their Table 2 (p. 1486), the values for causal power, mean causal judgments, and mean counterfactual judgments across the three values for *Pr(E/C)*, *Pr(E/not C)* used in the experiment are, respectively, <.875, 72.9, 76.8>, <.5, 48.9, 52.8>, <.5, 41.5, 75.0>. Assessed qualitatively, in the first two conditions, causal power and the other two judgments do not seem particularly far apart. It is only in the third case that the counterfactual judgment diverges substantially from causal power and from the causal judgment.

this is a misunderstanding.)[23] Some subsequent experiments attempted to refine the counterfactual question to eliminate possible misunderstanding, and subjects were also asked to provide subjective estimates of the probabilities *Pr(E/C)* and *Pr(E/not C)*. Subjects' probability estimates were quite accurate (from which Perales and Shanks conclude that the experimental results cannot be explained by mis-estimation of these conditional probabilities even though information about these was presented in the more demanding sequential format). Again subject estimates of *Pr(E/C)* were, Perales and Shanks claim, a better predictor of counterfactual judgments than either power PC or Δp (2008, 1486).

In a third set of experiments, in which the hypothetical scenario involved the effect of fertilizer on plants, Perales and Shanks (2008) attempted to replicate the format in which data were presented in Cheng's experiments more closely—the relevant frequency data were presented "simultaneously" rather than sequentially but in two different formats. In one format information about the response of fertilized and unfertilized plants were interleaved; in the second format data from all of the fertilized plants were grouped separately from data from the unfertilized plants. (In another words, treatment and control data were physically separated.) The first experiments produced results that were similar to the previous experiments, while the second produced rather different results. Here *mean* counterfactual judgments approximately matched causal power. However, at an individual level none of participants judgments were fully in accordance with causal power, with individual judgments deviating in different ways from causal power across two conditions in which causal power was equal.[24]

Perales and Shanks interpret their results as casting doubt on the argument from Buehner et al. 2003 that deviations from the predictions of the power PC model are in part due to misperception of the conditional probabilities *Pr(E/C)* and *Pr(E/not C)*. They also conclude that the counterfactual question employed by Buehner et al. does not closely track causal power, contrary to what these authors claim. Instead subjects behave non-normatively, with their answer to this question being strongly influenced by *Pr(E/C)* perhaps because the probe question focuses on instances of *C* and *E*. Finally, they also infer (as is certainly

[23] Notice, however, that the subjects are apparently behaving in accord with Adams's thesis, according to which the assertability of an indicative conditional is given by the conditional probability of the consequent given the antecedent.

[24] This pattern, in which mean judgments or responses in an experiment are well approximated by the researcher's model even though the model does not fit the responses of most or any individuals in the experiment, seems to be present in a number of experiments in psychology. For critical discussion, see Eberhardt and Danks 2011. There are ways of making sense of this pattern that appeal to the idea that individual responses reflect a sampling process of some kind from an underlying normative model, but I will not pursue this idea here.

illustrated by their experiments) that the format in which data is presented can strongly influence causal judgment.[25]

This back-and-forth between Cheng et al. and Shanks et al. (as well as related points in Griffiths and Tenenbaum) raises a number of interesting questions. One result that emerges from all of the experiments is that judgments of causal strength depend on the particular verbal probe employed. In some respects, use of different verbal probes for different purposes seems completely unproblematic. As noted earlier, it is an assumption of virtually all current theories of causal strength judgments that there is a difference between judgments of how strongly the evidence supports a claim that some causal relationship exists and judgments of how strong that relationship is, assuming that it exists.[26] It thus makes sense to employ different verbal probes to elicit these two kinds of judgments (e.g., "How confident are you that some causal relationship exists" versus some probe to elicit strength) and to take additional steps that the verbal probes employed do not result in conflation of these two kinds of judgment.[27] Here an analytical distinction between two different sorts of considerations (support or confidence vs. strength) suggests the appropriateness of employing different verbal probes.

But even if support and strength questions are separated in this way, the same strength question (i.e., the same verbal question) may be interpreted by subjects in different ways, as the first experiment in Buehner et al. 2003 illustrates. It is uncontroversial (again as an analytical or normative point) that there is a big difference between asking (i) "What difference does *c* make to *e* in the current learning context?" and (ii) "What difference would *c* make if no other causes of *e* were present?" The evidence from this experiment is fairly compelling that some subjects interpreted the verbal probe used in this experiment along the lines of (i) and others in accord with (ii), so that the original causal strength question was in this respect "ambiguous." This suggests replacing it with another, less ambiguous probe. On the other hand, one might worry that the alternative counterfactual question Buehner et al. employed in subsequent experiments is very closely tailored to the power PC theory. In effect, subjects are explicitly asked to

25 Unsurprisingly, Cheng is not convinced by the Perales and Shanks results. Among other considerations, she is concerned that it may not have been made clear to the subjects in these experiments that the hypothetical scenarios with which they were presented involved random assignment. Based on current work, she believes that when this is made clear and other problems in experimental design are addressed, results are obtained that are more in accord with the power PC model than any of its competitors (based on a telephone conversation, March 1, 2017).

26 Our reasons for thinking that these judgments are logically or conceptually independent is presumably based on some kind of normative/conceptual analysis, although one that is accepted by virtually all difference-making accounts of causation. However it is a further, empirical question whether subjects recognize such a distinction in their judgments.

27 For example, by asking the strength question only of those subjects who have previously indicated some degree of confidence that a causal relation exists, as Lombrozo et al. do in experiments described subsequently.

estimate just the quantity that the causal power theory attempts to capture and so (one might think) it is not surprising if the resulting judgments conform well to that theory. On the other hand, recall that Perales and Shanks 2008 claim that even when subjects are asked this counterfactual question, their judgments do not conform to the power PC theory. I will return to this issue later.

Note also that Buehner et al. agree that if a causal strength question had been employed that corresponds to interpretation (i), this would elicit results that, as an empirical matter, are better described by the Δp model; and that, moreover, this model gives the normatively appropriate response to this question. However, presumably Buehner et al. would not take this to show that Δp is after all the correct account of causal strength judgment since they think that a different verbal probe is more appropriate for eliciting such judgments. (They think that the right verbal probe corresponds to (ii)—a claim that is tied to their conceptualization of causal power.) This raises a number of questions, one of which is this: once one separates out judgments of support, is there some single verbal probe for causal strength judgments that is best or most appropriate or at least "neutral" between competing models of such judgments? If not, what should we conclude when different verbal probes seem to yield different results?

Without trying to settle these questions, let me make some additional observations. The first is simply that, as the preceding discussion illustrates, analytical, normative work is essential in answering them. It is the analytical/normative distinction between the issue of how strong the evidence is that a causal relationship exists and the issue of how strong that relationship is, that suggests the need for distinct verbal probes for support and strength. Similarly, it is an analytical distinction that draws our attention to the difference between (i) and (ii). This illustrates how in interpreting both armchair judgments about cases and experimental results there is a need for such analytical work.[28]

To extend this point, it is worth reflecting on the particular assumptions about causal structure that characterize both the Buehner et al. experiments and the Shanks et al. experiments. In all of these experiments subjects are asked to make judgments about a causal structure in which both cause C and effect E are "binary events" and some alternative cause or causes may be present that also affect E. The information available to subjects when they are asked to make strength judgments is information about the values of $Pr(C)$, $Pr(E)$, $Pr(E/C)$, and so on, so of course it is natural (perhaps virtually guaranteed) that unless subjects are totally confused, they will base their strength judgments on some function of these

[28] This is work that armchair philosophers may be in a good position to perform, although I again emphasize that it may be highly nontransparent when a verbal probe or question is ambiguous if one relies just on a priori reflection. Empirical results can suggest non-obvious ambiguities. It is also worth adding that one way of discovering that a probe is ambiguous is to ask subjects how they understand the probe—their answers are not guaranteed to make sense, but they may.

quantities. In experiments of this sort, conceptualizing strength as connected in some way to a judgment of how big a boost the occurrence of C provides to the probability E in comparison with some alternative seems very natural, and of course this is reflected in both the Δp and power PC models.

However, once one moves outside of this paradigm and subjects are provided with other sorts of information, the possible bases for strength judgments and the possible conceptualizations of strength change. For example, in experiments conducted by Lombrozo and colleagues that are described in Section 7.4, subjects are provided with information about stability/invariance of relationships across changes in background conditions (background invariance), in this case the presence of moderating variables. As we shall see, it is a matter of mathematics that background invariance can vary for relationships that have the same values for Δp and causal power. Moreover, while both Δp and causal power are defined only for binary variables, the notion of background invariance also applies to variables that take more than two values. It is an empirical discovery, described in Section 7.4, that some of the most commonly employed verbal probes for causal strength when applied to relationships involving binary variables also reflect the influence of background-invariance-related considerations, with more invariant relationships being rated as "stronger."[29] One question this raises, adumbrated in Section 7.1, is whether, in view of these considerations, we should further distinguish among strength judgments, designing different verbal probes for different aspects or dimensions of strength. That is, given that analytically/normatively the question of how much C boosts the probability of E (in a situation in which C is introduced by an intervention when other causes of E are absent) is distinct from the question of how stable the $C \rightarrow E$ relationship is across background changes, does it make sense to employ a single verbal probe for causal strength that tracks both of these considerations, or should we instead look for distinct verbal probes for different dimensions or aspects of strength? I will argue in what follows that there seem to be advantages and disadvantages to both strategies.

As the preceding discussion illustrates, normative analysis can help to make us aware of possible ambiguities and unclarities in verbal probes, and suggest replacements that are more satisfactory. However, as noted earlier in connection with the counterfactual probe used by Buehner et al. (2003), when a normative model of causal judgment is invoked to support the use of a particular verbal probe, and the results of that probe are then used to support the descriptive adequacy of the normative model, there is an obvious worry that this procedure

[29] I take it that this suggests that at least in this context, subjects are interpreting these strength probes as meaning something generic like "How good or paradigmatic is this causal relationship?" rather than just in terms of something having to do specifically with probability raising. The title of Vasilyeva et al.'s 2018 paper, "Stable Causal Relationships Are Better Causal Relationships," seems to reflect this idea.

may be question-begging or at least that it may not amount to a test of what we think we are testing.[30] Suppose we have a theory according to which (judgments of) the "goodness" or "appropriateness" of a causal claim is affected by the presence of feature F. (E.g., F might be degree of background invariance.) If we adopt a verbal probe that is too closely tied to whether or not F is present, we may end up only testing whether subjects can recognize whether F is present, with the connection with the "goodness" of the causal claim being lost. For example, rather than asking about the appropriateness of asserting the claim that *C* causes *E*, we might instead ask subjects a more focused question about the extent to which the $C \rightarrow E$ relation is background invariant, thereby making it likely that we have a probe that is responsive to this feature, assuming subjects can recognize its presence. However, responses to this question will not tell us anything about whether subjects regard more background-invariant causal relationships as "better" causal relationships. To the extent that we are interested in this latter issue, it is arguable that it is better to employ a more generic, less targeted probe that is thought to be connected to judgments of how good or paradigmatic the causal relationship is. There thus seem to be arguments both for and against the use of more specific probes.

What then should we do when there is disagreement about which of several different verbal probes is most appropriate for measuring some target, as in the experiments described earlier? A number of possibilities suggest themselves. One is to investigate empirically the relationships between different probes and also the factors that influence the results associated with each. Do different probes for causal strength (or other causal judgment questions of interest) give substantially different results in a large range of different circumstance or only under special conditions? Are results from different probes related in some systematic way (e.g., does probe 1 usually produce stronger strength ratings than probe 2?). When probes give different results, can we discover at least in part why they do so? If we think that one or more of the probes are defective, can we identify some non-question-begging source of error in it? If, say, probe 1 requires subjects to estimate some quantity that we know on independent grounds they are not good at estimating and probe 2 does not require this and that is why they diverge, this may provide some grounds for preferring probe 2. In any case it seems desirable to invest more effort in explicit thinking about which probes are most appropriate for a given research question and how different probes may differentially influence subject responses.[31]

[30] In other words, responses to the probe may just be tests of whether subjects understand the probe and can recognize the presence or absence of the feature the probe is designed to track rather than a test of whether subjects' judgments of the goodness of the causal relationships is influenced by the presence or absence of the feature in question.

[31] As argued in what follows, similar points hold for X-phi research and for the questions that armchair philosophers ask themselves when they make judgments about cases.

A second strategy that may be preferable is, when possible, to reduce reliance on verbal probes altogether. One way of putting some of the problems posed by the use of verbal probes is that they often involve a loss of control by the experimenter. The experimenter uses certain words in the probe but cannot control how those words are interpreted by the subjects, so that subjects may be answering a different question or engaged in a different task than what the experimenter intended. The use of outcome measures that do not involve verbal probes may help ameliorate this problem when the experimental task and these nonverbal measures are sufficiently unambiguous. (This has been done in a number of the experiments described elsewhere in this book.)

Is it possible to develop nonverbal measures of subject assessments of causal strength (and not just measures of judgments about whether a causal relationship is present) or at least measures that don't consist in asking subjects directly about causal strength? Vasilyeva et al. 2018 took some steps in this direction in experiments described subsequently when they asked subjects to choose among different causes of some desirable effect where the relationships between these causes and the effect varied in background invariance. (However, the choices were still hypothetical and reported verbally, rather than involving an actual choice expressed in behavior.) They found that subjects chose more invariant causal relationships and, moreover, that their choices tracked their verbally expressed strength judgments, which arguably provides some support for the claim that the verbal measures are tracking something of more general significance. Of course it also would be worthwhile to see whether their choices expressed in actual behavior track strength judgments. Similarly, in experiments like Cheng's one might present subjects with information about the frequency of headaches with or without various medicines in the format she employed and then ask subjects to imagine that they want to avoid developing a headache and to choose a medicine to accomplish this.[32] Would subjects make choices based on Δp, causal power, or something else? One can also ask what verbal probe of causal strength best correlates with the choices they make.[33]

[32] Another possibility is to ask subjects to predict how a cause will behave in some new context based on information provided in some original context and then compare the prediction with strength judgments elicited by verbal probes in the original context. The prediction is of course a verbal report, but presumably it speaks in favor of a particular strength probe that it produces results that track such predictions, assuming that strength judgments reflect, among other considerations, information relevant to such predictions. Cheng has employed this general strategy in experiments discussed in Chapter 8 but without asking subjects to make strength judgments.

[33] As Thomas Blanchard has pointed out to me, one can think of this general strategy as motivated by the functional perspective on causation advocated in Chapter 1: a good indicator of whether subjects' verbal responses are genuinely aspects of their causal cognition is whether these responses are associated with characteristic goals of causal cognition having to do with manipulation, prediction, and control. If subjects who give stronger verbally expressed strength judgments to certain relationships in response to some verbal probe than to other relationships also privilege those relationships in tasks involving manipulation, prediction, and other tasks involved in causal learning and reasoning, this provides reason to think that the verbal probe is really tracking features that

Of course a skeptic might deny that a subject's nonverbal actions or choices in such cases have anything to do with their judgments of "causal strength," but one might hope that in many cases a suitably designed nonverbal task is less likely to be interpreted in ways that are different from what the experimenter intends and that researchers are less likely to be faced with the problem that small variations in the task lead to different results than when verbal probes are used.

Finally let me return to another issue previously mentioned in passing. This is the difficult question of what to make of subject responses that are (apparently) normatively incorrect and, relatedly, whether an apparently normatively incorrect response is really that or just an indication that the subject is interpreting the question differently than intended, and responding in a normatively appropriate way, given that different interpretation.[34] As noted previously (Chapter 3), many (and arguably all well-designed) experiments on causal cognition exclude subjects who give evidence of failing to understand important aspects of the scenarios or tasks with which they are presented. Obviously, however, this standard can be used in a question-begging way to exclude subjects who do not behave in accord with experimenter's preferred model. Presumably (at the least) one wants the criteria for excluding subjects to be independent of or not presuppose the correctness of the very theory or model under test—ordinary comprehension checks on verbally presented scenarios usually conform to this requirement.

There are, however, gray areas. For example, what should we think about the results in Perales and Shanks 2008 in which it is claimed that a number of subjects who attempt to answer the counterfactual query devised by Buehner et al. 2003 respond by citing *Pr(E/C)*? On the one hand, it would not be surprising if some subjects indeed do this since what this amounts to is (something like) a failure to distinguish *Pr(E/C)* from *Pr(E/intervene on C)*—something one finds even in econometric textbooks, as Pearl (2000) documents. But if subjects fail to make this distinction, what exactly follows? It would certainly follow that the counterfactual question is not a good one for getting at what Buehner et al. were hoping to measure. However, it is not clear that it follows that such subjects think and judge about causation in a way that reflects no difference between causal power

play an important role in causal cognition. For example, the results of Cheng et al. (2007) in which subjects infer the existence of causal relationships or causal interactions apparently on the basis of causal power-based considerations go well beyond merely fitting some particular verbal probe of strength.

[34] As noted earlier, this is an issue that arises elsewhere in cognitive psychology. When subjects are asked, in Kahneman's and Tversky's well-known study, whether it is more probable that (i) Linda is a feminist bank teller or (ii) she is bank teller, do they interpret the question as asking whether they should assign a higher probability to a conjunction than one of its conjuncts, and answer that question in the affirmative (and hence in a normatively inappropriate way)? Or do they interpret the question in some other way (e.g., as asking, given that Linda is a bank teller, is it more likely that she is a feminist than not) in which case their answers may well be normatively appropriate?

and *Pr*(*E*/*C*) or in a way that identifies causal power and *Pr*(*E*/*C*) or even that they identify the answer to the counterfactual probe employed by Buehner et al. with *Pr*(*E*/*C*). For one thing, that people recognize some difference between causal power and *Pr*(*E*/*C*) might be shown by the discovery of some other measure that better tracks causal power and distinguishes it from *Pr*(*E*/*C*). Second, to the extent that subjects respond to the counterfactual probe by providing an estimate of *Pr*(*E*/*C*), it is arguable that they are confused by or do not understand the counterfactual probe. (Similarly, if when asked for an estimate of *Pr*(*E*/*intervene on* C), they respond with an estimate of *Pr*(*E*/*C*), it might be argued that they lack a clear understanding of *Pr*(*E*/*intervene on* C) or its relation to *Pr*(*E*/*C*).) If people do not understand the counterfactual probe or what is meant by *Pr*(*E*/*intervene on* C), it seems misleading to describe them as identifying these quantities with *Pr*(*E*/*C*) or as treating them as measured by *Pr*(*E*/*C*).[35] Note that these questions about how to interpret non-normative responses are completely general, arising whether the responses take the form of judgments about cases from the armchair, judgments elicited by surveys, or judgments in experiments.

Another general issue raised by the experiments described previously has to do with the extent to which it is possible and fruitful to separate off those processes and representations that have to do with causal cognition from those having to do with other, distinct cognitive and perceptual processes that serve as an input to causal cognition or may otherwise influence it. For example, if people tend to misestimate probabilities from frequency data when these are presented in certain formats and (because those probabilities are used as input when people make causal strength judgments) they then make causal strength judgments that do not reflect the true frequency data in the way predicted by some model of causal judgment, does that show that the model is mistaken? It is arguable (and Buehner et al. 2003 do argue) that the answer to this question is no—that if the misestimation hypothesis is correct, this shows us something about how people estimate probabilities but not necessarily anything about causal cognition. (This of course assumes that the processes involved in probability estimation are in relevant respects distinct from those involved in causal judgment, which is a claim that Buehner et al. explicitly make.) The claim that what is going on in such cases is that people are misestimating probabilities need not be objectionably ad hoc if there is some independent way of showing that such misestimation occurs. For example, it may be possible to obtain information from other sources about subject estimates of frequencies and show that these diverge from the true frequencies. If subjects' estimates are plugged into the researcher's model of causal

[35] Consider someone who is asked to estimate *Pr*(*E*/*C*) and replies with an accurate estimate of *Pr*(*E*). It seems problematic to describe this subject as thinking that these two quantities are identical or that the latter is a good estimate of the former. It seems at least as accurate to describe the subject as confused about what *Pr*(*E*/*C*) represents or what *Pr*(*E*) represents or both.

cognition and the result tracks the subject's causal judgments, this can be thought of as providing support for the model. (In fact this is what Buehner et al. attempt to show.)

Another possibility is to look for formats for the presentation of frequency information that are known to lead subjects to make more accurate probability estimates. Here one would expect improved fit between the causal cognition model and judgments of strength (or whatever the dependent variable is). To the extent that people can do probability estimation tasks in contexts not directly involving causal judgments (as they clearly can) and one can observe how various formats affect reliability, one may be able to separate out the effects of probability estimation per se from other factors that affect causal judgment.

As Buehner et al. suggest, we can think of these issues in terms of a contrast between two different strategies or methodologies for modeling causal cognition. One strategy begins by adopting some particular measure (e.g., of causal strength, operationalized in a certain way) as its target explanandum and then tries to account for patterns in that measure in terms of a causal cognition model. A second strategy takes seriously the possibility that responses to the measure may be the upshot of a number of different processes, only some of which are directly relevant to causal cognition per se. For example, responses may reflect the influence of memory limitations, as discussed earlier. Or responses may reflect varying processes of semantic interpretation of the probe. To the extent this is so, trying to provide a unified model or explanation of such responses may be a mistake. Instead, those following the second strategy will look for new procedures (more perspicuous formats for the presentation of data that make fewer demands on memory, improved verbal probes) that better isolate the particular cognitive processes that are of interest from other extraneous factors. This is a matter of designing better experiments and measures (generating new data) that more effectively isolate the phenomenon of interest rather than just trying to model the data generated by whatever probe happens to be in common use.[36] Buehner et al. make it clear their preference for the second strategy:

> Our goal was to study people's natural capability to discover causal relations, with as little interference as possible from other mental processes. Demands on mental processes that could influence performance but are not directly relevant to causal inference would add ambiguity and statistical noise to our results. To reduce demands on comprehension, we used a visual format wherever we thought it would aid the clarity of presentation and reduce potential errors in comprehension. To reduce demands on memory, we presented all trials

[36] The second procedure is tied to the discovery of strategies for isolating phenomena in the sense of Bogen and Woodward 1988 and trying to explain phenomena rather than data.

> relevant to each candidate cause simultaneously, on the same screen. Finally, to keep boredom and fatigue to a minimum, we included as few conditions as possible that would still attain our goals. (1129)

They add:

> The choice between finding good fits to existing data and conducting better controlled experiments can be viewed in a broader context. Any pattern of psychological results is potentially the product of a multitude of distinct psychological processes, many of which are extraneous to the central process in question. These extraneous processes may disguise or mask the output of the process of central concern. If the goal of psychological research is to fit various patterns of data, irrespective of the extraneous psychological processes likely to give rise to those patterns, then models replete with parameters will no doubt have an advantage, as they have additional degrees of freedom. However, fitting arbitrary parameters to fit an unspecified mixture produced by multiple psychological processes will not deepen the understanding of any psychological process.
>
> If the goal of psychological research is instead to understand particular processes that are evidently important, then a more fruitful research strategy is to attempt to isolate the operation of the process under study. Achieving this goal may involve minimizing the influences of extraneous processes that normally would operate under typical everyday situations. The researcher's strategy, just like that of an everyday causal reasoner operating in accord with the power PC theory, is to control for and reduce the influence of alternative causes of the effect in question, so that the candidate causal process can manifest itself as clearly as possible. Thus, contrived situations in which alternative causes are eliminated or controlled can in fact be most informative. What justifies the creation of such situations is the simple and coherent explanation of a complex pattern of findings. (1139)

7.5. Some Implications for X-phi and for Appeals to Intuition in Traditional Philosophical Methodology

I turn next to an exploration of the implications of some of the issues canvassed previously for the roles assigned to intuitive judgment in philosophy, both in X-phi and in more traditional forms of philosophical methodology. My basic point is simply that issues about possible ambiguities in verbal probes, subject heterogeneity in interpreting such probes, how to deal with non-normative responses, and so on that arise in experimental psychology are also likely to arise in both X-phi and in connection with armchair methods or appeals to intuition. All of these methodologies face a common set of problems.

In an experimental context in which a question is posed (e.g., about causal strength or whether a causal relationship is present) and subjects are asked for a verbal response, it is obvious that one needs to worry that different subjects may interpret the question differently from one another or differently from what the experimenter intends. If so, the verbal probe may not measure what the experimenter thinks it is measuring or may not measure the same variable for all subjects. Similarly some subjects may be confused in various ways, producing verbal responses that, while semantically meaningful, may be difficult to interpret when taken as sources of information about their cognition.

As previous chapters have suggested, it is natural to wonder whether the same things sometimes happen when philosophers elicit intuitions/judgments about cases, either from themselves or others. Suppose Philo describes a case and reports having such and such an intuition/judgment about it. Cleanthes reports the same or a different intuition about the case. Each is in effect asking himself a question ("What is my intuition or judgment about whether this is a case of X?"). When (and how) can we be confident that they are asking themselves the same question or even that they are asking a coherent, intelligible question? Suppose the example is one in which Philo and Cleanthes both report their intuitions about the strength of the causal relationship in a certain scenario. In the experiments described earlier, it seems clear that different subjects sometimes interpret this request for causal strength differently, even when the same verbal probe is used. The same subject may also respond differently when different verbal probes (probes that antecedently one might think are not very different from one another) are used. It would seem unduly optimistic to assume that the same thing can't happen to Philo and Cleanthes. Indeed, going further, shouldn't we also be concerned about the possibility that Philo himself may be unaware of possible ambiguities in the question he is asking himself? That is, he may report his intuition about causal strength (or whatever), without recognizing that the question he asks himself may be unclear or that when he asks himself what he thinks are versions of the same question, expressed slightly differently, he is actually asking himself different questions to which different answers are appropriate.

Moreover, if this can happen for causal strength, why can't it also happen for intuitions/judgments about whether various scenarios are cases of causation at all? As a possible illustration (cf. Chapter 2) consider a case in which some subjects who are asked whether a causal relationship is present interpret the question as about whether a nonzero total cause is present and others interpret it as a question about causation along a route (contributing causation).[37] If you are tempted to respond that trained philosophers would immediately recognize such an ambiguity, consider that in the early philosophical discussion surrounding Hesslow's (1976) well-known example of birth control pills, in which there is cancellation along two different routes with the result that there is causation along a route but a zero total cause, few if any discussants seemed to recognize this ambiguity.

In raising these questions about the interpretation of verbal probes, I don't mean to suggest that they are unsurmountable in principle for judgments about cases, any more than they are so in psychology experiments.[38] Additional discussion between Philo and Cleanthes (as well as additional examples and analysis of these) may help to clarify how both understand key terms.[39] Distinctions based on normative theories can help to make us aware of possible ambiguities and unclarities in verbal probes used in eliciting intuitions, just as they can for verbal probes used in experiments. My point is simply that whatever role they assign to judgments about cases, if parallel results from experimental psychology are any guide, both armchair philosophers and those conducting X-phi research need to be concerned about whether their verbal probes are understood uniformly and unambiguously, both by themselves and by others. And just as it is desirable in experimental psychology to combine verbal measures with measures of non-verbal behavior, a similar strategy might be usefully employed in X-phi.

Next consider the previous suggestion from Buehner et al. (2003) that in psychology experiments subjects' responses (verbal and otherwise) are often the upshot of lots of different psychological processes within individual subjects, to which we may add the further complicating possibility that the relative contributions of these processes may vary across subjects. It is an obvious possibility that this may be true of philosopher's intuitive judgments (or some of them) as well.[40] To the extent that we treat such judgments simply as possible sources of information about how the philosopher and others judge, this may not be a problem since for these purposes what matters is just how reliable the judgments

[37] The example of the fielder and the window in Chapter 3 provides another illustration of this possibility.

[38] Of course it is also an empirical question to what extent such ambiguities and other interpretive problems are actually present in the verbal probes used by philosophers and others. I don't claim to know that such problems are widespread since I think that they haven't been adequately investigated.

[39] This is a point at which the armchair philosopher (or better, several of them in dialogue) may have an advantage over both survey research and at least some experiments. Dialogue and reflection may help to disclose ambiguities and unclarities in questions that may not be apparent when a subject is asked for a one-off answer (Is this a case of causation or free will?) with no further probing of how the question is understood.

[40] This is another reason why thinking of intuitions/judgments about cases as explained in a unified way by the possession of a concept may be misguided—even if we think of the concept as playing some role in generating the judgment, the judgment will reflect much else as well. Moreover (what we call), the concept may itself be the reflection in judgment of a number of disunified processes.

are about these targets. On the other hand, to the extent that these judgments are taken to be what we are trying to explain (so that we try to produce a theory that appropriately systematizes these or describes how they are generated), we do need to consider the possibility that whatever underlies these judgments may be very complicated, reflecting the impact of many disparate processes that may not lend themselves to a unified explanation. In some cases better-chosen scenarios and verbal probes that focus on more specific phenomena associated with causal cognition and attempt to account for them may help with this problem.

7.6. Psychological Studies of Omissions, Preemption, and Double Prevention: The Role of Invariance

I turn now to a rather different set of psychological experiments that concern causal judgments about omission and double-prevention scenarios. Here too considerations having to do with invariance will be central.

I begin with Walsh and Sloman (2011), who presented subjects with a series of scenarios. In one, a coin stands unstably on edge, about to fall tails. Billy and Suzy roll marbles in such a way that, if nothing interferes, each will strike the coin so that it lands heads. Billy's marble strikes first. In this scenario, 74 percent of subjects judged that Billy's marble caused the coin to land heads.[41] In a second scenario, an unstable coin is again on edge, about to land heads. A third party rolls a marble in such a way that if it strikes the coin, it will land tails. However, a book blocks the path of the marble. Frank removes the book, but if he had not, Jane would have. The marble strikes the coin and it lands tails. In this scenario only a minority of subjects (38 percent) judge that Frank caused the coin to land tails, and virtually no one judges that Jane caused this effect.

Walsh and Sloman's Billy/Suzy scenario is a case of causal preemption in which Billy's roll causes the coin to fall heads but in which this effect is not counterfactually dependent on the cause in the simple, straightforward captured by the counterfactuals "if *c*, then *e*" and "if not *c*, then not *e*." However, transmission or a connecting process is present between the cause and the effect. In the second scenario, the relation of Frank's action to the coin's falling tails is like double prevention, but with the (important) added complication that because of how Jane would have acted, the coin's falling tails is not counterfactually

[41] As remarked in Chapter 3 this raises the question of what we are to make of the 26 percent of subjects who did not endorse this judgment. Of course it is possible that some of these subjects were operating with a (highly nonstandard) understanding of causation that supported their responses. However, a natural thought is that some of the subjects either did not understand the question they were asked or were not motivated to seriously try to answer it. If so, there are arguably good reasons for excluding these subjects from the experiment.

dependent on Frank's action, unlike the standard cases of double prevention. Walsh and Sloman do not report any results about experiments with a standard double-prevention structure without a second, preempted preventer (i.e., the same scenario as before with Jane absent).[42] I assume that they think, regarding such a Jane-absent case, that although more subjects might judge that Frank's action was a cause than in the case in which Jane is also present, more still would judge Billy was a cause in the first scenario. If this expectation is correct and if similar results were obtained in other experiments, this would seem to support the conclusion that as far as folk thinking about causation (or at least actual causation) goes, cases in which a "connecting process" is present are more likely to be judged as causal than cases of double prevention and further that most people do not regard double-prevention relations as causal. Indeed, these are among the conclusions Walsh and Sloman draw. Of course similar conclusions are advocated by a number of philosophers on the basis of more intuition-like considerations.

I will return to this claim later, but I want first to describe some additional experiments, due to Lombrozo (2010) and to Vasilyeva et al. (2018), that complicate matters in an interesting way and which relate directly to the discussion of double-prevention relations in Chapter 6 and also to the discussion of invariance in Chapter 5. Lombrozo's (2010) experiments explored people's causal judgments regarding pairs of double-prevention scenarios. One relationship in the pair involved either intentional action, an artifact with a designed function, or a biological adaptation. The other relationship was a similar double-prevention scenario but without these features. (Her assumption, discussed in more detail in what follows, was that the first relationship in each pair would be seen as more invariant, for reasons of the sort discussed in Chapter 6.) For example, in one pair, subjects were presented with the following scenario:

> The diet of a certain kind of Australian shrimp consists of three kinds of foods: alphaplankton, bacterioplankton, and cromplankton. Alphaplankton contain chemical A, which triggers a reaction that changes the shrimp's skin to make it reflect high frequencies of ultraviolet light. Bacterioplankton contain chemical B, which neutralizes chemical A and thereby prevents the shrimp from reflecting high frequencies of ultraviolet light. However, cromplankton contain chemical C, which binds to chemical B and thereby prevents chemical B from preventing chemical A from making the shrimp reflect high

[42] In other words, if Walsh and Sloman's intention was to do an experiment that pits (i) a case in which simple dependence of the effect on the cause is absent but a connecting process is present against (ii) a case in which simple dependence is present but a connecting process is absent, their experiment fails to accomplish this. This illustrates how a more adequate analytical understanding of differences in causal structure can be relevant to experimental design.

frequencies of ultraviolet light. Because of these interactions, a shrimp that has eaten alphaplankton, bacterioplankton, and cromplankton will reflect high frequencies of UV light. (2010)

Thus chemical C is in a double prevention relation with respect to reflection of ultraviolet light. Some subjects were presented only with the preceding description with no indication that a biological adaptation is involved. Other subjects were presented with additional material that describes the ability to reflect high-frequency UV light as an adaptation:

> Reflecting high frequencies of UV light is biologically important, as it aids the shrimp in regulating its temperature by reflecting the frequencies of light with the most energy. In fact, while eating bacterioplankton is important for nutritional reasons, Australian shrimp have evolved to eat alphaplankton and cromplankton because these foods result in the reflection of high frequencies of UV light and thereby improve temperature regulation. (2010)

The two groups were then asked for judgments about the causal status or strength of the double-prevention relations in the scenarios. In particular, they were asked to answer questions about "how appropriate" (as judged on a six-point scale) it is to claim the relations in question are causal. Other scenarios were similar but, as described previously, had to do with whether the double-prevention relation involved a designed artifact or an intentional action (as opposed to an action that produced the same outcome as an unintended byproduct) as the manipulated variable.

Although (in additional experiments that I will not describe here) Lombrozo found, in agreement with Walsh and Sloman, that subjects judged it more appropriate to describe cases (or at least the cases she considers) involving the presence of a connecting process as causal than (at least some) cases involving double prevention, she also found that subjects distinguish among cases of double prevention, judging it more appropriate to describe cases involving intentional action, designed function, and biological adaptation as more causal than cases of double prevention not involving these features. For example, in the pair of scenarios described earlier, subjects were more willing to judge that the double-prevention structure is causal when they were told that this light-reflecting tendency was a biological adaptation than when it had no adaptive significance. Similarly for parallel examples involving effects that were intended or the result of a designed function. Since none of Lombrozo's cases of double prevention involve an uninterrupted connecting process, or transfer of energy/momentum from the putative cause to its effect, it cannot be the presence or absence of these features that accounts for subjects' willingness to regard some cases as more paradigmatically causal than others.

Following a suggestion in a paper of mine (Woodward 2006) along the lines described in Chapter 6, Lombrozo 2010 proposed that the underlying unified explanation of this pattern of judgment is that double-prevention relations are judged to be more appropriately described as causal or as having higher causal strength to the extent that they are more insensitive or background invariant. In other words, the feature that the experimental subjects are differentially responding to when they assign different measures of causal strength to different double-prevention relations is the difference in background invariance of such relations. As argued in Chapter 6, when double-prevention relationships obtain as a result of a biological adaptation, they tend to be such that they are buffered against normal environmental variations, and hence less sensitive/more background invariant than otherwise similar double-prevention relations that are not the results of biological adaptations. Similarly for designed artifacts involving double-prevention relations: to the extent that these are products of design, they will be designed in such a way that the double-prevention relation on which they rely will continue to operate under a range of conditions and hence will be relatively insensitive. Similarly also for cases in which some outcome O is associated with behavior B via a double-prevention relation, with O being intended by the agent, when these are contrasted with otherwise similar cases in which O is not intended—other things being equal, double-prevention relations between B and O when O is intended are more background invariant and hence judged to be more appropriately described as causal.[43]

Lombrozo's analysis is thus similar to the suggestion I made in Chapter 6: that a unified explanation of why subjects judge it more appropriate to describe some double-prevention relations as causal in comparison with other double-prevention relations can be found in the fact that these subjects are tracking the extent to which the relations are sensitive or background invariant: the more invariant relations are those that the subjects find it more appropriate to describe as causal. As argued in Chapters 5 and 6, there is a normative rationale for this pattern of judgment: to the extent that subjects value the identification of invariant relationships, it makes sense for them to make the distinctions they do among double-prevention relationships. Lombrozo (2010, 327) expresses the point this way:

[43] Recall the discussion from Chapter 6 concerning why the relation between behavior B undertaken with the intention of causing O is likely to be more background invariant than the relation between B and O when there is no such intention.

> Causal ascriptions are valuable insofar as they identify relationships that are sufficiently stable or invariant across situations to be useful in prediction and intervention. This proposal mirrors a hypothesis in Lombrozo and Carey (2006) about the function of explanation, called "Explanation for Export." According to Explanation for Export, explanations identify factors that are "exportable" in the sense that they are likely to subserve future prediction and intervention. If the function of isolating parts of causal structure, be it in an explanation or a causal claim, is to subserve future prediction and intervention, then "good" causes should be those that reflect stable, invariant relationships that are exportable to relevant situations.

Returning to Walsh and Sloman's results, we are now in a better position to assess what they show. Assuming that their subjects are representative (and ignoring the complication discussed in footnote 41), their experiment does show that their subjects are more willing to judge the relation present in one example involving a connecting process as causal than they are willing to judge one example of a double-prevention relation as causal. Obviously, though, it does not follow that people judge all relations as causal as long as a connecting process is present or all relations involving dependence as non-causal as long as a connecting process is absent or that people make no distinctions regarding causal status among processes of the latter sort. Nor does it follow that what drives people's causal judgments is just whether a connecting process is present rather than other features of the scenarios presented, such as level of background invariance, the latter being what Lombrozo et al. in effect claim. To repeat a point made earlier, to establish conclusions about what causally explains subjects' judgments in scenarios like those used by Walsh and Sloman, one needs to control for alternative explanations for what is influencing those judgments, including explanations invoking background invariance. Walsh and Sloman's study does not do this.[44]

Lombrozo's experimental results arguably provide evidence that subjects value (and differentially respond to the presence of) insensitivity /background invariance in causal relationships. However, there are still worries about confounding: perhaps in the experiments already described, subjects are responding to some feature of the relationships other than their background invariance—a feature that is correlated with variations in background invariance. In particular, consider the following possibility: suppose subjects have a joint probability distribution over E, C, and B_i for all possible values of these variables. They then use this to compute the average effect of C on E, averaging over all of the possible values of the background variables weighted by their

[44] Of course a similar conclusion holds if a philosopher were to consult her intuitions about scenarios like those described by Walsh and Sloman and judge as their subjects do.

probability. When C and E are binary variables, this amounts to comparing situations in which C is present with situations in which it is absent, so that they compute $\Delta p = Pr(E/C) - Pr(E/not\ C)$ or some function thereof where $Pr(E/C) = \Sigma\ Pr(E/C.B_i)Pr(B_i/C)$ and $Pr(E/not\ C) = \Sigma\ Pr(E/not\ C.B_i)Pr(B_i/not\ C)$. It seems plausible that in many real-life cases, relatively background invariant relationships will also be relationships for which Δp tends to be high since background invariance means, in calculating Δp, that one averages over many background circumstances in each of which that relationship obtains.[45] If there is a relatively large difference between $Pr(E/C.B_i)$ and $Pr(E/not\ C.B_i)$ in many of these B_i, Δp will be relatively large. In the Australian shrimp example, to the extent that the relation between eating alphaplankton and reflecting ultraviolet light holds in many rather than few background circumstances, one might expect, averaging over those background circumstances, the overall probability boost provided to reflecting ultraviolet light by eating alphaplankton to be relatively higher in comparison with a situation in which the relation is relatively non-invariant.

Thus, in the preceding examples, the subject's apparent preferences for more background-invariant relationships might just reflect the fact that these are relationships for which Δp is higher (than Δp for the less stable relationships). For example, if causal relationships that reflect a biological adaptation hold in background circumstances B_i for which $Pr(E/C.B_i)$ is large while $Pr(E/C.B_i)$ is small for non-adaptive counterparts, then Δp may be higher for the former for this reason. We thus need to disentangle the role of Δp from the role of background invariance in causal judgment. This has been done in a series of more recent experiments reported in Vasilyeva et al. 2018. These authors constructed scenarios in which background invariance (which in this paper they describe as a matter of "stability")[46] was varied while ΔP was held fixed. They found an independent effect of the former on causal judgment.[47] To understand how such independent variation is possible, recall first that in probabilistic contexts like those under discussion, what is relevant to the assessment of background invariance is the invariance of some form of the probabilistic relation involving C, E, and B_i—for example, whether $Pr(E/C.B_i)$ remains (close to) the same for different B_i. Second, recall that in general the assessment of background invariance does not just reflect the frequency with which those background circumstances occur but rather the number or range of background circumstances in which the relationship holds, independently of whether these are common or rare. For

[45] Of course, as noted earlier, when a cause raises the probability of its effect by only a small amount, it is possible for Δp to be low and yet for this probability raising to occur across different background circumstances so that this causal relation is relatively background invariant. As far as I know there have been no experimental investigations of how subjects judge in this sort of situation.

[46] "Stability" is also used in Woodward 2010 to describe background invariance or insensitivity.

[47] A similar independent effect of background invariance on responsibility judgments is found in Grinfeld et al. 2020.

example, in a probabilistic context the extent to which $Pr(E/C.B_i)$ remains the same for different B_i regardless of the actual frequency of the B_i is relevant to the assessment of background invariance. Recall that, conceptually, background invariance is understood counterfactually in such a way that even background circumstances that never actually occur are relevant to background invariance as long as it is true that if those circumstances were to occur, the relationship of interest would hold in those circumstances. (Of course if $Pr(B_i) = 0$, the corresponding conditional probabilities will not be well defined, but this just brings out the distinctness of background invariance from quantities that can be defined in terms of conditional probabilities.)

These observations enable us to understand how it is possible to manipulate background invariance independently of Δp and related quantities. As an illustration, in one set of experiments conducted by Vasilyeva et al., subjects were presented with scenarios in which they were told that a scientist was investigating the hypothesis that for certain fictional animals, zelmos, eating yona plants is causally related to developing sore antennas. Participants were first presented with results of an experiment described as performed by the scientist that took the form of a two-by-two covariation table cross-classifying zelmos based on whether they ate yonas or not (the participants were told there was random assignment) and whether they developed sore antennas or not. Participants were then told that a second experiment was performed, again with random assignment of diet but in this experiment half of the zelmos were given salty water and the other half were given fresh water (both were described as normally occurring within their environment), with this variable and the diet variable described as varying orthogonally. Contingency information concerning the results of this experiment was again presented to participants. Some participants saw contingency information in which the ingestion of salt water acted as a moderator variable for the effect of diet on sore antennas, with the eating yona → sore antenna relationship being strong for those ingesting salt water (i.e., the probability of sore antennas after eating yonas was high for those ingesting salt water) and non-existent for those zelmos ingesting freshwater. Other participants saw contingency information which suggested that the eating yona → sore antenna relationship was not moderated by the kind of water ingested—it was the same for those ingesting salt water and for fresh water. Thus the unmoderated relationship was more background invariant than the moderated relationship with the former but not the latter holding to the same extent across both the saltwater and freshwater conditions. The frequencies in the contingency tables were chosen so that value of Δp was the same across both the moderated and unmoderated relationship and was also the same as the value for the table in the first experiment in which results of the moderator were not presented.

Subjects were then asked how much they agreed (on a seven-point scale) with a series of statements about the relationship between eating yonas and sore antennas without mentioning the kind of water that the zelmos drank—these included both type- and token-level causal statements ("For zelmos, eating yonas causes their antennas to become sore," "For some particular zelmo, Timmy, eating yonas caused Timmy's antennas to become sore") and what Vasilyeva et al. called "explanation" judgments ("How much do you agree with the following explanation of why zelmos' antennas become sore?: For zelmos, antennas become sore because of eating yonas"). The result of this experiment was that participants were significantly more likely to agree with both the causal and explanatory claims, for both type and token judgments, when these were unmoderated (greater background invariance), despite the fact that Δp was equated across the two sorts of claims. For example, on a seven-point scale, the mean judgment of agreement was above 5 for the unmoderated causal claims and around 3.5 for the moderated claims. Interestingly, there was little difference in the results for the causal and explanatory claims.[48]

In this experiment, the portion of the population in the moderating condition for which salt water functioned as an enabler of sore antennas was relatively small—50 percent of the total population—thus raising the possibility that subjects responded to this feature (which Vasilyeva et al. call "actual scope") rather than background invariance per se. (Recall that degree of background invariance is at least in part independent of the actual frequency with which background conditions occur, since background invariance also tracks whether a relationship would hold were non-actual background conditions realized.) However, for this reason, actual scope can also be manipulated independently of background invariance, and, moreover, this can be done in a way that involves holding Δp constant. When this was done in a second experiment (by increasing the proportion of the population in which salt water functions as an enabler of

[48] On the other hand, Vasilyeva and Lombrozo report, in another paper (2015), evidence for a "dissociation" between causal strength judgments and judgments of explanatory goodness. In particular they report results that they interpret as showing that explanatory judgments are more sensitive to what they call "mechanistic" information and less sensitive to "covariational" information than causal strength judgments. For reasons of space I will not discuss this paper here except to say that I think it raises a number of issues about the interpretation of verbal probes that are similar to those highlighted elsewhere in this chapter. As a more general point, issues concerning the relation between causal claims and causally explanatory claims (and probes for such claims) deserve further exploration, both empirical and analytical. As noted earlier, a number of philosophers have claimed, on various grounds, that there are fundamental differences (other than the obvious ones) between causal claims and causal explanations and that this is reflected in ordinary practices of causal judgment. They have then used this distinction to argue for various philosophical theses—for example, that although absences cannot be causes, they can figure in causal explanations. In assessing these claims it would be desirable to have a more systematic understanding of whether people make distinctions of the sort described, the circumstances in which they deploy such distinctions, and their normative significance.

sore antennas to well above 50 percent while holding Δp constant) the result again was that background/invariance had an independent effect (i.e., independent of actual scope) on judgment for both causal and explanatory claims and for both type and token judgments. Finally, in a third experiment, participants were given an opportunity to choose among different means for producing some desired end, where the means/ends relationships varied in background invariance (again holding Δp constant). Subjects tended to choose the more background-invariant relationships for those with the same Δp. Experiments conducted by other researchers show a broadly similar pattern (see Grinfield et al. 2020).

Yet another possibility is that the participants were making their causal judgments by computing causal power in the sense of Cheng 1997. As explained previously, Cheng's measure ($p_c = \Delta p / 1 - Pr(e/not\ c)$ (7.4)) is an invariance-based measure in the sense that it incorporates requirements having to do with the invariance of the causal power of a candidate cause *c* to produce *e* across changes in the frequency with which *c* occurs as well as across variations in whether other causes of *e* are operative. However, background invariance in the sense described previously involves a somewhat different invariance requirement from those imposed by Cheng. As noted earlier, conceptually Cheng's measure corresponds to a question like "How much would *c* increase the probability of *e* if it were introduced into a situation in which no other causes of *e* were present?"[49] Obviously this is a different question from the question of whether *c* increases the probability of *e* relatively uniformly across different background circumstances, including those that may be different from or unrealized in the original context in which p_c is calculated. In other words Cheng's measure (7.4) does a good job of capturing some dimensions of invariance but is not designed to capture others. Another way of bringing this out is to suppose that we simply apply measure (7.4) to estimate the causal power of eating yonas for sore antennas to a scenario in which the kind of water ingested plays a moderating role versus a scenario in which water does not play such a role. We would then find that the causal power measure p_c does not track degree of background invariance, in part because p_c (like Δp) is a measure of average causal strength that tracks the overall impact of *c* on *e* in a population and not the way in which the impact of *c* on *e* may or may not vary across the population. In fact, Vasilyeva et al. find in their experiments that when causal power in the sense of p_c as defined earlier is equated across moderated and non-moderated conditions, background invariance still has an independent effect on causal judgment. Let me add, however, as

[49] Note also the difference between this question (and the associated counterfactual probe Buehner et al. recommend) and the probe that Vasilyeva et al. employ. It would be odd if the latter probe generated the same judgments as Cheng's probe. Thus to the extent that the two probes are distinct, one would not expect strength judgments based on Vasilyeva's probe (which apparently are sensitive to background invariance) to fit strength judgments based on causal power.

Vasilyeva et al. recognize, there are other reasons why it is not clear that p_c is the appropriate measure of causal power to use for scenarios like those used in their experiments: for one thing, in the moderated condition, an interaction effect is present that violates one of Cheng's conditions for the applicability of the causal power measure. When interactions are present, Cheng proposes a more complex measure of causal power. However, Vasilyeva claim that their data are also inconsistent with the possibility that the participants are using this measure or other causal power-based possibilities.

7.7. Conclusion

Rather than trying the reader's patience by further exploring these experiments, I want to turn, by way of conclusion to this chapter, to some more general reflections (since my goal is, after all, the overall significance of invariance in causal judgment, both normatively and descriptively, rather than the ins and outs of these particular experiments). The experiments described in this chapter seem to show that, as an empirical matter, features of causal relationships having to do with invariance influence ordinary subjects' judgments (as expressed in responses to various causal strength probes) about how "strong" or "good" those relationships are or about the extent to which the subjects find it appropriate to describe those relationships as causal. In fact, several distinguishable invariance-linked features seem to influence causal judgment, including those incorporated into Cheng's causal power model and the considerations having to do with background invariance explored by Vasilyeva, Lombrozo, and colleagues. (I expect other invariance-based features of the sort described in Chapter 5 also to be influential.) Of course, as contended throughout this book, showing these empirical influences on causal judgment does not establish that they are normatively justified or appropriate. However, as argued in Chapter 5, there are strong normative reasons why various sorts of invariance considerations should matter in causal judgment. Thus to the extent that we find invariance considerations influencing people's judgments and behavior, we have reason to think that they are judging and behaving in normatively appropriate ways—that they are in these respects "rational" in their causal cognition, thus providing support for the rational model approach advocated in earlier chapters.

Returning to the role of intuition/judgment about cases in establishing conclusions about either causal cognition or "causation itself," I note that many accounts of causation developed by philosophers do not assign an explicit role to invariance-based considerations (at least under that description) either in how we think or how we ought to think about causation. Suppose that a traditional philosopher consulting her intuitions/judgments about various cases either

(i) reports that invariance-based considerations do not influence her judgments or, alternatively, (ii) simply fails to report that they do (perhaps because the possible relevance of invariance does not occur to her). If the philosopher is like the subjects in the experiments described in this chapter, we have reasons to suppose that in (i) she is mistaken and that in (ii) she has missed something important that influences her judgments. If the philosopher claims then that she is not like ordinary subjects and that in virtue of her philosophical training her different, invariance-neglecting judgments are more reliable or more informative than those of ordinary subjects, we can remind her that there is a normative theory supporting the role of invariance in the ordinary judgments, thus casting doubt on her claims of special expertise. In this way a combination of empirical results and normative theory can play a role in circumscribing and critically assessing claims deriving from intuition.

8
Proportionality

8.1. Introduction

This chapter is about the role of proportionality in causal judgment.[1] This notion was introduced by Yablo (1992) and has been the topic of a considerable amount of subsequent discussion. I see proportionality as one criterion among several having to do with the choice of variables for causal analysis.[2] The notion is applicable when (and as I will formulate it, only when) we face a choice among different candidates for the cause variable for a fixed effect or explanandum and where the candidate cause variables are themselves related in a specific way: they stand in what might loosely be described as a non-causal[3] hierarchy of abstractness or more or less fine-grainedness with respect to one another. (The determinate/determinable relation invoked by Yablo—see the subsequent discussion—is one way in which variables or properties in such a hierarchy may be related, but it may not be the only way.) Illustrations are provided by two extensively discussed cases to which the notion of proportionality has been applied: in the first, a variable representing the presence or absence of a particular shade of red such as scarlet stands lower in such a hierarchical relation to a more generic, less specific variable representing the presence or absence of the color red, which in turn is more specific (less abstract) than an even higher-level variable that just records whether some object is some color or other (rather than transparent). As a second illustration, one might think that if, for a variable whose values represent mental states, the same mental state can be multiply realized by different neural states, the mental states might be viewed as abstractions from (or coarse-grainings of) these neural states.

[1] Note for readers familiar with my previous discussions of proportionality in Woodward 2008 and Woodward 2010: I now think that these previous treatments are inadequate in various respects, both in failing to restrict the application of the notion of proportionality in certain necessary ways (described subsequently) and in understanding proportionality as imposing the requirement that the functional relation between cause and effect must be one to one. The formulation I currently prefer—(P) in Section 8.4—abandons the one-to-one requirement and is intended to replace what I have said earlier about proportionality. For further discussion, see Woodward 2018.

[2] As argued earlier, invariance is another criterion that legitimately influences variable choice.

[3] It is crucial to what follows that when variables or properties at one level are more fine-grained realizations of variables or properties at a "higher" level, the relations between these two sets of properties should not be thought of as causal. The relationship is rather of some other kind characterized by notions like supervenience or realization.

Causation with a Human Face. James Woodward, Oxford University Press. © Oxford University Press 2021.
DOI: 10.1093/oso/9780197585412.003.0009

The background to this is the observation that both in ordinary life and in science, causal claims can be formulated at different "levels" or "grains." It is widely (and in my view) correctly believed that some choices of "level" (or of variables associated with certain levels) are better than others for purposes of causal analysis. Moreover, although there is some tendency in philosophical discussion to suppose otherwise, there are strong reasons for thinking that the best choice of level is not always the most specific or fine-grained one.[4] Equally it is not always the most abstract or general level that is better or more appropriate. Moreover, as an empirical matter, people do not behave as if they believe that either of these choices is always best. Instead, people seem to believe that which choice or choices is best or better depends on the empirical details of the behavior of the system they are trying to understand—an assessment that is supported by normative analysis. This is also reflected in the fact that in science such choices are regarded as nontrivial and as requiring careful thought. Proportionality can be understood as one possible consideration that provides guidance for a good or appropriate choice of level. As such, it addresses a problem that is ubiquitous in causal thinking, both lay and scientific. Thus, as with the other considerations influencing variable choice discussed in this book, proportionality has both a normative and an empirical side: there is the question of what normative rationale, if any, supports the use of proportionality (and how to formulate a proportionality condition that has a suitable normative rationale), and there is also the empirical question of the extent to which people conform to such a condition.

Let me add that in what follows I will largely but not entirely focus on the role of proportionality in the assessment of type-level causal claims. Although I will not try to systematically explore this issue, my guess is that, as an empirical matter, proportionality considerations play a somewhat different role in the assessment of actual cause or token-level claims than in the assessment of type-causal claims.[5] In particular, failures to fully satisfy proportionality are (as an empirical matter) sometimes (and perhaps often) regarded as more acceptable for actual cause claims than for type-level claims. An appendix to this chapter describes a possible normative rationale for this difference.

[4] Some writers deny that abstract properties can ever be causally efficacious, claiming that strictly speaking only maximally specific properties can be causes—for discussion see, for example, Crane 2008.

[5] Much of the philosophical literature, including Yablo's original discussion, focuses on the role of proportionality in token-causal claims.

8.2. Yablo on Proportionality

In one of Yablo's illustrations[6] a pigeon is trained to peck at targets of any shade of red and only such targets. The pigeon is presented with scarlet targets and pecks. Consider the following two claims (understood, for the present, as type-level claims).

The red color of the target causes the pigeon to peck. (8.1)

The scarlet color of the target causes the pigeon to peck. (8.2)

Yablo claims (8.1) is superior or "preferable" to (8.2) on the grounds that (8.2) is "overly specific" and fails to be "proportional" to its effect. We will consider Yablo's more precise characterization of proportionality as well as some alternative characterizations shortly, but his intuitive idea is that causes are proportional to their effects when they do not "contain too little" (where "too little" means being inappropriately narrow and omitting crucial elements) *and* do not contain "too much" (where this means being overly broad and containing irrelevant or superfluous elements). The cause cited in (8.2) is overly narrow or too fine-grained (since the pigeon will peck in response to non-scarlet shades of red). Claim (8.1) does not have this defect, given the facts stipulated in the example and thus satisfies proportionality. By contrast, if we were to describe the cause as the target's having some color or other (as opposed to being transparent), it would be overly broad.

Yablo's more precise characterization of proportionality is as follows: first, proportionality is understood as applying to properties that are related as determinates (e.g., scarlet) and determinables (e.g., red), where for our purposes we can think of determinate properties as more specific ways of realizing the corresponding determinable properties or as standing to those properties in a subordinate to superordinate relation. In general the presence of a determinate property necessitates the presence of the corresponding determinable property, but not conversely. Yablo's view is that cause C is proportional to effect E if and only if C is "required" for E and C is "enough" for E. C is required for E iff none of its determinables screens it off and C is enough for E iff it screens off all of its determinates. In this context C_1 screens C_2 off from E iff, had

[6] I use this example because it has been widely used in discussions of proportionality. For some illustrations of the use of proportionality that are more scientifically serious, see Kendler 2005 and Woodward 2010.

C_1 occurred without C_2, E would still have occurred (Yablo 1997).[7] Applied to the pigeon example, the idea is that the presentation of the red target is proportional to the effect of pecking because all of the determinables of red (such as the target's being some color or other) fail to screen off red from pecking (the counterfactual "If the target had not been red but had been colored, the pigeon would have pecked" is false). Moreover, for any determinate of red, such as the target's being scarlet, if that determinate had not obtained but the target was still red, pecking would have obtained. The presentation of a scarlet target is not proportional to the effect (pecking) because there is a determinable of scarlet (namely red) such that if that determinable had obtained, the pigeon would have pecked.[8]

Although Yablo does *not* take satisfaction of a proportionality condition to be a strictly necessary condition for the truth of causal claims (1992, 277), he does suggest that (8.1) and (8.2) "compete" with respect to truth and that since (as he supposes) (8.1) is true, (8.2) must be false. (In other words, the preferability of (8.1) to (8.2) is understood in terms of their differing truth values.) A number of other writers either interpret Yablo as claiming that satisfaction of some proportionality condition is necessary for a causal claim to be true (e.g., Shapiro and Sober 2012—see the subsequent discussion) or advocate this position themselves (e.g., List and Menzies 2009).

These claims raise a number of interesting descriptive and normative issues. First, although Yablo's proposal is normative (he thinks we *ought* to prefer proportional to non-proportional causal claims), it also seems natural to interpret him as suggesting[9] that, as an empirical matter, subjects exhibit a preference of some kind for causal judgments that satisfy a proportionality requirement (on some suitable characterization of that requirement). Is this descriptive claim correct? If so, precisely what characterization of a proportionality requirement best captures people's judgments—Yablo's or some alternative? Relatedly, if subjects do (as an empirical matter) exhibit a preference for causal claims that

[7] "Screening off" in this context is thus understood in terms of counterfactuals rather than in terms of conditional independence. For a more precise definition see footnote 42.

[8] There are many other similar examples in the literature. Suppose a kind of platform will collapse if more than 2,000 kg are placed on it (Woodward 2008). Weights of 3,173 kg are placed on a series of such platforms and each collapses. The claim that placing weights of 3,173 kg on the platforms causes collapse does not satisfy proportionality or satisfies it less well than some alternatives, while the claim that placing a weight of more than 2,000 kg causes collapse does satisfy proportionality. Similarly, to take an example from the psychological literature (Lien and Cheng 2002, although they do not use the word "proportionality"), "Smoking Virginia Slims causes lung cancer" and "Inhaling fumes causes lung cancer" are respectively too narrow and too broad in the characterizations of their causes, while "Smoking causes lung cancer" does much better with respect to proportionality.

[9] Obviously Yablo expects his readers to respond to examples like (8.1) and (8.2) in a way that is consistent with the judgments that he thinks are normatively appropriate. That is, he expects that readers will, as an empirical matter, share his assessments of (8.1) and (8.2). This is one of many cases in which cases or examples are used by philosophers against the background of the assumption that judgments about those cases will be shared, so that, as claimed in Chapter 3, the philosopher uses her own responses to predict how others will judge.

satisfy some version of proportionality, what is the nature of this preference? Do subjects think that causal claims that fail to satisfy proportionality are (for that reason) always false? Or do they think that such claims can sometimes be true (assuming other appropriate requirements on causation are satisfied) but are deficient (or comparatively deficient) in some other way—for example, misleading, unsatisfactory for explanatory purposes, or less informative than they might be? Will a treatment of proportionality that, as a descriptive matter, best captures people's judgments take the form of a dichotomous, all-or-nothing condition (as Yablo's condition does), so that a causal claim either satisfies this condition or not, with failure to satisfy the condition being associated with falsity? Or should we think of proportionality as a more graded requirement/condition that can be satisfied to a greater or lesser degree? Note that this last possibility seems to fit better with the notion that comparative failures of proportionality are associated with failures along some other dimension besides truth/falsity, since the latter dimension is "binary."

To the best of my knowledge there has been relatively little discussion of these empirical issues in the psychological literature. The published research that most directly bears on these issues is Lien and Cheng 2002, which is discussed in Section 8.8.[10] These authors find that their subjects do judge and make causal inferences in a way that suggests that they prefer causal claims that satisfy a proportionality-like condition. However, Lien and Cheng do not directly address the issue of whether non-proportional (or less proportional claims) are judged to be false. (They also don't use the word "proportionality" or cite the philosophical literature on this notion.) The issue of whether subjects judge that claims like (8.2) are false might be addressed, of course, just by asking them a yes/no question, although there are more sophisticated possibilities. It would also

[10] Issues having to do with granularity of variables and proportionality are discussed in Soo 2019. After the drafting this chapter, an anonymous referee for Oxford University Press drew my attention to Johnson and Keil 2014. Like Lien and Cheng, this paper also explores how causal judgments are influenced by level-based considerations. However, Johnson and Keil use examples of levels involving what they call partonomic relations. These are cases in which candidate causes and effects at different levels are related by part/whole or component relations—for example, superordinate chemical reactions having subordinate reactions as parts or subprocesses. This contrasts with the Lien and Cheng's examples as well as the cases involving proportionality discussed in the philosophical literature, which involve what Johnson and Keil call "taxonomic" relations—relations like that between "red" and "scarlet." Nonetheless Johnson and Keil find that subjects assign stronger causal ratings to variables at the same level or with matching grain. One can think of this as also reflecting a proportionality-like consideration. Their paper is very interesting on the subject of the role of superordinate/subordinate relations and strategies of matching levels in restricting hypothesis spaces and guiding causal inference. Finally, as I was revising the final version of this manuscript, my attention was drawn to Blanchard et al. forthcoming. This is an experimental study of ordinary judgments about causal exclusion—it has to do with the extent to which causal claims formulated in terms of hierarchically related lower- and higher-level variables "exclude" one another. The authors find, in agreement with the views defended later in this chapter, that ordinary judgment does not conform to the exclusionist intuitions advocated by many philosophers.

be interesting to investigate more systematically whether, as I expect, subjects would give higher causal strength ratings (for some appropriate verbal probe) to claims like (8.1) in comparison with claims like (8.2), given the empirical facts stipulated in the example. There is also the issue, raised previously, of whether proportionality is weighted differently in connection with type-level judgments in comparison with singular or actual cause judgments and, if so, why this is the case.

In addition to these empirical questions, appeals to proportionality raise a number of important normative issues. Here a comparison with invariance is instructive. Although many accounts of causation pay little attention to invariance-based considerations, few researchers have explicitly argued against the idea that invariance matters normatively in causal assessment. By contrast, a number of philosophers have argued that proportionality is a misguided or normatively indefensible requirement. (For various forms of such arguments, see Shapiro and Sober 2012; Bontly 2005; McDonnell 2017.) Thus the question of whether there is a normative rationale for some version of a proportionality condition and, if so, what that rationale is a matter of considerable disagreement.

Obviously whether one thinks that a proportionality condition is normatively defensible is going to depend on just what one takes that condition to involve. Much of the criticism directed against the condition is directed against versions that claim that satisfaction of proportionality is necessary for a causal claim to be true (e.g., Sober and Shapiro 2012). Of course, as suggested previously, one can agree that proportionality is not necessary for truth while still thinking that, other things being equal, causal claims that do better in terms of proportionality are in some way preferable (because they are more informative, provide better explanations, or conduce more to other goals we have) and thus that in this sense proportionality is a normative desideratum. This is my own view, as explained in what follows.

Writers who deny that proportionality is necessary for the truth of causal claims sometimes conclude that proportionality is a "merely pragmatic" virtue or perhaps a condition relevant to the assessment of *causal explanations* (with explanatoriness itself understood as a pragmatic virtue) but not to causation itself. For example, Bontly (2005, 332) writes: "Proportionality is not in fact a constraint on the causal relation but rather a pragmatic feature of our use of causal language, derived from general principle of language use. If so, it turns out that proportionality has little if anything to do with the nature of causation itself." In a related vein McLaughlin (2007, 15) writes: "For the record, I myself think that rather than a constraint on causation, proportionality is a pragmatic constraint on explanation" (quoted in Harbecke, forthcoming).

Here the reader is urged to recall the remarks on "pragmatic" considerations in Chapter 1. If "pragmatic" is associated with means/ends justification, then

I fully agree that to the extent that a proportionality condition has a normative justification, this will take the form of arguments showing that satisfaction of this condition conduces to various ends or goals associated with causal thinking. This is the same form of argument that (according to me) can be used to provide a normative justification for other features like invariance that are relevant to the assessment of causal claims. In this respect (conduciveness to goals associated with causal thinking) I agree that proportionality is a pragmatic virtue. Note, however, that this notion of pragmatic does not support Bontly's claim that proportionality "derives from general features of language use" (presumably features having to do with the "pragmatics" of language such as those captured by Gricean maxims for conversational relevance). My own view instead sees proportionality as justified in terms of considerations and goals that are distinctively associated with causal thinking rather than with language use in general. For similar reasons, I would resist the invidious contrast (between causation and causal explanation) implied by the claim that proportionality has nothing to do with the "nature" of causation, although perhaps it is connected to causal explanation. In part this is a matter of my misgivings, expressed in previous chapters, about the uses to which philosophers have put the supposed contrast between what does and does not belong to the nature of causation. Putting aside this general concern, my view (recall Chapter 1) is that proportionality has to do with what I have called distinctions among causal relationships, in the sense that some true causal claims (or causal relationships) will satisfy proportionality conditions to some considerable extent and other true causal claims will not. Thus if the nature of causation has to do with necessary conditions for a causal claim to be true, then I agree that proportionality is not such a necessary condition. However, on my view, it is also true that a concern with proportionality is closely bound up with goals and functions associated with causal thinking rather than being, as it were, an external add-on to these.

In the remark quoted earlier, McLaughlin explicitly contrasts "causation" with "causal explanation," with the suggestion that "pragmatic" concerns play a large role in the latter. Of course there are obvious differences that are denied by no one—causation is a relation in the world, while "causal explanations" (as well as causal claims, whether or not these are used to explain) involve devices—words, graphs, equations, etc.—that are used to *represent* causal relationships. It is unclear, however, that it follows from this that when causal language is used to provide causal explanations, "cause" in such explanations somehow behaves very differently from (is governed by different rules than) "cause" when it is used in other contexts ("plain vanilla cause"/"cause with no implication of explanation"?). At the very least such a view requires considerably more spelling out than it has hitherto received. As an empirical matter, it seems to me, as suggested in Chapter 7, that it is at best unclear whether ordinary subjects distinguish between "cause" and "causally explains" in the manner suggested.[11] I will add that from a normative perspective, if one thinks of causal explanation along the lines that I favor, with this understood as providing information that can be used to answer what-if-things-had-been-different questions, then it is natural to think that one explains causally simply by providing causal information that can be used to answer such questions. Of course one can provide causal information that fails to answer such questions[12] or fails to answer them very well, but on this picture there is no sharp contrast between causal claims and causal explanations, at least of the sort some philosophers claim exists.[13] On the basis of these considerations I think we should resist the suggestion that proportionality is a condition that applies only to causal explanation but not to causal claims more generally.[14]

Of course this is not to say that proportionality has nothing to do with explanatory goodness—on the contrary. Rather my point is that this connection is not captured very well by adoption of a framework that distinguishes sharply between causal claims and causal explanations in the way Bontly and McLaughlin do. Thus when I say that when a causal claim is defective along the dimension of proportionality, the deficiency in question should be understood to pertain to

[11] Even if, as suggested in footnote 47, Chapter 7 explanation probes are more sensitive to mechanistic information than to covariational information, it certainly doesn't follow that proportionality has to do with causal explanation rather than causation. Indeed, proportionality considerations often seem to abstract away from mechanistic information and, as observed subsequently, to track covariation information.

[12] To take a trivial example, "The cause of *e* is the cause of *e*" is (assuming that *e* has a cause) a true causal claim, but (one would think) not explanatory. But I don't think that we need a sharp distinction between "cause" and "causally explains" to make sense of why this claim is unexplanatory.

[13] Robb and Heil (2018) claim that causation is a "metaphysical" notion and causal explanation an "epistemological" notion and that appeals to what they call "explanatory practice" in discussions of mental causation conflate these two notions. But even if we accept these characterizations, it does not follow, as they seem to suppose, that causal explanation and causation have nothing to do with one another or that one cannot learn something about how people think (and ought to think) about causation by considering how they think (and ought to think) about causal explanation and conversely. Robb and Heil's discussion thus strikes me as a good example of the misuse of the epistemology/metaphysics distinction that I complain about in the introduction.

There are other prominent philosophical accounts that distinguish between causal claims and causal explanations, but in my view these too are problematic. For example, according to Lewis (1986), while causal claims cite causes, a causal explanation of some event *e* can work by citing any causal information about the causal history of *e*, including the information that *e* had no causes, so that the latter counts as a causal explanation even though it does not work by citing causes. I doubt, however, that many people would regard "*e* had no causes" as a causal explanation of why *e* occurred and it is hard to see what the normative justification would be for judging otherwise.

[14] On the other hand, suppose that for some reason one is convinced that proportionality has to do (just) with causal explanation and not with causation. Then one can understand the discussion that follows as having to do with the role of proportionality in causal explanation. However, I would still want to resist any implication that because proportionality is associated with explanation, it is for this reason uninteresting or unworthy or attention or "pragmatic" in a way that implies it is "subjective." In this connection it is worth noting that in the quotations from Bontly and Mclaughlin, the association of proportionality with pragmatic considerations or explanation functions as a rhetorical device for not considering it further. It is yet another example of what I called a device of dismissal in the introduction.

the causal claim itself and should be understood in terms of goals associated with causal thinking rather than in terms of a separate category of causal explanation, governed by different rules and/or embedded within the framework of a pragmatic theory of explanation.

8.3. A Closer Look at Proportionality

With this as background, let me now turn to a more detailed exploration of what proportionality involves or how it might be understood. To do this, let's return to (8.1)–(8.2). To represent (8.1) and (8.2) within an interventionist framework, we need to express them in terms of claims about variables. Neither (8.1) nor (8.2) is explicit about how we are to think about the variables figuring in them. However, focusing first on (8.1), we naturally interpret it as employing a binary cause variable *RED* that can take either of two values, {red, not-red}, and a binary effect variable *PECK* capable of taking the values {peck, not peck}. Claim (8.1) might then be represented as

RED causes *PECK* (8.3)

where this is unpacked as implying that

An intervention that sets *RED* = *red* is followed by *PECK* = *peck* (8.3a)

An intervention that sets *RED* = *not red* is followed by *PECK* = *not peck*. (8.3b)

Thus we are interpreting (8.1) as implying both that an intervention that sets the target's being red leads to pecking and that an intervention that sets the target to a non-red color leads to the pigeon not pecking. Put more simply (and incorporating the idea that causal claims are intrinsically contrastive, which has a natural interventionist motivation; see Woodward 2003): we are interpreting (8.1) as the claim that the target's being red rather than not red causes the pigeon to peck rather than not peck. When interpreted in this way, (8.1) is true, given the facts specified in the example.

Reasoning in a parallel way, we can interpret (8.2) in terms of a cause variable *SCARLET* that takes the values {scarlet, not scarlet} as well as the *PECKS* variable:

An intervention that sets *SCARLET* = *scarlet* is followed by *PECK* = *peck* (8.4a)

An intervention that sets *SCARLET* = *not scarlet* is followed by *PECK* = *not peck*. (8.4b)

Statement (8.4a) is true. How about (8.4b)? Here we face a choice point. If we require that for (8.4b) to be true, *all* interventions that set *SCARLET* = *not scarlet* must be followed by *PECK* = *not peck*, then (8.4b) is false (since interventions that set the color of the target to non-scarlet shades of red are followed by pecking). Thus on this understanding of what it implies, (8.2) is false.

On the other hand, for (8.4b) to be true we might require only that *some* interventions that set *SCARLET* = *not scarlet* be followed by non-pecking. Since there are some ways of setting the target to a non-scarlet color (e.g., blue) that are followed by non-pecking, on this interpretation (8.4b) and hence (8.2) are true.[15]

To the best of my knowledge, there have been no systematic empirical investigations of whether most subjects interpret (8.4b) in terms of the requirement that all interventions that set *SCARLET* = *not scarlet* must be followed by *PECK* = *not peck*, or instead in terms of the requirement that some interventions that set *SCARLET* = *not scarlet* are followed by *PECK* = *not peck* and accordingly whether (8.2) is generally judged to be true (although perhaps less than fully perspicuous in some way) or to be false.[16]

[15] Interventions that set *SCARLET* = *not scarlet* are ambiguous interventions in the sense of Spirtes and Scheines (2004). However, the case is different from their total cholesterol case (discussed in Chapter 5) in several ways: first, in the total cholesterol case, interventions are ambiguous for all values of the total cholesterol variable. In the pigeon case, one of the two possible interventions—the *SCARLET* = *scarlet* intervention—is not ambiguous. Moreover, in the cholesterol case, unlike the pigeon case, the candidate cause variable cannot take the value "absent." Perhaps in the case of binary variables we have a higher tolerance for ambiguous interventions involving setting the variable to the value "absent" as long as interventions are not ambiguous for the value "present."

[16] Some (perhaps many) readers may find it natural to regard (8.2) as false and hence to think that the "some interventions" interpretation according to which it is regarded as true is far-fetched. It is worth noting, however, that influential theories of causation apparently judge it to be true. For example, Lewis (1986) tells us at one point that in evaluating a counterfactual of form, "If c had not occurred . . . ," we should imagine a possible world like the actual world in which c is "wholly excised" (rather than replaced with some event that is similar to c). This can be interpreted as implying that in assessing the counterfactual "If the target had not been scarlet . . . ," we are to imagine a situation in which no target is presented or something similar, in which case the pigeon will not peck. Under this interpretation, (8.2) is true. In addition, considerations of charity seem to lead us to regard causal claims that are overly specific in the way that (8.2) is as true. To use an example of Glymour's (1986), suppose that it is claimed that (S) "Shlomo's smoking four packs a day caused his lung cancer." Here we tend to take the contrast state to be one in which Shlomo does not smoke at all (or very little—his smoking is "wholly excised") and, on the assumption that in this case, Shlomo would not have

Of course if (8.1) is true and (8.2) is false (when these are interpreted along standard interventionist lines), there is no puzzle about why (8.1) is preferable to (8.2)—apparently we don't need to bring proportionality considerations into the picture to explain this preference. Suppose, on the other hand, that (8.2) is regarded as true (because, for example, it is interpreted in terms of the requirement that some interventions that set *SCARLET* = *not scarlet* are followed by *PECK* = *not peck*). In this case, we cannot appeal to the falsity of (8.2) to explain why (8.1) is preferable to (8.2), but (I contend) it still seems plausible that (8.1) is preferable to (8.2). One way of motivating this assessment is to note that within an interventionist framework, there are at least two different ways in which a causal claim might be deficient (or at least limited in various ways or lacking some desirable feature). At this point I will describe these roughly and imprecisely, just to provide a motivating intuition, with a more careful statement (P) provided later:

(8.5a) A causal claim might falsely claim that some (intervention-supporting) dependency relationship is present when it is not. Call this *falsity*. Violations of this requirement are ruled out by interventionist requirements like (M).

(8.5b) A causal claim might fail to represent one or more dependency relations that are present in the system of interest and that should be represented. Call this *omission*.

I assume that not all failures to represent dependencies along the lines of (8.5b) involve problematic deficiencies or limitations but that some such failures do—failures to represent what *should* be represented. (This is explained in more detail subsequently.) To anticipate, I see failures of proportionality as having to do with certain sorts of failures of this sort.

To apply (8.5a)–(8.5b) to (8.1)–(8.2), when (8.2) is interpreted in accord with the "all interventions" interpretation, it falsely claims that a dependency relationship is present when it is not—hence violating *falsity* (8.5a). When (8.2) is interpreted in terms of the "some interventions" interpretation, it does not violate *falsity* but, as I will interpret *omission* (8.5b), it does violate this condition in virtue of failing to represent or convey the information that there are other shades of red besides scarlet that lead to pecking. Again this failure seems to be a defect or limitation of some kind. By contrast, (8.1) respects both *falsity* and *omission*. This gives us reason to prefer (8.1) to (8.2) even if (8.2) is regarded as

developed lung cancer, we judge (S) to be true. It seems uncharitable to judge (S) to be false on the grounds that if Shlomo had not smoked 4 packs but 3.5 packs instead, he still would have got lung cancer. The interpretation that allows S to be true corresponds to the "some interventions" reading of (8.2).

true. Here we are supposing that even if true, a causal claim can be defective because of information it fails to provide, at least when there is a readily available alternative that does provide that information.

If we confine attention just to a comparison of (8.1) and (8.2), it may seem simpler to explain our preference for (8.1) in terms of (8.1) being true and (8.2) being false. In other cases, however, this option is not available. Suppose, following Franklin-Hall (2016), that the causal facts involving the pigeon are as described earlier. Consider a new binary variable C, which has two values, scarlet and cyan, with the pigeon pecking when the target is scarlet but not when it is cyan. Now consider

C causes *PECKS* (8.6)

where this is understood as implying that

An intervention that sets C = *scarlet* is followed by P = *pecking*
An intervention that sets C = *cyan* is followed by P = *not pecking*.

Putting aside misgivings about whether the variable C is it itself objectionable in some way,[17] it seems plausible that (8.6) is true—or at least we can't judge it to be false on the grounds that we judged (8.2) (under the "all interventions" interpretation) to be false. (8.6) does not falsely claim that dependency relations exist when they do not. On the other hand, there does seem to a deficiency in (8.6) along the dimension captured by *omission*. In comparison with (8.1), (8.6) fails to represent many facts about the dependency relations present in the example: (8.6) does not tell us that non-scarlet but red targets lead to pecking and that non-red non-cyan targets are followed by non-pecking. I take this (as well as other considerations described later) to suggest that we need something along the lines of *omission* to make the kinds of discriminations among causal claims that we want (and need) to make. A properly formulated proportionality condition should accomplish this.

[17] C is what Franklin-Hall calls a "non-exhaustive" variable in the sense that it fails to fully span or exhaust what we intuitively think of as the relevant space of possible values for a "good" cause variable—in the present case, this is the full range of possible colors. One might think C is an objectionable variable for that reason. It is not clear to me, however, how to formulate a requirement that cause variables be exhaustive in a way that is clear, normatively defensible, and distinct from the proportionality requirement (P) formulated subsequently. In one sense every variable seems "exhaustive" of its possible values since variables are defined in part in terms of their possible values. If every variable is exhaustive, what is really meant by the complaint of non-exhaustivity is that we should be employing a different variable.

8.4. Proportionality Formulated

Turning now to the task of providing such a formulation, let me re-emphasize the following: we may think of all of the examples discussed earlier (8.1, 8.2, 8.6) as having to do with *variable choice*—in particular, choices concerning variables representing causes. Some choices of variables seem better (more appropriate, perspicuous, informative, etc.) than others in formulating causal relationships, given the empirical facts holding for the system that we want to characterize, and it is this idea that we want to capture. For example, *RED* seems a better choice for the cause variable in the pigeon example than *SCARLET* or *C*. I take proportionality to have to do with a principle governing variable choice that might be characterized as follows:[18]

(**P** = Proportionality) Suppose we are considering several different causal claims/explanations formulated in terms of different candidate cause variables $V_1 \ldots V_n$ that are members of a set **V**. Each of these can be used to represent different claims about patterns of dependency relations involving some target effect or explanandum *E*, which is fixed or pre-specified. The variables V_i stand in non-causal hierarchal relations (e.g., realization) to one another. Thus we are choosing among pairs one of which is a candidate cause variable and the other of which is an associated dependency relation linking that cause variable to the specified effect, with each such pair at a different level in the hierarchy. For example, we are to choose between the pair {*RED* and (8.1)} versus {*SCARLET* and (8.2)}. Then a choice of variable V_i (and of the dependency claims regarding *E* in which V_i figures) satisfies proportionality better than an alternative choice from **V** to the extent that those dependency claims avoid *falsity* (8.5a) and *omission* (8.5b)—that is, to the extent that (i) nonexistent dependency relations involving *E* are not falsely represented as present (as noted earlier, this can be understood in terms of satisfaction of **M**) and to the extent that (ii) existing dependency relations (from among the variables in **V**) involving *E* that ought to be represented are represented.

[18] Two caveats and a clarification: First, as already intimated, proportionality is a ceteris paribus desideratum—a cause variable that satisfies proportionality can be defective in other ways and thus a bad choice for such a variable. Second, although (**P**) is intended to apply to certain situations in which we choose among candidate cause variables that vary in their level of abstractness, I don't claim (**P**) applies to all cases in which candidate variables differ in abstractness—see Section 8.1 for the kinds of cases to which I take (**P**) to apply. Finally, as remarked earlier, (**P**) is intended to replace the characterization of proportionality—also called (**P**)—in Woodward 2010. As I said at the beginning of this chapter, I now regard the formulation in Woodward 2010 as mistaken for several reasons, not the least of which is that it wrongly imposes the requirement that fully proportional relations must be injective.

Obviously (P) requires additional explication. Applying it requires that we first fix or pre-specify the effect variable (e.g., *PECKS*) and then choose among different candidate variables for characterizing the dependency relations governing that effect. The motivation for this requirement is that without such a prior specification of the effect variable, the problem of choosing among different cause variables becomes completely indeterminate and unconstrained.[19] In other words, what we are interested in is choosing among representations of dependency relations (and associated candidate cause variables) governing pecking and not in the representation of dependency relations governing other possible effects or explananda—for example, whether the pigeon coos or blinks. Moreover we are interested just in the conditions under which pecking vs non-pecking occurs and not in, say, fine-grained variations in pecking rate.

Second, in applying (P) we are interested just in the problem of choosing among cause variables that bear certain non-causal hierarchical relationships to one another. This includes variables that are related as *SCARLET* is to *RED*—that is, via supervenience, coarse-graining, realization, and determinate/determinable relationships, among others. (Again (P) is *not* intended to apply to cases in which we are choosing among variables that do not stand in such relations. See subsequent discussion.) When variables are related in such ways, there will be pairs of their values (pairs involving one value for each variable) that, for non-causal reasons, cannot hold together, for the same individual or unit. "Non-causal reasons" is meant to capture constraints that exclude possibilities holding for logical, conceptual, supervenience-based, or similar relations but not for causal reasons. When this is the case, I will say that this combination of values is not *compossible*. For example, the variable *SCARLET* cannot take the value = *scarlet* for a particular target while the *RED* variable takes the value = *not red* for that same target. This pair of values is not compossible, although not because the target's being scarlet *causally* excludes its being not red.[20]

Contrast such failures of compossibility with the following: suppose (following Franklin-Hall 2016) that whether the pigeon will peck or not depends

[19] If we count any failure to represent causally relevant information about any possible effect (e.g., whether the pigeon blinks or coos) rather than a prespecified one as a failure of proportionality, we will render that requirement effectively empty. Every causal claim (other than a theory of everything) will exhibit arbitrarily large failures of proportionality.

[20] This reflects the generally accepted requirement that variables can only stand in causal relations when they are "fully distinct" as well as a normative criterion for when variables are "fully distinct" that I and others have defended elsewhere (see Woodward 2015b, 2016a, forthcoming b; Hitchcock 2012). Briefly this requires that we distinguish between variables and their values. While fully distinct variables must have values all of which are compossible, different values of the same variable always exclude one another in the sense that the same individual cannot take two distinct values for the same variable. Thus if the values for a color variable include red and blue, the same target cannot take both values. This gives us a criterion for when we should employ causal representations with distinct variables rather than collapsing those variables into a single variable with distinct values. For an application of this idea to Franklin-Hall's (2016) discussion of proportionality, see Woodward 2018a.

(causally) not just on the color of the target but also on whether or not its chin is tickled, represented by a variable *TICKLES*. Each value of *TICKLES is* compossible with each value of *RED*—the same pigeon can be tickled and presented with a red target, not tickled and presented with a red target, and so on. Similarly, all possible values of *RED* are compossible with the possible values of *PECKS*. (Recall that failures of compossibility have to do with impossibilities that obtain for non-causal reasons.) As this example illustrates, when variables are compossible, they are candidates for variables that can stand in causal relationships; when variables are not compossible, they are not candidates for a cause/effect relationship: an object's being scarlet cannot cause it to be red.

The reason for this detour into compossibility is that the proportionality requirement (P) is to be understood as applying only to choices among variables that are related hierarchically in the way described previously and the values of which are not fully compossible. In other words, in the pigeon example, proportionality has to do with the choice among variables like *RED*, *SCARLET*, and *C* in describing the dependency relations bearing on *PECKS*. Proportionality does not have to do with whether or not we should include the *TICKLES* variable along with, say, *RED* in accounting for *PECKS*. *TICKLES* and *RED* do not stand in the kind of hierarchical relationship to which (P) is intended to apply—we excluded such variables in the formulation of (P). There may or may not be good reasons for including the *TICKLES* variable in an account of pecking, but they will not have to do with proportionality.[21] Thus there is a deep difference between the way in which *RED* and *SCARLET* are related and the considerations that are relevant to choosing between them as cause variables and the way in which *RED* and *TICKLES* are related—again (P) just bears on the former.[22]

Finally, let me mention an additional constraint that may seem so commonsensical as to be unnecessary[23] but that is sometimes rejected in philosophical discussion: the alternative variables (and the dependency relationships in which they stand) must be such that we are aware of them and can formulate them; we are to choose among *known* alternative variables that we know how to connect via dependency relations we can exhibit to the effect of interest. Philosophers with a taste for "what must be possible in principle" arguments may contend that there "exists" an explanation Q in terms of fundamental physics (e.g., quantum

[21] For additional discussion and motivation for restricting (P) in this way see Woodward 2018a and also Blanchard 2020. (Although Blanchard does not endorse (P), he endorses a similar restriction on the application of a proportionality requirement.)

[22] To use language that is sometimes employed in discussions of proportionality, we need to distinguish the "horizontal" problem of choosing among distinct variables from the "vertical" problem of choosing among variables that are hierarchically related. (P) only applies to the latter.

[23] In formulating a proportionality-like condition, Lien and Cheng (2002) describe the requirement that the subjects choose among known alternative variables as "obvious" and treat it as not requiring additional discussion for this reason.

field theory) of the pigeon's pecking. They may then worry that (P) recommends that we always prefer explanations in terms of Q.[24] However, unless we can explicitly formulate the Q-explanation, we are not faced with the problem of choosing between Q and an explanation in terms of a variable like *RED*. (P) is not intended to address the *Q* versus *RED* problem but rather the problem of choosing among variables like *RED* and *SCARLET* and associated dependency relations that are known or that we can exhibit. I will add that to the extent that we are interested in the role of proportionality in the empirical psychology of causal cognition, this constraint will seem trivial—people's reasoning won't reflect the influence of variables and dependency relations that they are unaware of or are unable to formulate.[25]

Once we understand the proportionality condition to be restricted in the way described, its application is straightforward: ceteris paribus, we should choose the variable from the set V that correctly represents more rather than fewer of the dependency relations concerning the effect or explanandum that are present, up to the point at which we have specified dependency relations governing all possible values of *E*. The greater the extent to which such dependency relations are represented, the better proportionality is satisfied. Thus *RED* allows for a fuller (indeed as full as possible, given the way in which the effect is specified) representation of the causal relationships present in the pigeon example than either of the other two variables *SCARLET* and *C* (again, assuming that we are confining ourselves to variables that are hierarchically related to red) and hence better satisfies proportionality.

Several other points about (P) are worth noting. First, (P) is obviously nonbinary: it can be satisfied to a greater or lesser degree by a causal claim depending on the extent to which various existing dependency relations are represented. However, if the represented dependency relations bearing on the effect fully specify conditions under which all values of the effect occur, proportionality will be fully satisfied and we can speak of the cause and effect being proportional simpliciter. I will suggest later that this is not as difficult to satisfy as some may suppose. By contrast, whether a causal claim is true or not, as assessed by, say, (M) is a binary matter. As the preceding discussion illustrates, a causal claim can

[24] Even supposing that we are able to formulate an explanation in terms of Q, it is not clear that (P) judges it superior to more upper-level explanations for reasons discussed in Section 8.6.

[25] Both Franklin-Hall (2016) and Weslake (2010) use examples involving the in-principle possibility (where this notion is not further explained or characterized) of formulating causal claims or explanations in terms of "fundamental physics" (as with Q earlier) to object to previous formulations of proportionality by me and others. They seem to think that it should not matter that we cannot actually state or write down such explanations, presumably because from the point of view of the metaphysician, all that matters is in-principle possibility. But if we are interested in methodology or empirical psychology, the distinction between causal and explanatory claims that we have some realistic possibility of formulating and those that we do not clearly does matter. Here too my remarks in the introduction are relevant.

be true, in the sense of satisfying (M) (or (8.5a)) and yet not do particularly well (in comparison with alternatives) with respect to proportionality. The extent to which a causal claim satisfies proportionality (and in particular (8.5b)) is thus a distinct dimension of causal assessment that goes beyond assessment of whether the claim is true. When thus understood, proportionality should not be regarded as a candidate for a necessary condition for the truth of causal claims, contrary to the way it is often treated in the literature.

Second, note that (P) has to do with "objective" features of causal claims that track how matters stand in the world. That the claim (8.1) formulated in terms of the *RED* variable does a better job of satisfying proportionality than the alternative claims (8.2) and (8.6) formulated in terms of *SCARLET* and *C* is a reflection of the objective fact that the pigeon pecks at red and only red targets and that (8.1) fully captures these facts about the dependency relations present in the example and (8.2) and (8.6) do not. Finally, note also that on this understanding of proportionality there is no mystery about why proportionality is, ceteris paribus, a virtue or desirable feature of causal claims—its normative justification is straightforward. To the extent that a causal claim satisfies proportionality understood as (P), it will provide more information about dependency relations governing the effect than alternatives that satisfy proportionality to a lesser degree: it will do a better job at providing information that is associated with such distinctive aims of causal thinking as manipulation and control.[26] Although proportionality is thus a pragmatic virtue in the sense that it has a means/ends rationale or justification, it is not "pragmatic" in the sense that it depends on the idiosyncrasies of particular people's interests or other similarly "subjective" factors. Moreover, although, as noted earlier, a causal claim can be true without satisfying (P) to any very great extent, the considerations that go into the assessment of proportionality are truth-based in the sense that they have to do with the extent to which a causal claim captures certain truths about dependency relationships.

[26] At the risk of repeating the discussion in previous chapters it may be useful to contrast this rationale or justification for adoption of a proportionality requirement with a more intuition and conceptual analysis-based justification, which is nicely described although not endorsed in Dowe 2010. In Dowe's words,

> [This] is the view that philosophy is conceptual analysis, the aim being to give a theory which best captures the concept of causation that we all share, and so evidence for and against a theory should be the intuitions that any competent user of English will have concerning the truth of sentences of the form "this causes that." Folk intuitions are that sentences like "being presented with a red object caused Sophie to peck" and "my deciding to lift my arm is the cause of my doing so" are true; i.e. ultimately such intuitions are what justify the proportionality principle. (447)

The rationale I have described does not claim that a proportionality requirement is built into the concept of causation, does not claim that intuitions concerning proportionality reflect our mastery of the concept of cause, and does not claim that such intuitions are what justify proportionality. Rather, as I have explained, the justification for proportionality has to do with the desirable information that causal claims better satisfying proportionality provide.

Proportionality is thus not based on considerations that have nothing to do with truth or with what nature is like.

Now consider another case—this one a purported counterexample due to Shapiro and Sober 2011. Their immediate target is the following characterization of proportionality (P*) (which is obviously different from (P)):

> (P*) A statement of the form "C caused E" obeys the constraint of proportionality precisely when C says no more than what is necessary to bring about E. (89)

They assume that (P*) is intended as a candidate for a necessary condition for a causal claim to be true. In Shapiro and Sober's example, real-valued variables *X* and *Y* are related by some non-monotonic function F that maps two different values of *X*—for example, 3 and 22—into the same value of *Y* ($y = 6$), with other values of *X* being mapped into different values of *Y*. (i.e., the function is not injective or one to one.) There is an obvious sense in which $X = 3$ is not "necessary" (or is not "required") for $Y = 6$ since $X = 22$ also yields $Y = 6$. So proportionality, understood as (P*), is violated in this case. If we take (P*) to be necessary for a causal claim to be true, it follows that F does not truly describe a causal relationship, even if it correctly describes how *Y* responds to interventions on *X* and qualifies as true according to the interventionist criterion (M). For similar reasons, (P*) implies that it is false that $X = 3$ causes $Y = 6$.

I assume that many readers will regard these assessments (of the earlier causal claims as false) as "unintuitive"—this certainly seems to be Shapiro and Sober's assessment and one that they expect others to share. More specifically, the intuitive judgments of many of us are that the preceding causal claims are not false (and arguably are not defective in other ways) merely because they involve non-injective relationships. Of course the methodology advocated in this book does not license rejecting (P*) merely because it leads to unintuitive conclusions. On the other hand, given that (P*) has such consequences, one may well wonder whether there is any defensible normative basis for (P*), understood either as a necessary condition for the truth of causal claims or, more weakly, as capturing a desideratum of some kind—that being that for some reason it is desirable to avoid causal relationships that are not one to one or injective.

What does (P) (my preferred explication of proportionality) imply about these examples? That depends on what is included in the cause or explanans side of the examples. Consider the following possibility: in addition to the actual values of *X* and *Y*, we are given a full specification of the function F, relating *X* and *Y* so that we are told for each possible value of *X* what the corresponding value of *Y* will be. This information—call it *A*—seems to fully satisfy (P). Ex hypothesi, the information *A* does not claim the existence of any dependency relations that do not exist, and (in the relevant sense) it fully describes those that do exist. This accords with our judgment that the injective character of F does not by itself show that there is anything defective about F (or *A*) as a characterization of the causal relationship in the example. This also shows that Shapiro and Sober's (P*) is not equivalent to (P), even putting aside the consideration that the former is put forward as a necessary condition for causation and the latter is not. (I say more about this shortly.) It also follows that certain informal characterizations according to which a causal claim satisfies proportionality to the extent that the cause contains only what is "necessary" or "required" for the effect or according to which causes must be just "enough" for their effects (and no more) are not equivalent to (P). We can agree that these characterizations in terms of what is required or enough are inadequate while endorsing (P).

Suppose, on the other hand, that rather than being given the information in *A*, the causal claim with which we are presented is just

$$X = 3 \text{ causes } Y = 6 \tag{8.7}$$

and nothing more. That is, we are not given the rest of the information contained in F, the function relating *X* and *Y*, beyond (8.7). Then there is an obvious sense in which relevant dependency information has been omitted. Although we haven't been told anything that is false, we haven't been told what the value of *Y* will be for conditions other than $X = 3$. So in this respect (P) is far from being fully satisfied. However, it is not clear that there is anything wrong with the assessment that, ceteris paribus, it would be preferable to replace (8.7) with a claim that provides more information about the dependency relations governing *Y*, such as that provided by *A*. That (P) supports this assessment seems a consideration in support of, rather than against, (P). Note also that this assessment does *not* depend on the idea that there is something unsatisfactory about causal relationships that are characterized by non-injective functions.

In Yablo's original discussion of proportionality as well as a number of subsequent discussions in the philosophical literature, the causal relationships to which this notion was applied were assumed to be binary. Under this assumption, if the relationship between cause and effect is described by a nontrivial function,[27] the kind of case considered by Shapiro and Sober, which involves a non-injective function, cannot arise. However, many variables employed in

27 Nontrivial means that the function is not a constant function; that it maps some different values of *X* into different values of *Y*. That *Y* will be different for at least two different values of *X* is required by the interventionist criterion (M). However, this does not imply that just any nontrivial functional relationship in the binary case will satisfy (P) well. The relationship (8.6) in Franklin-Hall's *CYAN* example is such a nontrivial function, but it does not satisfy (P) very well.

ordinary causal cognition and in science are nonbinary, and it is certainly desirable to extend the notion of proportionality to cover such variables. In my view it is an important attraction of (P) that it does this.

(P) differs from Yablo's understanding of proportionality in several other respects as well. As explained earlier, (P) is a graded notion and is not proposed as a necessary condition for a causal claim to be true. In addition, to anticipate the more extended discussion to follow, Yablo's understanding does seem to be that non-one-to-one relationships violate proportionality, while, as we have noted, (P) does not have this implication.

It is also worth noting that (P) has some straightforward implications for the kinds of functional relationships that will fully satisfy proportionality—one can thus think of (P) as associated with desiderata in the choice of functions to represent causal relationships, just as certain invariance conditions are. First, in the binary case, if the mapping between cause and effect is not a function, then some functional (i.e., deterministic) relationship (assuming there is one that truly describes the dependencies present in the example) will better satisfy (P). For example, the relationship between *SCARLET* and *PECKS* is not a function, since the non-scarlet value of *SCARLET* is mapped into different values of *PECKS* (depending on whether that non-scarlet value is or is not a shade of red). By contrast, the relationship between *RED* and *PECKS* is (described by) a function. When the variables involved are nonbinary, (P) will again be best satisfied by some functional relationship (again if one exists that truly describes the dependency relations). As we have seen, such a function need not be one to one. However, to best satisfy (P), the function should be onto or surjective: every value of the effect variable should be the image of some value of the cause variable. For example, given an effect variable Y with 3 possible values, y_1, y_2, y_3, a cause variable X with two values x_1 and x_2 and an associated dependency relationship F for which $F(x_1) = y_1$ and $F(x_2)= y_2$, (P) would be better satisfied by an alternative variable X^* (and an associated functional relationship F^*) that correctly specifies the values of X^* that are mapped into y_3, assuming there are such. The variable/functional relationship F^*/X^* tells us more about what the values of Y depend on than does F/X. Note also that when the relationship between cause and effect is described by a function, full satisfaction of (P) requires that there be at least as many possible values or states of the cause variable as there are states or values of the effect variable.

Suppose next that the candidate cause-and-effect variables are both binary but that some or all of the candidate cause/effect relationships are probabilistic (with probabilities strictly between 0 and 1) rather than deterministic. There are delicate issues about how to understand probabilities (and when ascriptions of probabilities are "correct") in situations of this sort. However, in what follows I propose to put these aside and assume that we are choosing among candidate causes related hierarchically that bear different probabilistic relationships to the effect, where each of these relationships is correct as far as they go, but may differ in informativeness. As an example, suppose that the true probability of pecking given that the target is scarlet is .3 and the true probability of pecking given the target is red is .6.[28] What understanding of proportionality best fits such contexts? If we think of (P) as motivated by the idea that we should prefer cause variables that allow for the formulation of relationships that tell us more rather than less about the conditions under which various values of the effect occur, then a natural thought is that we should prefer the variable *RED*. Similarly, if we have a choice between a variable *C* that permits the formulation of a deterministic relationship and an alternative variable *C** that permits only the formulation of a probabilistic relationship, we should prefer the former—this provides more information in the relevant sense. Going further, we might consider generalizing this to a preference for candidate cause variables that assign more extreme rather than less extreme probability values to the effect. That is, we might prefer causes that assign probability values that are as close as attainable to 1 and 0, where "attainable" means that these probability values are empirically correct. If we follow this proposal, then, in cases in which we are choosing among hypotheses each postulating single causes[29] (each satisfying the interventionist condition (M)) at different levels of the hierarchy, proportionality will be better satisfied to the extent that the choice of the cause variable *C* is such as to maximize $\Delta p = Pr(E/C) - Pr(E/-C)$. In other words, in cases of this sort, among the candidate cause variables we should choose the one for which Δp is maximal, subject to the constraint that the assigned probability values are correct. As we shall see, this is also the proportionality-like normative requirement advocated by Lien and Cheng (2002) for such cases. It also is the rule that best describes the behavior of the subjects in their experiments.[30]

[28] This can happen if non-scarlet shades of red are stronger causes of pecking than scarlet and if such non-scarlet shades occur more frequently. On the other hand, it might also be argued that because under these circumstances the variable *RED* is heterogeneous (it confers different probabilities of pecking depending on the shade of red, it should be split into several variables and that there is no well-defined probability *simpliciter* of pecking given red. This would mean that the possibility envisioned in the main text cannot arise.

[29] The restriction to cases in which at each level there is a single candidate cause for *E* is important: unlike the situation for which Cheng's causal power measure is designed, we are assuming that at each level there are no "other causes" of *E* that operate independently of *C*. Again, we are comparing candidate causes *RED* and *SCARLET* that are at different levels of abstraction In such cases, causal power reduces to Δp.

[30] We can illustrate this idea by returning to the pigeon example. The contrast Δp between the probability of pecking given that the target is red and the probability of pecking given that the target is not red is maximal—it is equal to 1 – 0 = 1. By contrast, although *Pr*(*Pecks/target is scarlet*) = *1*, *Pr*(*Pecks/target is not scarlet*) is, assuming it is well defined, greater than zero but less than one. Thus Δp when the cause is characterized as "red" is greater than Δp when the cause is characterized as "scarlet," and on this basis "red" is preferable.

Proportionality is sometimes presented as the requirement that cause and effect be "commensurate" or that both have a "grain" that allows them to "fit" appropriately with one another. Although this way of putting matters can be misleading,[31] we should also be able to see from (P) what is right about this idea. Full satisfaction of (P) requires (among other things) that the "grain" of the cause (understood as the number of possible values of the cause variable) should line up or fit with grain of the effect variable in the sense that the cause variable should have enough values to provide information about the conditions under which each of the possible values of the effect variable will occur, with the ideal case being one in which the value of the cause will determine the value of the effect for all values of the latter. Moreover, even when a candidate cause variable has enough states or values, it can still carve things up in a deficient way, as illustrated by (8.2) and (8.6), which we can again think of as involving a failure of appropriate graining.

8.5. (P) versus (P+)

So far we have focused mainly on cases in which some candidate cause variable fails to fully satisfy (P) and another choice of variable would constitute an improvement with respect to (P). There is, however, another possible set of cases (with Shapiro's and Sober's example providing one illustration) to be considered which, so to speak, are the other side of the coin. Suppose that we have a cause variable *Z* that fully satisfies (P) with respect to effect *E*. What should we make of alternative causal claims that also fully satisfy (P) but employ a cause variable *X* that is more fine grained than *Z*—that is, that makes distinctions in values or states that are unnecessary from the point of view of accounting for *E*? In other words, *X* goes "beyond" (P) in the sense of making distinctions that are not required by (P).

Consider the following idealized illustration, which resembles a number of examples that have been discussed in the philosophical literature.[32] The effect is a variable *R* that takes one of three possible values (*monkey reaches right* = r_1, *monkey reaches left* = r_2, *monkey does nothing* = r_3). One candidate cause variable for *R* is an "intention" variable *I* which also takes one of three possible values i_1, i_2, i_3 corresponding to whether the monkey intends to reach right, left or do

[31] For example, the variable *C* in (8.6) has (in one relevant respect) the same grain as the explanandum *PECKS* (since both have two possible states or values), but (8.6) is defective from the point of view of (P).

[32] See, e.g., Woodward 2008.

nothing. The monkey always reaches in the direction he intends to reach, so that the relation $R = F(I)$ between I and R is deterministic. Now suppose that each of these values for the intention variable I is multiply realized by the values of a neurological variable N with values n_{ij}: whenever either n_{11}, n_{12} or n_{13} occurs, value i_1 of I is realized and correspondingly for the values i_2 (realized by n_{21}, n_{22}, or n_{23}) and I_3 (n_{31}, n_{32}, n_{33}). These are the only possible values of N and N is the only neurological variable whose values are realizers for I. The relation $R = G(N)$ between N and R is thus also deterministic. As before, assume that the realization relationship is not a causal relationship: the values of N don't cause the values of I that they realize. Thus the choice between N and I is one of those choices among hierarchically related variables to which (P) applies but we will not need to be any more precise than this about what realization involves.

The variable I and the associated relationship $R = F(I)$ satisfy (P) but so do N and $R = G(N)$, given the way we have interpreted (P). The difference between F and G is simply that the former is injective and the latter is not and our assessment previously was that the non-injective character of G was no barrier to its satisfying (P). So if we are comparing F and G, (P) does not prefer one to the other: G and N are not penalized for making unnecessary distinctions. As far as (P) goes, G and N are just as good as F and I.

Suppose, however, that we change the comparison: in some actual case, we are given just the information that

$$N = n_{11} \text{ causes } R = r_1 \tag{8.8}$$

where this information is correct and is all the information that is conveyed. This claim does leave out a lot of dependency information relevant to conditions under which R takes its possible values and in this respect falls short of fully satisfying (P). (In this respect it is parallel to the claim that $X = 3$ causes $Y = 6$ in the Shapiro/Sober example.) So (P) will judge alternatives to (8.8), such as the psychological causal claim that presents all of the information in F as well as the value taken by I as superior.[33]

[33] Consider yet another comparison: the claim that (8.9) I causes $R = r_1$ (where this claim is true and it is all the information we are given). This also leaves out dependency information about some of the conditions under which R takes its possible values (it does not tell us the conditions under which $R = r_2$ or $R = r_3$) and in this respect falls short of fully satisfying (P). However, it might be argued that (8.8) does worse than (8.9) with respect to (P): (8.8) tells us only what will happen with respect to R for one of the nine possible values of N, while (8.9) is comparatively more informative in telling us what will happen under one of the three possible values of I. Furthermore, in the one case in which

One of the primary motivations for the original introduction of the notion of proportionality by Yablo was to defend the *superiority* of upper-level explanations (in particular explanations that appeal to mental causes like intentions) over lower-level explanations (in particular explanations that appeal to physical or neurological variables). Although it is arguable, as noted earlier, that (P) supports a preference for the psychological causal claim that presents all of the information in *F* and *I* over the neurobiological claim (8.8), and similarly a preference for (8.9) over (8.8)—see footnote 34—we also noted that (P) does not tell us to prefer an explanation of *R* that appeals to the psychological variable *I* and the accompanying generalization *F* to the alternative explanation that appeals instead to the neurobiological variable *N* and the generalization *G*. Some (including perhaps Yablo) will think that in this second case as well the psychological explanation in terms of *I* and *F is* preferable to the explanation in terms of *G* and *N* and hence that it is a limitation of (P) that it does not yield this result. Let us explore this.

I noted earlier that one way of thinking about the relationship between the causal claim involving *F* and the claim involving *G* is that the former makes distinctions among values of variables that are unnecessary in accounting for *R*, while the latter does not do this. (P) does not treat such unnecessary distinctions as a defect, but it might seem plausible that they should be so regarded—this would yield a preference for explanations appealing to *F*.[34]

In adopting this latter view, we are in effect going beyond (P) to adopt a strengthened conception of proportionality that I will call (P+): not only do we want variables and generalizations that fully conform to (P) but we also don't want variables and generalizations that make distinctions or that appeal to information that goes beyond what is required by (P). Fully satisfying (P+) thus requires use of a cause variable that has exactly the same number of values or states (exactly the same grain) as the effect variable so that in cases in which the cause-effect relationship is described by a function, that function must be one

N is informative about *R*, the corresponding value of *I* conveys the same information about *R*, so that in this sense (8.8) does not tell us anything about *R* that we are not told by (8.9). Thus in this case it is arguable that (P) tells us to prefer (8.9) to (8.8).

[34] It is also sometimes argued that the causal claim formulated in terms of *G* and *N* is defective in failing to show what the various states of *N* that lead to the same state of *R* have in common, while the causal claim formulated in terms of *F* and *I* does not have this defect. As formulated, this claim is mistaken. Although it is true that if the claim involving *G* just consists of the information described here, it does not tell us what the various values of *N* leading to the same state of *I* have in common, the claim involving *F* and connecting *I* to *R* suffers from the same deficiency, assuming that we mean something non-trivial by "in common." *F asserts* that i_1 leads to r_1 and so on, but it does not explain *why* this happens. Nor does *F* tell us what the various neural realizers of i_1 etc. have in common. A nontrivial way of showing what all of the values of *N* that lead to the same value of *R* have in common would be to identify some more abstract common neurobiological characterization of those neural states.

to one. When we find ourselves operating with a cause variable *V* for which two or more values v_1, v_n are mapped into the same value of the effect, then as far as proportionality goes (there may of course be other considerations) it would be an improvement (according to (P+)) to replace *V* with another variable *V** that collapses all such values of *V* into a single value *v** of *V**.[35] (In effect, we can think of the values of *V** as obtained via a partition of the values of *V* into equivalence classes, where values of *V* belong to the same equivalence class if they lead to the same value of *R*.)

To further illustrate the difference between (P) and (P+), return to the contrast among the *RED* (8.1), *SCARLET* (8.2), and *CYAN* (8.6) claims about pecking. To explain the superiority of (8.1) over these alternatives, we require only (P). By contrast, to motivate the claim that an upper-level psychological explanation that appeals to *F* is superior to the more fine-grained neurobiological explanation that appeals to *G*, we require the stronger (P+). Put differently, the considerations that support the superiority of the (8.1) *RED* causes *PECKS* claim are distinct from the considerations that might be claimed to support the superiority of *F* over *G* in accounting for *R*. One can agree about the superiority of (8.1) without agreeing that the psychological explanation appealing to *F* is superior to the neurobiological explanation appealing to *G*—the superiority of (8.1) doesn't support the latter claim.

One way of describing the difference between (P) and (P+) is that the former is more permissive: (P) allows us to use more fine-grained variables and accompanying generalizations when this involves no loss of dependency information but it also does not require this unless there is a gain in dependency information. By contrast (P+) says that the use of the less fine-grained variables is preferable as long as there is no loss in dependency information. Adopting (P) rather than (P+) allows us to avoid taking on the burden of arguing that the upper-level explanation is preferable even when there is no loss of dependency information.

8.6. Proportionality and Conditional Independence

There is no doubt more that might be said about the comparative merits of (P) and (P+).[36] But rather than spending additional space on this,[37] I want to turn

[35] Paul Griffiths has drawn my attention to Pocheville et al. 2017 which yields an account of proportionality that is, given some modeling assumptions, equivalent to P+ for categorical variables. This paper also makes the important point that proportionality for effects raises very different issues than proportionality for causes, as I acknowledge in appendix 2 to this chapter.

[36] For more on conditional independence and related themes, see Woodward, forthcoming, a, b, and c.

[37] Nonetheless I can't resist some additional brief remarks. It might be claimed that (P+) is superior to (P) on the grounds that (P) allows (but (P+) does not) causal claims that involve redundancies or that (P) allows claims that are less simple and compact than the claims permitted by (P).

instead to a related consideration that bears on the motivation for adopting some version of proportionality and which also suggests that, in practical terms, the contrast between (P) and (P+) may, in many cases, be less consequential than one might suppose. As an empirical matter, there are many cases in science in which we are unable (for various reasons having to do with computational and epistemic constraints) to formulate explanations or true causal claims involving upper-level effects that appeal to fine-grained, lower-level variables like the neurobiological variable *N*. In such cases, (P) (and for that matter, (P+)) can provide a kind of reassurance: if we can find an upper-level variable (or variables) fully satisfying this condition, we can appeal to these instead (rather than the more fine-grained variables), knowing that as far as proportionality considerations go, we lose nothing by doing so. Suppose (as seems likely) that in many cases we have no idea what fine-grained neural variables and accompanying generalizations are involved in the causation of behavior, but we are in possession of a psychological theory that fully satisfies (P) (or does as well with respect to (P) as any neurobiological theory that we are able to construct would). Then there is no puzzle about why we use (and think that we are justified in using) the psychological theory. I take this to suggest that, insofar as we are interested in understanding why it is reasonable to use the psychological theory, constraints like (P+) are unnecessarily strong; we don't need them to explain (and justify) aspects of our use

For example, when *G* and *N* versus *F* and *I* are compared as causal accounts of *R*, *F* and *I* may seem simpler on the grounds that they require only that we keep track of three different stares and how they are related to *R*, as opposed to nine different states required if we employ *G* and *N*. The account in terms of *I* thus gives us a more compact summary of what is relevant to *R*. Note, however, that this looks like superiority along the dimension of what is sometimes called descriptive simplicity (different descriptions, differing in complexity, of the same alternative) as opposed to the kind of simplicity that is supposedly relevant to choosing among different empirical alternatives. That is, as far as *R* goes, *G* and *N* and *F* and *I* seem to convey effectively the same empirical information. If this is right the explanation that appeals to *F* and *I* is "simpler" than the explanation that appeals to *G* and *N* in something like the way in which the use of polar rather than Cartesian coordinates might be simpler for a certain problem, even though either choice of coordinates conveys the same information. If so, although causal claims conforming to (P+) may be easier to use, we don't need to think of them as superior to those conforming to (P) along the dimension of conveying dependency information relevant to the explanandum. Of course it is possible that there is some additional dimension that is relevant to causal assessment that supports the use of (P+) and does not reduce to ease of use, but so far no candidates have been produced.

A second relevant observation is due to Hoffman-Kloss (2014). She criticizes previous formulations of proportionality including mine by noting their awkward implications for common cause structures. To use her example, suppose one has a heat source that is used to melt chocolate in a water bath which is also attached to a thermometer, so that the temperature of the heat source is a common cause of the melting and the readings. The chocolate will melt when the temperature is greater than 35 degrees C, so if the dependent variable measuring melting is two-valued *E* = {melts, not melts}, the cause variable that best satisfies (P+) and does not make unnecessary distinctions will similarly have just two values corresponding to temperatures above and below 35 degrees. However, the thermometer readings *R* register the water temperature in increments of a single degree, so satisfying (P+) with respect to *R* requires a much more fine-grained cause variable. Such a fine-grained variable will of course violate (P+) with respect to *M*. Similar problems will arise in many other cases in which variables have multiple effects. (P) avoids this difficulty.

of upper-level causal claims and aspects of our practices of abstraction, since (P) plus the nonavailability of the lower-level theory provides all the justification that is needed. On the other hand, this consideration also suggests that in many realistic cases, we may not be presented with a sharp conflict between (P) and (P+): the causal claims that are able to produce or construct may, as in the case of the psychological theory that appeals to *F*, satisfy or come close to satisfying (P+) as well as (P), simply because the many-one claims that would violate (P+) are not ones that we are able to formulate. Of course this line of argument requires that it be true, as an empirical matter, that in some substantial range of cases upper-level generalizations like the psychological explanation appealing to *F* do as well in satisfying (P) as lower-level generalizations (of a kind that we are able to produce).

To further motivate this claim, let me put it in a more general context and introduce some additional apparatus. I claim that it is a striking empirical fact that the difference-making features cited in many lower-level theories sometimes can (see Woodward 2020b, forthcoming a) be absorbed into variables that figure in upper-level theories without a significant loss of difference-making information with respect to many of the effects explained by those upper-level theories. (Of course this is not always possible, but it is more common than many suppose.) In other words, in many cases with respect to a range of possible explananda, the upper-level theory will do just as well as the lower-level theory in satisfying (P). The relationship between the psychological and neurological theories of reaching behavior envisioned earlier illustrates this possibility. Given the values of the psychological variable *I*, variations of the values of the lower-level variable *N* make no further difference to the values of *R* (as reflected in the generalization *F*), so that from the point of view of providing difference-making information about *R*, *F* and *I* do just as well as *N* and *G*. As a more realistic illustration, it is almost but not quite true that, given the values of various thermodynamic variables like temperature that are difference-makers for various aspects of the macroscopic behavior of a gas, further variations in the microscopic state of the gas as described by, for example, the positions and momenta of the individual molecules making it up are irrelevant to its macroscopic behavior. This is why if we wish just to explain features of the macroscopic behavior of the gas, it is often sufficient to just appeal to the value of the thermodynamic variables and the generalizations in which these figure.

To further spell this out in a more general way, let us say, following (M), that a set of variables X_i is *unconditionally causally relevant* (alternatively, irrelevant or independent) to *E* if there are some (no) changes in the values of each X_i when produced by interventions that are associated with changes in *Y*.[38] A set of variables Y_k is irrelevant to variable *E conditional* on additional variables X_i if

[38] Ideas that are similar to the account developed here are developed in a machine-learning context in considerably more formal detail in Chalupka, Perona, and Eberhardt 2015. This work has

the X_i are unconditionally relevant to E, the Y_k are unconditionally relevant to E, and, conditional on the values of X_i, changes in the value of Y_k produced by interventions and consistent with these values for X_i are (unconditionally) irrelevant to E.[39] In other words, changes in the X_i are causally relevant to E in the sense captured by (M) and, conditional on the values taken by X_i, further variations in the Y_k make no difference to E. We can think of this as a kind of generalization of the "screening-off" idea used by Yablo to characterize proportionality, as described in Section 8.2: the X_i screen off the Y_k from E in the sense just described.[40] The idea that we are considering is that in such a case, we may appeal just to the X_i to explain the values of E—the Y_k (and the generalizations in which they figure) provide no further causal or explanatory information relevant to E. One might of course go on to say that the explanation in terms of the X_i is preferable to the explanation in terms of the Y_k (or even that the claims in terms of the Y_k are false—this would be to adopt (P+) as opposed to (P)). However, as we have noted, it is not necessary to claim this to vindicate the use of the X_i instead of the Y_k—all that is required (in accordance with (P)) is that the explanation in terms of the X_i be as good as the explanation in terms of the Y_k. In other words, we can think of (P) as reflecting or fitting with the ideas about conditional irrelevance just described. Put in terms of proportionality requirements, the idea is that we can often find upper-level variables, related to lower-level variables, typically by some kind of coarse-graining operation, that satisfy the requirements of proportionality just as well the lower-level variables and thus (as far as proportionality is concerned) are equally satisfactory form the point of view of explanation

influenced my interpretation of proportionality and conditional independence. Similar ideas can also be found in Ellis 2016 in the context of a discussion of downward causation.

[39] Some additional observations: first, as originally employed by writers like Reichenbach (1956) and Salmon (1971), "screening off" is a statistical notion that is characterized in terms of conditional probabilistic (in)dependence. In the present context, I follow Yablo in understanding this notion (and the accompanying notions of relevance/irrelevance) in terms of counterfactuals describing what happens under interventions rather than in terms of statistical dependence—for example, if Y is conditionally irrelevant to E, given X, then, if (i) one intervenes to fix X at some value, (ii) further variations in Y due to interventions consistent with (i) will not change E. In other words, we are considering nested counterfactuals the antecedents of which involve reference to two interventions: X is fixed at some value via an intervention and then Y is set via a separate intervention at some value that is consistent with the value to which X has been set. Note that while there is no inconsistency in setting X to some value and then setting Y to a value that is consistent with that value of X, we can't do it the other way around: once we have set Y to a particular value, this determines a value for X; we cannot coherently set X to some different value. Second, note that conditional irrelevance is much stronger than what is described in the philosophical literature as "multiple realizability." Taken literally, the latter requires only that *some* different values of the same or different micro-variable(s) realize the same value of a macro-variable. Conditional irrelevance requires that *all* variations at the micro level consistent with the value of the macro-variables that such variations realize make no difference to E.

[40] Note that, as with (P*), relevance and irrelevance are always defined relative to an effect or explanandum E. Y may be irrelevant to E conditional on X but this may not be true for some alternative explanandum E^*.

and causal analysis. Although if we accept (P), there is nothing wrong with the use of variables that make finer-grained distinctions, there may also be no motivation for employing such variables—no further gain from the point of view of causal analysis or explanation.

8.7. Some Empirical Questions

As so far described, (P), (P+), and the accompanying ideas about conditional irrelevance are normative proposals. However, one can also ask about the extent to which, as an empirical matter, various groups of subjects conform to these principles, and this in turn suggests a number of experimental questions.

8.7.1. Causal/Explanatory Strength Judgments

Suppose subjects are presented with information about two sets of variables *X* and *Y* that are related to one another via a non-causal coarsening or hierarchical relation (*X* is a coarsening of *Y*, so that in this sense it is at a higher "level" than *Y*). *X* and *Y* are both unconditionally relevant to *E* and conditional on *X*, *Y* is irrelevant to *E*. Subjects are then presented with candidate causal/explanatory claims relating *X* to *E* and relating *Y* to *E*. Is it the case, as (P) and the conditional irrelevance idea would suggest, that they would regard causal and explanatory claims about *E* that are framed in terms of the coarser-grained variable *X* as just as satisfactory as those framed in terms of *Y*? That is, would the subjects give strength ratings (for some appropriate verbal probe) to causal and explanatory claims appealing to *X* that are as high as the ratings given to causal and explanatory claims appealing to *Y*? Alternatively, would they regard causal and explanatory claims appealing to *X* as *better* than those appealing to *Y*, on the grounds that *Y* makes unnecessary distinctions from the point of view of accounting for *E*—that is, do their causal and explanatory judgments follow (P+) rather than (P)? Or would they judge that causal claims appealing to the more fine-grained variable *Y* are superior on the grounds that more detail is always better (causes need to be maximally specific)?

8.7.2. Learning Variables and Relationships at Particular Levels of Abstraction

In Section 8.7.1 we supposed that subjects were presented with alternative causal claims formulated at different "levels" and asked to rate them. That is, they are not asked to discover or formulate these claims themselves but just to rate them once they are provided. Another possibility is not to provide subjects with explicit formulations of these alternative hypotheses but rather to explore whether subjects can discover them, given appropriate dependency information. If subjects can do this, do they then exhibit preferences for hypotheses at certain levels, and, if so, what principles guide their choices? Suppose, for example, subjects are given information about patterns of dependence that allow for the discovery of relationships between *X* and *E* and between *Y* and *E* where again *X* is related to *Y* via some coarse-graining or similar operation. (Subjects are not told, however, that *X*, *Y*, or alternative variables are appropriate ones to use.) When asked questions like "What causes (or explains *E*)?" do subjects prefer to cite one of these variables rather than the other? Can they discover these variables (and causal hypotheses formulated in terms of them) from information about dependency or contingency information even if they are not explicitly told about these variables and relationships? To what extent can they discover and to what extent do they prefer to cite variables and accompanying causal claims that satisfy (P) (or (P+)) better than alternatives? When subjects learn about dependency relations in one set of circumstances, which variables and levels of abstraction do they prefer when asked to generalize to new situations?[41]

Of course it is natural to expect that many of these questions will receive similar results—that is, that subjects will assign higher strength ratings to those causal claims that they are most willing to spontaneously endorse, to generalize and so on. This is what is found in an empirical study by Lien and Cheng discussed in the following section. However, it is of course an empirical question whether this pattern holds generally.

[41] An illustration: Suppose subjects are given data about the pecking behavior of a number of pigeons that show that the pigeons peck in response to targets with various particular shades of red and not in response to colors that are non-red. However, they are not told the pigeons peck in response to red and only red targets. Do they exhibit a preference for causal claims formulated in terms of the cause variable *RED*? (One but not the only form such a preference might take is higher strength assignments to *RED* causes *PECKING* than to *SCARLET* causes *PECKING*.) Do subjects spontaneously discover that the *RED* variable is the most appropriate one? (Presumably the variable *RED* is "available" to them in the sense that they have this word in their vocabulary. We might also ask about cases in which the variable that will best satisfy (P) is a new, unfamiliar variable that they need to discover.) Relatedly we might also ask about which variables and level of abstraction guides their willingness to generalize about new situations. For example, suppose that in the setup just described, subjects are not explicitly encouraged to formulate a generalization containing the variable *RED* but are instead told that the pigeon has been presented with a new target that is a shade of red that has not been previously seen—for example, maroon. Will subjects judge that the presentation of this target will cause the pigeon to peck and will they describe the redness of the target as the cause of the pecking? (This would be another form that a preference for *RED* might take.)

8.8. Lien and Cheng on Levels of Abstractness

I turn next to some experimental results due to Lien and Cheng (2002) bearing on the extent to subjects are guided by proportionality-like considerations in causal judgment and inference. As noted earlier, Lien and Cheng do not use the word "proportionality" in describing their results, but as far as I am aware, theirs is the empirical study that bears most directly on this notion.

In broad outline Lien and Cheng presented subjects with hypothetical soil ingredients that (they were told) were candidates for causes of plant blooming as well as contingency information linking whether plants were fed these ingredients and the rate at which they bloomed. These ingredients were described as varying in ways that fell into hierarchical structures or classes of increasing abstractness, so that these classes were related in the way described by (P). The experimental questions then had to do with whether the subjects preferentially learned and endorsed claims about the causes of blooming formulated at one of these levels and what principles appeared to be guiding this choice of levels.

In more detail, the experiment (which had a rather complicated design) proceeded as follows: in the *learning* phase, subjects were given information about different hypothetical substances *s* that were fed to groups of a particular kind of plant as well as the frequency of blooming in each group. They were told that plants were kept in the same constant environmental conditions, so that any variation in their blooming would be attributable to the substances they were fed. Subjects were also given information showing that most plants that were not fed with any substance did not bloom. Each substance was described at varying levels of abstractness. For example, the substances varied as to color, and these variations were represented at each of three levels of abstractness—particular shades of color (e.g., pine-green), general type of color (green), and (at a very general level) whether the color was "warm or cool." Similarly, the substances varied in shape, and subjects were given information that represented these shapes both in highly specific ways (e.g., as a triangle of certain dimensions), at an intermediate level of specificity (e.g., a type of shape such as triangular), or in a still more abstract way (regular in the sense of rotationally symmetrical versus irregular).

In this phase of the experiment, although each subject was shown information about the covariation between each of the substances and blooming, the frequency of blooming was manipulated across two groups of subjects. In each such group, the patterns of covariation were chosen in such a way that the contrast $\Delta p = Pr(E/C) - Pr(E/-C)$ was maximized when the candidate cause C was at the most abstract category in the hierarchy in which it fell. However the causally relevant variable varied across the two groups: for one group shape was the relevant variable, with the irregular shapes in contrast to regular shapes producing the maximum contrast. In the other group color was the relevant variable, with Δp being maximized when the color of s was warm rather than cool. (Recall that choosing the cause variable for which Δp is maximal is a natural interpretation of (P) for binary probabilistic causes, assuming that there is just one cause of the effect.) In contrast, for each group, blooming versus non-blooming covaried less than maximally with more specific descriptions of the soil ingredients. That is, Δp was smaller when the substances were described as blue rather than red or as a triangle with an irregular shape rather than a regular one. Then in a "dynamic learning phase" subjects were asked to predict the frequency of blooming for various groups of plants fed various substances on the basis of the information that they had previously received, and were given feedback about the correctness of their answers.

Next, the *test* phase: first, subjects in both groups were given a categorization task in which they were asked to sort substances in accord with what caused blooming. The goal here was to determine the level of abstractness of the causal relationship the subjects recognized and employed in this task. Most subjects gave the highest causal ratings to the most abstract causal relationships. That is, they categorized in accord with the irregular/regular shape contrast or the warm/cool color contrast, depending on the data they had seen.

Second, subjects in both of the groups previously established were told that the plants were placed in a novel environment and were fed a novel substance s^* that was irregularly shaped and warm colored and that most of the plants bloomed. Subjects were then asked whether s^* caused the plants to bloom. Notice that in this case because both the environment and s^* are new (and perfectly correlated), the covariational evidence from this new experiment alone is ambiguous about whether s^* or the new environment caused blooming. However, Lien and Cheng hypothesized that the subjects who had learned (during the learning phase) that irregular shapes caused blooming would be more willing to judge that in this new situation s^* caused the blooming (since s^* is irregular), in comparison with the subjects who had learned in the test phase that cool colors cause blooming. This is because s^* is warm colored and hence the claim that s^* caused blooming was "inconsistent" with what the latter group had previously learned. This prediction was borne out. Subjects in the former group give a higher causal rating to the claim that s^* caused the blooming.

The general pattern in the experiment, then, is that when given a choice among causal relationships that might be formulated at different levels (where these levels stand in a hierarchical relationship), subjects appeared to learn or induce causal relationships from the data they were given at the level that maximized Δp and that they were guided by relationships at this level in their willingness to infer new causal relationships. In the particular case explored in Lien and Cheng's experiment, the level or choice of variable that maximized Δp was the most abstract level, but of course the covariational data might have been chosen in such a way that the strategy of choosing the variable or level that maximizes Δp leads instead to the choice of an intermediate or maximally specific level. Lien and Cheng did not do such experiments, but in such cases one would expect subjects to prefer these intermediate levels instead. As noted earlier, in the case of binary variables related in the way described in Lien and Cheng's experiment, the choice of variables that maximizes Δp will also be the choice that best satisfies (P), so that we can also think of their experiment as showing that subjects choose in accord with (P) in their experiment. (Their experiment does not discriminate between (P) and (P+).)

These experiments provide an additional illustration of the interplay between descriptive and normative considerations in causal cognition that is one of the main themes of this book. First, in agreement with the claims of Yablo and others, the experiments show that subjects do sometimes prefer (learn and make use of) causal relationships or descriptions of causal relationships that are not characterized in a maximally specific or detailed way. Instead, they sometimes prefer more abstract characterizations. More specifically, as an empirical matter subjects choose in a way that maximizes the contrast $\Delta p = Pr(E/C) - Pr(E/-C)$. To the extent that the generalization that subjects choose in a way that maximizes Δp holds more generally, this implies that which level of abstractness subjects prefer for the choice of causal variable will depend on the empirical details of the case. Sometimes they will prefer more abstract descriptions and sometimes more specific descriptions. More specifically, at least in these experiments, subjects infer and judge in a way that conforms to a proportionality requirement.

Of course the empirical fact that subjects do this does not by itself show that it is justifiable or normatively correct for them to do so. However, as Lien and Cheng argue, and is also suggested by our earlier discussion of (P), there is an obvious normative rationale for this behavior. This suggests that subject preference for variables at a more abstract level in these experiments is not a mistake or due to confusion. Subject preference for characterizations of cause that maximize the contrast $\Delta p = Pr(E/C) - Pr(E/-C)$ makes sense because this is the level that is most informative about the conditions under which the effect will and will not occur. This raises the following question for those who (typically on metaphysical grounds) claim that more specific cause variables are always better: in what sense, if any, are those who in the appropriate circumstances prefer more abstract or coarser cause variables that better satisfy proportionality to more specific cause variables making a mistake? Why should we be bound by some supposed metaphysical requirement to provide maximally specific descriptions of causes if proportional descriptions are often more informative about matters that we care about?[42]

8.9. Conclusion

A convention of book writing is that books should not end abruptly, but instead should have a substantial conclusion as a kind of gentle leave-taking. But this is already a relatively long book and I see little point in rehashing what I have said earlier. My overall message has been that there is a fruitful collaborative project concerning understanding causation and causal cognition—one that draws together results from philosophy, psychology, and other disciplines like statistics and machine learning. I hope to have convinced readers that progress on these topics can be made by cooperation across different disciplines rather than by boundary policing and strategies of dismissal.

Appendix 1: Proportionality as Applied to Actual Cause Claims

I noted in Section 8.1 that a proportionality requirement often seems more plausible or natural (where this is understood as an empirical observation about how people judge) when applied to type as opposed to token (or actual cause claims). For example, (i) "The impact of the rock thrown by Suzy caused the window to break" may seem completely appropriate even though (i) apparently does less well along the dimension of proportionality than (ii) "The impact of a rock caused the window to break," assuming that given the impact, it does not matter whether the thrower is Suzy. Here is a conjecture that attempts to make normative sense of this observation: The basic idea is that, as suggested in Chapter 2's appendix, actual cause claims—particularly those involving human actions, are commonly used to ascribe responsibility. (This does *not* require that "cause" in such cases just means "responsible" but merely that the causal claim is used in the way described.) Ascriptions of responsibility may require non-proportional characterizations of causes because this is the only available way of distinguishing among candidate causes for purposes of ascribing responsibility. For example, if the impact of a rock causes a window to break, then, to the extent

[42] At the risk of belaboring a point made earlier, again note the difference between asking, on the one hand, whether proportionality is part of the "nature of causation" or whether it is part of "our concept" of causation and, on the other hand, asking instead whether there is a normative rationale for preferring causal claims that better satisfy proportionality. Again issues of the latter sort tend to be more tractable.

that we are interested in attributing responsibility, we may be very interested in whether it was the rock thrown by Suzy or the rock thrown by Billy that caused the breaking. If it was Suzy's rock, we accept (i) and regard (ii) as less satisfactory because it provides no information about the responsible party. More generally, recall from Chapter 2's appendix that in making actual cause claims, we are often interested in discriminating among various candidates for the actual cause of some outcome. In a number of cases, the only characterizations of these candidate causes that we have available and that can be used to distinguish among them may involve some features that are causally irrelevant to the effect. If the candidate causes for the explosion are a gas leak and Jones's mixing "those yellow chemicals" together in the laboratory (this is the only description we have available), we may settle on the latter as the actual cause despite the fact that it involves a characterization of the cause that does not do very well with respect to proportionality.

We can get some additional insight into what is going on in these examples by considering the variables used to represent them. In the Billy/Suzy rock-throwing example, we should employ two variables, one, *B*, representing whether Billy's rock impacts the bottle or not, and the other, *S*, representing whether Suzy's rock impacts the bottle. Notice that *B* and *S* are fully distinct variables with compossible values. (We assume it is possible for both rocks or neither to hit the bottle.) This makes it clear that in the Billy/Suzy causal attribution example, we face a very different problem from the problem that proportionality is designed to address. In the Billy/Suzy case, we want to know which of the values of two fully distinct variables was the actual cause of the outcome. By contrast, as explained earlier, proportionality has to do with a choice among variables that stand in non-causal hierarchal relations, where the values of these different candidate variables are not fully compossible. It is not surprising that proportionality seems a more appropriate consideration when faced with a choice between variables standing in a non-causal hierarchical relation than in the Billy/Suzy example.

To reinforce this analysis, consider the following variant of the pigeon example. As before the pigeon has been trained to peck at red and only red targets, and this has involved exposing it to training targets of different shades of red. However, now Billy and Suzy are playing a game. Suzy possesses a scarlet target that she can present to the pigeon and Billy a maroon target. The first person to cause the pigeon to peck wins a prize. In this situation we are interested in whether the (iii) presentation of the scarlet target or (iv) the presentation of the maroon target caused the pigeon to peck since this is what will determine the winner of the game. If (iii) is true of what caused the pecking, we are not going to reject it, as a basis for determining the winner, on the grounds that it is less proportional than the claim that attributes the causation of the pecking to redness.

Appendix 2: Proportionality for Effects?

In the discussion of (P) in this chapter I restricted the application of proportionality to a situation in which the effect variable was fixed or pre-specified, so that proportionality was understood as a criterion for choosing among alternative cause variables that are hierarchically related. As I noted, this was a departure from a previous discussion of mine (Woodward 2010) in which I treated proportionality as a condition on effect variables as well. For example, in discussing a case of Kendler's that had to do with the best way of characterizing the effects of a particular allele a, I suggested that describing the allele as causing a tendency to take risks was superior to characterizing the allele as causing a tendency to engage in a number of distinct effects including skydiving, cliff jumping, and driving fast. Whatever one thinks about this suggestion, it is now clear to me that it addresses a different question than the question to which (P) is addressed. In Kendler's case, we have a fixed cause, the presence or absence of the allele a, and different candidates for the effect (or effects) variable(s). Moreover, rather than (as (P) requires) comparing single variables at different levels of abstractness, as in the case of *SCARLET* versus *RED*, we are comparing a single variable *RISKTAKING* with an alternative choice in which there are a number of distinct effect variables *SKYDIVING*, *FASTDRIVING*, and so on, values of which (for different variables) are fully compossible. (Note, though, that the values of the more abstract variable *RISKTAKING* and the more specific variables like *SKYDIVING* that realize it are not fully compossible and seem to be at different "levels.")

I noted previously that the problem of choosing variables when both the cause and the effect variables are allowed to vary in grain or abstractness seemed hopelessly indeterminate. However, one might wonder about the following: suppose that we take the cause variable to be fixed, as in the allele case, and think of ourselves as choosing among different effect variables for that cause, at different levels of abstraction. Is there some condition that might be used to guide such choices? I have no suggestions about this, but it is a worthwhile problem.

References

Aitkin, M., Larkin, M., and Dickinson, A. 2000. "Super-Learning of Causal Judgments." *Quarterly Journal of Experimental Psychology* 53B: 59–81.

Alexander, J., Mallon, R., and Weinberg, J. 2010. "Accentuate the Negative." *Review of Philosophy and Psychology* 1: 297–314.

Alexander, J. and Weinberg, J. 2006–7. "Analytic Epistemology and Experimental Philosophy." *Philosophy Compass* 2: 56–80.

Anderson, J. 1990. *The Adaptive Character of Thought*. Hillsdale, NJ: Lawrence Earlbaum.

Angrist, J. and Pischke, S. 2009. *Mostly Harmless Econometrics*. Princeton, NJ: Princeton University Press.

Austin, J. 1956–57. "A Plea for Excuses." *Proceedings of the Aristotelian Society*, new series, 57: 1–30.

Beck, S., Riggs, K., and Burns, P. 2011. "Multiple Developments in Counterfactual Thinking." In Hoerl, C., McCormack, T., and Beck, S., eds., *Understanding Counterfactuals, Understanding Causation*. Oxford: Oxford University Press, 110–122.

Beebee, H. 2004. "Causings and Nothingness." In Collins, J., Hall, N., and Paul, L. A., eds., *Causation and Counterfactuals*. Cambridge MA: MIT Press, 291–308.

Bennett, J. 1984. "Counterfactuals and Temporal Direction." *Philosophical Review* 93: 57–91.

Biernaskie, J., Walker, S., and Gegear, R. 2009. "Bumblebees Learn to Forage Like Bayesians." *American Naturalist* 74: 413–423.

Blaisdell, A., Sawa, K., Leising, K., and Waldmann, M. 2006. "Causal Reasoning in Rats." *Science* 311: 1020–1022.

Blanchard, T. 2020. "Explanatory Abstraction and the Goldilocks Problem: Interventionism Gets Things Just Right." *British Journal for the Philosophy of Science* 71: 633–663.

Blanchard, T., Murray, D. and Lombrozo, T. Forthcoming. "Experiments on Causal Exclusion." *Mind and Language*.

Bogen, J. 2004. "Analysing Causality: The Opposite of Counterfactual Is Factual." *International Studies in the Philosophy of Science* 18: 3–26.

Bogen, J. and Woodward, J. 1988. "Saving the Phenomena." *Philosophical Review* 97: 303–352.

Bonawitz, E., Ferranti, D., Saxe, R., Gopnik, A., Meltzoff, A., Woodward, J., and Schulz, L. 2010. "Just Do It? Investigating the Gap between Prediction and Action in Toddlers' Causal Inferences." *Cognition* 115: 104–117.

Bonawitz, E. and Schulz, L. 2007. "Serious Fun: Preschoolers Engage in More Exploratory Play When Evidence Is Confounded." *Developmental Psychology* 43: 1045–1050.

Bontly, T. 2005. "Proportionality, Causation, and Exclusion." *Philosophia* 32: 331–348.

Bowers, R. and Timberlake, W. 2017. "Do Rats Learn Conditional Independence?" *Royal Society Open Science* 4: 160994.

Briggs, R. 2012. "Interventionist Counterfactuals." *Philosophical Studies* 160: 139–166.

Buchsbaum, D., Bridgers, S., Weisberg, D. S., and Gopnik, A. 2012. "The Power of Possibility: Causal Learning, Counterfactual Reasoning, and Pretend Play." *Philosophical Transactions of the Royal Society B: Biological Sciences* 367: 2202–2212.

Buehner, M., Cheng, P., and Clifford, D. 2003. "From Covariation to Causation: A Test of the Assumption of Causal Power." *Journal of Experimental Psychology: Learning, Memory, and Cognition* 29: 1119–1140.

Burgess, A. and Plunkett, D. 2013. "Conceptual Ethics 1 and 2." *Philosophy Compass* https://doi.org/10.1111/phc3.12086.

Cappelen, H. 2012. *Philosophy without Intuitions*. Oxford: Oxford University Press.

Carroll, C. D. and Cheng, P. W. 2010. "The Induction of Hidden Causes: Causal Mediation and Violations of Independent Causal Influence." In Ohlsson, S. and Catrambone, R., eds., *Proceedings of the 32nd Annual Conference of the Cognitive Science Society*, Austin, TX: Cognitive Science Society, 913–918.

Cartwright, N. 1983. *How the Laws of Physics Lie*. Oxford: Oxford University Press.

Cartwright, N. 1989. *Nature's Capacities and Their Measurement*. Oxford: Oxford University Press.

Chalupka, K., Perona, P., and Eberhardt, F. 2015. "Visual Causal Feature Learning." In *Proceedings of the Thirty-First Conference on Uncertainty in Artificial Intelligence*. Corvallis, OR: AUAI Press, 181–190.

Cheng, P. W. 1997. "From Covariation to Causation: A Causal Power Theory." *Psychological Review* 104: 367–405.

Cheng, P. W. and Lu, H. 2017. "Causal Invariance as an Essential Constraint for Creating a Causal Representation of the World: Generalizing the Invariance of Causal Power." In Waldmann, M., ed., *The Oxford Handbook of Causal Reasoning*. New York: Oxford University Press.

Cheng, P. W., Novick, L. R., Liljeholm, M., and Ford, C. 2007. "Explaining Four Psychological Asymmetries in Causal Reasoning: Implications of Causal Assumptions for Coherence." In Rourke, M., ed., *Topics in Contemporary Philosophy: Explanation and Causation*. Cambridge, MA: MIT Press, 1–32.

Clarke, R., Shepherd, J., Stigall, J., Waller, R., and Zarpentine, C. 2015. "Causation, Norms, and Omissions: A Study of Causal Judgments." *Philosophical Psychology* 28: 279–293.

Clatterbaugh, K. 1999. *The Causation Debate in Modern Philosophy, 1637–1739*. New York: Routledge.

Cole, J. 2016. "Intuitional Instability." In Sytsma, J. and Buckwalter, W., eds., *A Companion to Experimental Philosophy*. Malden, MA: Wiley, 568–577.

Craig, E. 1990. *Knowledge and the State of Nature*. Oxford: Oxford University Press.

Crane, T. 2008. "Causation and Determinable Properties: On the Efficacy of Colour, Shape, and Size." In Hohwy, J. and Kallestrup, J., eds., *Being Reduced: New Essays on Reduction, Explanation and Causation*. Oxford: Oxford University Press, 176–195.

Danks, D. 2003. "Equilibria of the Rescorla-Wagner Model." *Journal of Mathematical Psychology* 47: 109–121.

Danks, D., Rose, D., and Machery, E. 2014. "Demoralizing Causation." *Philosophical Studies* 171: 251–277.

Davidson, E. et al. 2002. "A Genomic Regulatory Network for Development." *Science* 295: 1669–1678.

Davis, L. 1968. "And It Will Never Be Literature—the New Economic History: A Critique." *Explorations in Entrepreneurial History* 6: 75–92.

Dayan, P. and Balleine, B. 2002. "Reward, Motivation, and Reinforcement Learning." *Neuron* 36(2): 285–298.

Dennett, D. 1984. *Elbow Room: The Varieties of Free Will Worth Wanting*. Cambridge, MA: Bradford Books.

Dickinson, A. 2001. "Causal Learning: An Associative Analysis." *Quarterly Journal of Experimental Psychology*, Section B, 54(1b): 3–25.

Dickinson, A. 2012. "Associative Learning and Animal Cognition." *Philosophical Transactions of the Royal Society of London B: Biological Sciences* 367(1603): 2733–2742.

Dickinson, A. and Balleine, B. 2000. "Causal Cognition and Goal-Directed Action." In Heyes, C. and Huber, L., eds., *The Evolution of Cognition*. Cambridge, MA: MIT Press, 185–204.

Dickinson, A. and Burke, J. 1996. "Within-Compound Associations Mediate the Retrospective Revaluation of Causality Judgements." *Quarterly Journal of Experimental Psychology B* 49(1): 60–80. https://doi.org:10.1080/713932614.

Dickinson, A. and Shanks, D. 1995. "Instrumental Action and Causal Representation." In Sperber, D., Premack, D., and Premack, A., eds., *Causal Cognition*. Oxford: Oxford University Press, 5–25.

Dorr, C. 2016. "Against Counterfactual Miracles." *Philosophical Review* 125: 241–286.

Dowe, P. 2000. *Physical Causation*. Cambridge: Cambridge University Press.

Dowe, P. 2009. "Causal Process Theories." In Beebee, H., Hitchcock, C. and Menzies, P. eds., *The Oxford Handbook of Causation*. Oxford: Oxford University Press, 213–233.

Dowe, P. 2010. "Proportionality and Omissions." *Analysis* 70: 1–5.

Doya, K., Ishii, S. Pouget, A., and Rao, R. 2006. *Bayesian Brain: Probabilisitic Approaches to Neural Coding*. Cambridge, MA: MIT Press.

Eberhardt, F. 2014. "Direct Causes and the Trouble with Soft Interventions." *Erkenntnis* 79: 755–777.

Eberhardt, F., and Danks, D. 2011. "Confirmation in the Cognitive Sciences: The Problematic Case of Bayesian Models." *Minds and Machines* 21: 389–410.

Eberhardt, F. and Scheines, R. 2007. "Interventions and Causal Inference." *Philosophy of Science* 74: 981–995.

Edgington, D. 2011. "Causation First: Why Causation Is Prior to Counterfactuals." In Hoerl, C., McCormack, T., and Beck, S., eds., *Understanding Counterfactuals, Understanding Causation*. Oxford: Oxford University Press, 230–241.

Edwards, B., Rottman, B., and Santos, L. 2011. "The Evolutionary Origins of Causal Cognition." In McCormack, T., Hoerl, C., and Butterfill, S., eds., *Tool Use and Causal Cognition*. Oxford: Oxford University Press, 111–128.

Eells, E. 1991. *Probabilistic Causality*. Cambridge: Cambridge University Press.

Elga, A. 2001. "Statistical Mechanics and the Asymmetry of Counterfactual Dependence." *Philosophy of Science* 68: 313–324.

Ellis, G. 2016. *How Can Physics Underlie the Mind?* New York: Springer.

Fawcett, T., Skinner, A., and Goldsmith, A. 2002. "A Test of Imitative Learning in Starlings Using a Two-Action Method with an Enhanced Ghost Control." *Animal Behaviour* 64: 547–556.

Fenton-Glynn, L. 2016. "A Probabilistic Analysis of Causation." *British Journal for the Philosophy of Science* 63: 343–392.

Fodor, J. 1983. *Modularity of Mind*. Cambridge, MA: MIT Press.

Franklin-Hall, L. 2016. "High Level Explanation and the Interventionist's 'Variables Problem.'" *British Journal for the Philosophy of Science* 67: 553–577.

Frisch, M. 2014. *Causal Reasoning in Physics*. Cambridge: Cambridge University Press.

Frosch, C., McCormack, T., Lagnado, D., and Burns, P. 2012. "Are Causal Structure and Intervention Judgments Inextricably Linked? A Developmental Study." *Cognitive Science* 36: 261–285.

Gebharter, A. 2017. "Uncovering Constitutive Relevance Relations in Mechanisms." *Philosophical Studies* 174: 2645–2666.

Geisler, W. 2011. "Contributions of Ideal Observer Theory to Vision Research." *Vision Research* 51: 771–781.

Gendler, T. 2010. "Philosophical Thought Experiments, Intuitions and Cognitive Equilibrium." In *Intuition, Imagination and Philosophical Methodology*. Oxford: Oxford University Press, 116–134.

Gerstenberg, T., Goodman, N., Lagnado, D., and Tenenbaum, J. 2012. "Noisy Newtons: Unifying Process and Dependency Accounts of Causal Attribution" *Proceedings of the 34th Annual Conference of the Cognitive Science Society*. Austin, TX: Cognitive Science Society.

Gerstenberg, T., Peterson, M., Goodman, N., Lagnado, D., and Tenenbaum, J. 2017. "Eye-Tracking Causality." *Psychological Science* 28: 1731–1744.

Giere, R. 1999. *Science without Laws*. Chicago: University of Chicago Press.

Glymour, C. 1998. "Learning Causes: Psychological Explanations of Causal Explanation." *Minds and Machines* 8: 39–60.

Glymour, C. 2001. *The Mind's Arrows: Bayes Nets and Graphical Models in Psychology*. Cambridge, MA: Bradford Books.

Glymour, C. 2004. "We Believe in Freedom of the Will So That We Can Learn." *Behavioral and Brain Sciences* 27: 661–662.

Glymour, C. 2007. "When Is a Brain Like a Planet?" *Philosophy of Science* 74: 330–347.

Glymour, C., Danks, D., Glymour, B., Eberhardt, F., Ramsey, J., Scheines, R., Spirtes, P., Teng, M., and Zhang, J. 2010. "Actual Causation: A Stone Soup Essay." *Synthese* 175: 169–192.

Goddu, M. and Gopnik, A. 2020. "Learning What to Change: Young Children Use 'Difference-Making' to Identify Causally Relevant Variables." *Developmental Psychology* 56: 275–284.

Goldman, A. 2007. "Philosophical Intuitions: Their Target, Their Source, and Their Epistemic Status." *Grazer Philosophische Studien* 74: 1–26.

Goodman, N. 1955. *Fact, Fiction and Forecast*. Cambridge, MA: Harvard University Press.

Gopnik, A. 2009. *The Philosophical Baby*. New York: Farrar, Straus and Giroux.

Gopnik, A., Glymour, C., Sobel, D. M., Schulz, L., Kushnir, T., and Danks, D. 2004. "A Theory of Causal Learning in Children: Causal Maps and Bayes Nets." *Psychological Review* 111: 3–32.

Gopnik, A., Meltzoff, A., and Kohl, P. 1999. *The Scientist in the Crib: What Early Learning Tells Us about the Mind*. New York: William Morrow.

Gopnik, A. and Schulz, L., eds. 2007. *Causal Learning: Psychology, Philosophy and Computation*. New York: Oxford University Press.

Gopnik, A. and Sobel, D. M. 2000. "Detecting Blickets: How Young Children Use Information about Novel Causal Powers in Categorization and Induction." *Child Development* 71: 1205–1222.

Gopnik, A., Sobel, D. M., Schulz, L., and Glymour, C. 2001. "Causal Learning Mechanisms in Very Young Children: Two-, Three-, and Four-Year-Olds Infer Causal Relations from Patterns of Variation and Covariation." *Developmental Psychology* 37: 620–626.

Griffiths, T. L., Sobel, D. M., Tenenbaum, J., and Gopnik, A. 2011. "Bayes and Blickets: Effects of Knowledge on Causal Induction in Children and Adults." *Cognitive Science* 35: 1407–1455.
Griffiths, T. L., and Tenenbaum, J. B. 2005. "Structure and Strength in Causal Induction." *Cognitive Psychology* 51: 334–384.
Griffiths, T. L. and Tenenbaum, J. B. 2009. "Theory-Based Causal Induction." *Psychological Review* 116: 661–716.
Grinfeld, G., Lagnado, D., Gerstenberg, T., Woodward, J., and Usher, M. 2020. "Causal Responsibility and Robust Causation." *Frontiers in Psychology* 11. https://doi.org/10.3389/fpsyg.2020.01069.
Hagmeyer, Y. and Sloman, S. 2009. "Decision Makers Conceive of Their Choices as Interventions." *Journal of Experimental Psychology General* 138(1): 22–38.
Hall, N. 2004. "Two Concepts of Causation." In Collins, J., Hall, N., and Paul, L., eds., *Causation and Counterfactuals*. Cambridge, MA: MIT Press, 225–276.
Halpern, J. 2016. *Actual Causality*. Cambridge, MA: MIT Press.
Halpern, J. and Pearl, J. 2005. "Causes and Explanations: A Structural-Model Approach. Part I: Causes." *British Journal for the Philosophy of Science* 56: 843–887.
Harris, P. 2000. *The Work of the Imagination*. Oxford: Blackwell.
Haslanger, S. 2000. "Gender and Race: (What) Are They? (What) Do We Want Them to Be?" *Noûs* 34: 31–55.
Hausman, D. and Woodward, J. 1999. "Independence, Invariance, and the Causal Markov Condition." *British Journal for the Philosophy of Science* 50: 521–583.
Heckman, J. 2005. "The Scientific Model of Causality." *Sociological Methodology* 35(1): 1–97.
Henne, P., Niemi, L., Pinillos, Á., De Brigard, F., and Knobe, J. 2019. "A Counterfactual Explanation for the Action Effect." *Cognition* 190: 157–164.
Hertwig, R., and Gigerenzer, G. 1999. "The 'Conjunction Fallacy' Revisited: How Intelligent Inferences Look Like Reasoning Errors." *Journal of Behavioral Decision Making* 12: 275–305.
Hesslow, G. 1976. "Two Notes on the Probabilistic Approach to Causality." *Philosophy of Science* 43: 290–292.
Hitchcock, C. 1995. "Discussion: Salmon on Explanatory Relevance." *Philosophy of Science* 62: 304–320.
Hitchcock, C. 2012. "Events and Times: A Case Study in Means-ends Metaphysics." *Philosophical Studies* 160: 79–96.
Hitchcock, C. 2017. "Actual Causation: What's the Use?" In Beebee, H., Hitchcock, C., and Price, H., eds., *Making a Difference: Essays on the Philosophy of Causation*. Oxford: Oxford University Press.
Hitchcock, C. 2018. "Causal Models." *Stanford Encyclopedia of Philosophy*. Accessed 5/24/21.
Hitchcock, C. 2001. "The Intransitivity of Causation Revealed in Equations and Graphs." *Journal of Philosophy* 98: 273–299.
Hitchcock, C. and Knobe, J. 2009. "Cause and Norm." *Journal of Philosophy* 106: 587–612.
Hitchcock, C. and Woodward, J. 2003. "Explanatory Generalizations, Part II: Plumbing Explanatory Depth." *Noûs* 37: 181–199.
Hoffmann-Kloss, V., 2014. "Interventionism and Higher-Level Causation." *International Studies in the Philosophy of Science* 28: 49–64.

Holland, P. 1986. "Statistics and Causal Inference." *Journal of the American Statistical Association* 81: 945–960.

Holyoak, K. and Cheng, P. 2011. "Causal Learning and Inference as a Rational Process: The New Synthesis." *Annual Review of Psychology* 62: 135–163.

Hoover, K. 2001. *Causality in Macroeconomics*. Cambridge: Cambridge University Press.

Icard, T., Kominsky, J., and Knobe, J. 2017. "Normality and Actual Causal Strength." *Cognition* 161: 80–93.

Ismael, J. 2015. "How to Be Humean." In Loewer, B. and Schaffer, J., eds., *A Companion to David Lewis*. Malden, MA: Wiley Blackwell, 188–205.

Janzing, D., Mooij, J., Zhang, K., Lemeire, J., Zscheischler, J., Daniusis, D., Steudel, B., and Scholkopf, B. 2012. "Information-Geometric Approach to Inferring Causal Directions." *Artificial Intelligence* 182–183: 1–31.

Janzing, D. and Schollkopf, B. 2010. "Causal Inference Using the Algorithmic Markov Condition." *IEE Transactions on Information Theory* 56(10): 5168–5194.

Johnson, S. and Keil, F. 2014. "Causal Inference and the Hierarchical Structure of Experience." *Journal of Experimental Psychology: General* 143(6): 2223–2241.

Kahneman, D., and Tversky, A. 1982. "The Simulation Heuristic." In Kahneman, D., Slovic, P., and Tversky, A., eds., *Judgement under Uncertainty: Heuristics and Biases*. Cambridge: Cambridge University Press, 201–210.

Kamm, F. 1993. *Morality, Mortality*. Vol. 1. New York: Oxford University Press.

Kendler, K. 2005. "'A Gene for . . .': The Nature of Gene Action in Psychiatric Disorders." *American Journal of Psychiatry* 162: 1243–1252.

King, G., Keohane, R., and Verba, S. 1994. *Designing Social Inquiry: Scientific Inference in Qualitative Research*. Princeton, NJ: Princeton University Press.

Kistler, M. 2014. "Analysing Causation in Light of Intuitions, Causal Statements, and Science." In Copley, B. and Martin, F., eds., *Causation in Grammatical Structures*. Oxford: Oxford University Press, 6–99.

Kitcher, P. 2011. *The Ethical Project*. Cambridge, MA: Harvard University Press.

Knobe, J. 2003. "Intentional Action in Folk Psychology: An Experimental Investigation." *Philosophical Psychology* 16(2): 309–323.

Knobe, J. 2016. "Experimental Philosophy Is Cognitive Science." In Systma, J. and Buckwalter, W., eds., *A Companion to Experimental Philosophy*. Chichester: Wiley Blackwell, 37–52.

Kohler, W. 1927. *The Mentality of Apes*. 2nd ed. New York: Vintage Books.

Kominsky, J., Phillips, J., Gerstenberg, T., Lagnado, D., and Knobe, J. 2015. "Causal Superseding." *Cognition* 137: 196–209.

Kushnir, T. and Gopnik, A. 2005. "Young Children Infer Causal Strength from Probabilities and Interventions." *Psychological Science* 16: 678–683.

Kushnir, T. and Gopnik, A. 2007. "Conditional Probability versus Spatial Contingency in Causal Learning: Preschoolers Use New Contingency Evidence to Overcome Prior Spatial Assumptions." *Developmental Psychology* 43: 186–196.

Kushnir, T., Wellman, H., and Gelman, S. 2009. "A Self-agency Bias in Children's Causal Inferences." *Developmental Psychology* 45: 597–603.

Kutach, D. 2013. *Causation and Its Basis in Fundamental Physics*. Oxford: Oxford University Press.

Lagnado, D. and Sloman, S. A. 2004. "The Advantage of Timely Intervention." *Journal of Experimental Psychology: Learning, Memory & Cognition* 30: 856–876.

Lake, B., Ullman, T., Tenenbaum, J., and Gershman, S. 2017. "Building Machines That Learn and Think Like People." *Behavioral and Brain Sciences* 40: E253.

Le Pelly, M., Griffiths, O., and Beesley, T. 2017. "Associative Accounts of Causal Judgment." In Waldman, M., ed., *The Oxford Handbook of Causal Reasoning*. Oxford: Oxford University Press, 13–28.

Lewis. D. 1973a. "Causation." *Journal of Philosophy* 70: 556–567.

Lewis, D. 1973b. *Counterfactuals*. Cambridge, MA: Harvard University Press.

Lewis, D. 1979. "Counterfactual Dependence and Times Arrow." *Noûs* 13: 455–476.

Lewis, D. 1986. "Postscripts to 'Causation.'" In *Philosophical Papers*, vol. 2. Oxford: Oxford University Press.

Lewis, D. 2000. "Causation as Influence." *Journal of Philosophy* 97: 182–197.

Lieder, F. and Griffiths, T. 2018. "Resource-Rational Analysis: Understanding Human Cognition as the Optimal use of Limited Computational Resources." *Behavioral and Brain Sciences 43*. https://doi.org/10.1017/S0140525X1900061.

Lien, Y., and Cheng, P. 2000. "Distinguishing Genuine from Spurious Causes: A Coherence Hypothesis." *Cognitive Psychology* 40: 87–137.

Liljeholm, M. and Cheng, P. W. 2007. "When Is a Cause the 'Same'? Coherent Generalization across Contexts." *Psychological Science* 18: 1014–1021.

List, C. and Menzies, P. 2009. "Nonreductive Physicalism and the Limits of the Exclusion Principle." *Journal of Philosophy* 106: 475–502.

Lober, K., and Shanks, D. 2000. "Is Causal Induction Based on Causal Power? Critique of Cheng (1997)." *Psychological Review* 107: 195–212.

Loewer, B. 2012. "Two Accounts of Laws and Time." *Philosophical Studies* 160: 115–137.

Lombrozo, T. 2010. "Causal-Explanatory Pluralism: How Intentions, Functions, and Mechanisms Influence Causal Ascriptions." *Cognitive Psychology* 61: 303–332.

Lombrozo, T. 2011. "The Instrumental Value of Explanations." *Philosophy Compass* 6: 539–551.

Lombrozo, T. and Carey, S. 2006. "Functional Explanation and the Function of Explanation." *Cognition* 99: 167–204.

Machery, E. 2017. *Philosophy within Its Proper Bounds*. Oxford: Oxford University Press.

Machery, E., Mallon, R., Nichols, S., and Stich, S. 2004. "Semantics, Cross-Cultural Style." *Cognition* 92: B1–B12.

Mackie, J. 1974. *The Cement of the Universe*. Oxford: Oxford University Press.

Mandel, D. 2003. "Judgment Dissociation Theory: An Analysis of Differences in Causal, Counterfactual, and Covariational Reasoning." *Journal of Experimental Psychology: General* 132: 419–434.

Marcus, G. 2018. "Deep Learning: A Critical Appraisal." arXiv.org:1801.00631.

Maudlin, T. 2007. *The Metaphysics within Physics*. Oxford: Oxford University Press.

McDermott, M. 1995. "Redundant Causation." *British Journal for the Philosophy of Science* 46: 523–544.

McDonnell, N. 2017. "Causal Exclusion and the Limits of Proportionality." *Philosophical Studies* 174: 1459–1474.

McLaughlin, B. 2007. "Mental Causation and Shoemaker Realization." *Erkenntnis* 67: 149–172.

Meltzoff, A. 2007. "Infants' Causal Learning: Intervention, Observation, Imitation." In Gopnik, A. and Schulz, L., eds., *Causal Learning: Psychology, Philosophy, and Computation*. Oxford: Oxford University Press, 37–47.

Menzies, P. 2006. "Review of *Making Things Happen*." *Mind* 115: 821–826.

Menzies, P. and Price, H. 1993. "Causation as a Secondary Quality." *British Journal for the Philosophy of Science* 44: 187–203.
Morgan, S. L. and Winship, C. 2015. *Counterfactuals and Causal Inference.* New York: Cambridge University Press.
Muenter, P. and Bonawitz, E. 2017. "The Development of Causal Reasoning." In Waldmann, M., ed., *The Oxford Handbook of Causal Reasoning.* Oxford: Oxford University Press, 677–698.
Murray, D. and Lombrozo, T. 2017. "Effects of Manipulation on Attributions of Causation, Free Will and Moral Responsibility." *Cognitive Science* 41: 447–481.
Nagel, J. 2012. "Intuitions and Experiments: A Defense of the Case Method in Epistemology." *Philosophy and Phenomenological Research* 85: 495–527.
Nichols, S. and Knobe, J. 2007. "Moral Responsibility and Determinism: The Cognitive Science of Folk Intuitions." *Noûs* 41: 663–685.
Novick, L. R., and Cheng, P. W. 2004. "Assessing Interactive Causal Influence." *Psychological Review* 111: 455–485.
Nyrup, R. 2015. "How Explanatory Reasoning Justifies Pursuit: A Peircean View of IBE." *Philosophy of Science* 82: 749–760.
Oaksford, M. and Chater, N. 2000. "The Rational Analysis of Mind and Behavior." *Synthese* 122: 93–131.
Ober, D. 2017. "Causation and the Probability of Causal Conditionals." In Waldmann, M., ed., *The Oxford Handbook of Causal Reasoning.* Oxford: Oxford University Press, 307–326.
Paul, L. 2010. "A New Role for Experimental Work in Metaphysics." *Review of Philosophy and Psychology* 1: 461–476.
Paul, L. 2012. "Metaphysics as Modeling: The Handmaiden's Tale." *Philosophical Studies* 160: 1–29.
Paul, L. and Hall, N. 2013. *Causation: A User's Guide.* Oxford: Oxford University Press.
Pearl, J. 1988. *Probabilistic Reasoning in Intelligent Systems.* San Francisco: Morgan Kaufmann.
Pearl, J. 2000. *Causality: Models, Reasoning and Inference.* Cambridge: Cambridge University Press.
Pearl, J. 2001. "Direct and Indirect Effects." In *Proceedings of the Seventeenth Conference on Uncertainty in Artificial Intelligence.* San Francisco: Morgan Kaufman, 411–420.
Pearl, J. 2018. "Theoretical Impediments to Machine Learning with Seven Sparks from the Causal Revolution." UCLA Cognitive Systems Laboratory, Technical Report R-475.
Pearl, J. and Mackenzie, D. 2018. *The Book of Why.* New York: Basic Books.
Perales, J. C., and Shanks, D. R. 2008. "Driven by Power? Probe Question and Presentation Format Effects on Causal Judgment." *Journal of Experimental Psychology: Learning, Memory, and Cognition* 34: 1482–1494.
Philips, J. and Shaw, A. 2015. "Manipulating Morality: Third-Party Intentions Alter Moral Judgments by Changing Causal Reasoning." *Cognitive Science* 39(6): 1320–1347.
Pineno, O. 2010. *The Thinking Rat: The New Science of Animal Behavior.* N.p.: CreateSpace Independent Publishing Platform.
Pocheville, A., Griffiths, P., and Stotz, K. 2017. "Comparing Causes—an Information-Theoretic Approach to Specificity, Proportionality and Stability." In Leitgeb, H., Niinluoto, N., Sober, E., and Seppälä, P. eds. *Proceedings of the 15th Congress of Logic, Methodology and Philosophy of Science.* London: College Publications, 260–275.
Poldrack, R. 2020. "The Physics of Representation." July 11. PhilSci Archive.

Povinelli, D. 2000. *Folk Physics for Apes*. Oxford: Oxford University Press.

Premack, D., and Premack, A. 2003. *Original Intelligence*. New York: McGraw-Hill.

Price, H. 2017. "Causation, Intervention and Agency: Woodward on Menzies and Price." In Beebee, H., Hitchcock, C., and Price, H., eds., *Making a Difference*. Oxford: Oxford University Press.

Reichenbach, H. 1956. *The Direction of Time*. Berkeley-Los Angeles: University of California Press.

Rescorla, R. 1988. "Pavlovian Conditioning: It's Not What You Think It Is." *American Psychologist* 43: 151–160.

Robb, D. and Heil, J. 2018. "Mental Causation." In Zalta, E. N., ed., *Stanford Encyclopedia of Philosophy*. Spring 2021 ed.

Rose, D., Danks, D., and Machery, E. In preparation.

Rottman, B. 2017. "The Acquisition and Use of Causal Structural Knowledge." In Waldmann, M., ed., *The Oxford Handbook of Causal Reasoning*. Oxford: Oxford University Press, 85–114.

Rubin, D. 1974. "Estimating Causal Effects of Treatments in Randomized and Nonrandomized Studies." *Journal of Educational Psychology* 66: 688–701.

Rubin, D. 1986. "Which Ifs Have Causal Answers" *Journal of the American Statistical Association* 81: 961–962.

Rubin. D. 2013. "Discussion of 'Experimental Designs for Identifying Causal Mechanisms' by K. Imai, D. Tingley and T. Yamamoto." *Journal of the Royal Statistical Society-A* 176: 45.

Russell, B. 1912–13. "On the Notion of Cause." *Proceedings of the Aristotelian Society*, new series, 13: 1–26.

Saffran, J., Aslan, R., and Newport, E. 1996. "Statistical Learning by 8-Month-Old Infants." *Science* 274: 1926–1928.

Salmon, W. 1971. "Statistical Explanation." In Salmon, W., ed., *Statistical Explanation and Statistical Relevance*. Pittsburgh: University of Pittsburgh Press, 29–87.

Salmon, W. 1984. *Scientific Explanation and the Causal Structure of the World*. Princeton, NJ: Princeton University Press.

Sanborn, A., Mansinghka, V., and Griffiths, T. 2009. "A Bayesian Framework for Modeling Intuitive Dynamics." In *Proceedings of the 31st Annual Conference of the Cognitive Science Society*, 1–6.

Schaffer, J. 2000. "Causation by Disconnection." *Philosophy of Science* 67: 285–300.

Schaffer, J. 2016. "Grounding in the Image of Causation." *Philosophical Studies* 173: 49–100.

Schaffer, J. and Knobe, J. 2012. "Contrastivism Surveyed." *Noûs* 46: 675–708.

Schulte, O. 2017. "Formal Learning Theory." In Zalta, E. N., ed., *Stanford Encyclopedia of Philosophy*. Spring 2018 ed.

Schulz, L., Gopnik, A., and Glymour, C. 2007. "Preschool Children Learn about Causal Structure from Conditional Interventions." *Developmental Science* 10: 322–332.

Shanks, D. and Dickinson, A. 1988. "Associative Accounts of Causality Judgment." *Psychology of Learning and Motivation* 21: 229–261.

Shapiro, L. and Sober, E. 2012. "Against Proportionality." *Analysis* 72: 89–93.

Simon, H. (1957) *Models of Man: Social and Rational*. New York: John Wiley.

Sloman, S. and Lagnado, D. 2005. "Do We 'Do'?" *Cognitive Science* 29: 5–39.

Sobel, D. M. and Kushnir, T. 2003. "Interventions Do Not Solely Benefit Causal Learning: Being Told What to Do Results in Worse Learning Than Doing It Yourself." *Proceedings of the 2003 Meeting of the Cognitive Science Society, Boston, MA.*

Sobel, D. M., Tenenbaum, J. B., and Gopnik, A. 2004. "Children's Causal Inferences from Indirect Evidence: Backwards Blocking and Bayesian Reasoning in Preschoolers." *Cognitive Science* 28: 303–333.

Sommerville, J. 2007. "Detecting Structure in Action: Infants as Causal Agents." In Gopnik, A. and Schulz, L., eds., *Causal Learning*. Oxford: Oxford University Press.

Soo, K. 2019. "The Role of Granularity in Causal Learning." PhD dissertation, University of Pittsburgh.

Spirtes, P., Glymour, C., and Scheines, R. 1993/2000. *Causation, Prediction, and Search.* First edition, New York: Springer-Velag, Second edition. Cambridge, MA: MIT Press.

Spirtes, P. and Scheines, R. 2004. "Causal Inference of Ambiguous Manipulations." *Philosophy of Science* 71: 833–845.

Spohn, W. 2012. *Laws of Belief*. Oxford: Oxford University Press.

Stern, R. Forthcoming. "Causal Concepts and Temporal Ordering." *Synthese.*

Steyvers, M., Tenenbaum, J., Wagenmakers, E., and Blum, B. 2003. "Inferring Causal Networks from Observations and Interventions." *Cognitive Science* 27: 453–489.

Stich, S. and Tobia, K. 2016. "Experimental Philosophy and the Philosophical Tradition." In Systma, J. and Buckwalter, W., eds., *A Companion to Experimental Philosophy.* Malden, MA: Wiley.

Strevens, M. 2019. *Thinking off Your Feet: How Empirical Psychology Vindicates Armchair Philosophy*. Cambridge, MA: Harvard University Press.

Suppes, P. 1970. *A Probabilistic Theory of Causality*. Amsterdam: North Holland.

Swain, S., Alexander, J., and Weinberg, J. 2008. "The Instability of Philosophical Intuitions: Running Hot and Cold on Truetemp." *Philosophy and Phenomenological Research* 76: 138–155.

Sytsma, J. and Buckwalter, W. 2016. *A Companion to Experimental Philosophy*. Malden, MA: Wiley.

Tenenbaum, J. and Griffiths, T. 2001. "Structure Learning in Human Causal Induction." In Leen, T., Dietterich, T., and Tresp, V., eds., *Advances in Neural Processing Systems*, vol. 13. Cambridge, MA: MIT Press, 59–65.

Tetlock, P. 2005. *Expert Political Judgment*. Princeton, NJ: Princeton University Press.

Tomasello, M. and Call, J. 1997. *Primate Cognition*. New York: Oxford University Press.

Vargas, M. 2013. *Building Better Beings: A Theory of Moral Responsibility*. Oxford: Oxford University Press.

Vasilyeva, N., Blanchard, T., and Lombrozo, T. 2018. "Stable Causal Relationships Are Better Causal Relationships." *Cognitive Science* 42: 1265–1296.

Vasilyeva, N., and Lombrozo, T. 2015. "Explanations and Causal Judgments Are Differentially Sensitive to Covariation and Mechanism Information." In Noelle, D., Dale, R. Warlaumont, S., Yoshimi, J., Matlock, T., Jennings, C., and Maglio, P., eds., *Proceedings of the 37th Annual Meeting of the Cognitive Science Society* Austin, TX: Cognitive Science Society, 2475–2480.

Visalberghi, E. and Trinca, L. 1989. "Tool Use in Capuchin Monkeys: Distinguishing between Performance and Understanding." *Primates* 30: 511–521.

Waldmann, M. R., Cheng, P. W., Hagmayer, Y., and Blaisdell, A. P. 2008. "Causal Learning in Rats and Humans: A Minimal Rational Model." In Chater, N. and Oaksford, M., eds., *Rational Models of cognition*. Oxford: Oxford University Press, 453–484.

Waldmann, M. and Hagmayer, Y. 2005. "Seeing versus Doing: Two Modes of Accessing Causal Knowledge." *Journal of Experimental Psychology: Learning, Memory, and Cognition* 31: 216–227.

Waldman, M. and Holyoak, K. 1992. "Predictive and Diagnostic Learning within Causal Models: Asymmetries in Cue Competition." *Journal of Experimental Psychology: General* 121: 222–236.

Walsh, C., Sloman, S., Ahn, W., Kalish, C., Medin, D., and Gelman, S. 1995. "The Role of Covariation versus Mechanism Information in Causal Attribution." *Cognition* 54: 299–352.

Walsh, C. and Sloman, S. 2011. "The Meaning of Cause and Prevent: The role of Causal Mechanism." *Mind & Language* 26 (1): 21–52.

Waters, C. 2007. "Causes That Make a Difference." *Journal of Philosophy* 104: 551–579.

Wegner, D. 2002. *The Illusion of Conscious Will*. Cambridge, MA: MIT Press.

Weinberger, N., Williams, P., and Woodward, J. Forthcoming. "The Worldly Infrastructure of Causation."

Weslake, B. 2010. "Explanatory Depth." *Philosophy of Science* 77: 273–294.

Williamson, T. 2007. *The Philosophy of Philosophy*. Oxford: Blackwell.

Wilson, M. 2006. *Wandering Significance: An Essay on Conceptual Behavior*. Oxford: Oxford University Press.

Wilson, M. 2017. *Physics Avoidance: Essays in Conceptual Strategy*. Oxford: Oxford University Press.

Wilson, M., and Woodward, J. 2019. "Counterfactuals in the Real World." In Fillion, N., Corless, R., and Kotsireas, I., eds., *Algorithms and Complexity in Mathematics, Epistemology, and Science. Fields Institute Communications*, vol. 82. New York: Springer.

Wilson, T. 2002. *Strangers to Ourselves: Discovering the Adaptive Unconscious*. Cambridge, MA: Harvard University Press.

Wolff, P. and Thorstad, R. 2017. "Force Dynamics." In Waldmann, M., ed., *The Oxford Handbook of Causal Reasoning*. Oxford: Oxford University Press, 147–168.

Woodward, J. 1990. "Supervenience and Singular Causal Claims." In Knowles, D., ed., *Explanation and Its Limits*. Cambridge: Cambridge University Press, 211–246.

Woodward, J. 1995. "Causality and Explanation in Econometrics." In Little, D., ed., *On the Reliability of Economic Models: Essays in the Philosophy of Economics*. New York: Kluwer, 9–61.

Woodward, J. 1999. "Causal Interpretation in Systems of Equations." *Synthese* 121: 199–257.

Woodward, J. 2000. "Explanation and Invariance in the Special Sciences." *British Journal for the Philosophy of Science* 51: 197–254.

Woodward, J. 2003. *Making Things Happen: A Theory of Causal Explanation*. New York: Oxford University Press.

Woodward, J. 2006. "Sensitive and Insensitive Causation." *Philosophical Review* 115: 1–50.

Woodward, J. 2007. "Interventionist Theories of Causation in Psychological Perspective." In Gopnik, A. and Schulz, L., eds., *Causal Learning: Psychology, Philosophy and Computation*. New York: Oxford University Press, 19–36.

Woodward, J. 2008. "Mental Causation and Neural Mechanisms." In Hohwy, J. and Kallestrup, J., eds., *Being Reduced: New Essays on Reduction, Explanation, and Causation*. Oxford: Oxford University Press, 218–262.

Woodward, J. 2009. "Agency and Interventionist Theories of Causation." In Beebee, H., Hitchcock, C., and Menzies, P., eds., *The Oxford Handbook of Causation*. Oxford: Oxford University Press.

Woodward, J. 2010. "Causation in Biology: Stability, Specificity, and the Choice of Levels of Explanation." *Biology and Philosophy* 25: 287–318.

Woodward, J. 2011a. "Causal Perception and Causal Understanding." In Roesller, J. ed., *Causation, Perception, and Objectivity: Issues in Philosophy and Psychology*. Oxford: Oxford University Press, 229–263.

Woodward, J. 2011b. "Psychological Studies of Causal and Counterfactual Reasoning." In Hoerl, C., McCormack, T., and Beck, S. R., eds., *Understanding Counterfactuals, Understanding Causation*. Oxford: Oxford University Press, 16–53.

Woodward, J. 2013. "Mechanistic Explanation: Its Scope and Limits." *Proceedings of the Aristotelian Society Supplementary Volume* 87: 39–65.

Woodward, J. 2014. "A Functional Account of Causation." Philosophy of Science Association Presidential Address. *Philosophy of Science* 81: 691–713.

Woodward, J. 2015a. "Methodology, Ontology, and Interventionism." *Synthese* 192: 3577–3599.

Woodward, J. 2015b. "Interventionism and Causal Exclusion." *Philosophy and Phenomenological Research* 91: 303–347.

Woodward, J. 2016a. "The Problem of Variable Choice." *Synthese* 193: 1047–1072.

Woodward, J. 2016b. "Causation in Science." In Humphreys, P., ed., *Oxford Handbook of the Philosophy of Science*. Oxford: Oxford University Press, 163–184.

Woodward, J. 2016c. "Causation and Manipulability." In Zalta, E. N., ed., *Stanford Encyclopedia of Philosophy*. Winter 2016 ed.

Woodward, J. 2018a. "Explanatory Autonomy: The Role of Proportionality, Stability, and Conditional Independence." *Synthese*. https://doi.org/10.1007/s1122.

Woodward, J. 2018b. "Normative Theory and Descriptive Psychology in Understanding Causal Reasoning: The Role of Interventions and Invariance." In Gonzalez, W. J., ed., *Philosophy of Psychology: Causality and Psychological Subject*. Berlin: De Gruyter.

Woodward, J. 2018c. "Causal Cognition: Physical Connections, Proportionality, and the Role of Normative Theory." In Gonzalez, W. J., ed., *Philosophy of Psychology: Causality and Psychological Subject*. Berlin: De Gruyter.

Woodward, J. 2018d. "Laws: An Invariance-Based Account." In Patton, L. and Ott, W., eds., *Laws of Nature: Metaphysics and Philosophy of Science*. Oxford: Oxford University Press.

Woodward, J. 2019. "On Wolfgang Spohn's *Laws of Belief*." *Philosophy of Science* 86: 759–772.

Woodward, J. 2020a. "Causal Attribution, Counterfactuals, and Disease Interventions." In Eyal, N., Hurst, S., Murray, C., Schroeder, S., and Wikler, D., eds., *Measuring the Global Burden of Disease: Philosophical Dimensions*. Oxford: Oxford University Press.

Woodward, J. 2020b. "Causal Complexity, Conditional Independence and Downward Causation." *Philosophy of Science* 87(5): 857–867.

Woodward, J. Forthcoming a. "Downward Causation and Levels." To appear in a volume on hierarchy and levels of organization in the biological sciences. Konrad Lorenz Institute.

Woodward, J. Forthcoming b. "Flagpoles Anyone?" To appear in *Theoria*. Also available on PhilSci Archive.

Woodward, J. Forthcoming c. "Sketch of Some Themes for a Pragmatic Philosophy of Science." In Andersen, H., ed., *The Pragmatist Challenge*. Oxford: Oxford University Press.

Woodward, J. Forthcoming d. "Mysteries of Actual Causation: It's Complicated."

Woodward, J. Forthcoming e. "Levels, Kinds and Multiple Realizability: The Importance of What Does not Matter." In Shenker, O., Hemmo, M. Ionannidis, S., and Vishne, G., eds., *Levels of Reality in Science and Philosophy: Re-examining the Multi-level Structure of Reality*. Cham, Switzerland: Springer.

Woodward, J. and Hitchcock, C. 2003. "Explanatory Generalizations, Part I: A Counterfactual Account." *Noûs* 37: 1–24.

Wright, L. 1976. *Teleological Explanations: An Etiological Analysis of Goals and Functions*. Berkeley: University of California Press.

Yablo, S. 1992. "Mental Causation." *Philosophical Review* 101: 245–280.

Index

For the benefit of digital users, indexed terms that span two pages (e.g., 52–53) may, on occasion, appear on only one of those pages.